INSTRUCTOR'S MANUAL FOR MILLER'S

ENVIRONMENTAL SCIENCE
WORKING WITH THE EARTH

FIFTH EDITION

Jane Heinze-Fry

Wadsworth Publishing Company
I(T)P™ An International Thomson Publishing Company

Belmont • Albany • Bonn • Boston • Cincinnati • Detroit • London • Madrid • Melbourne
Mexico City • New York • Paris • San Francisco • Singapore • Tokyo • Toronto • Washington

CONTENTS

PREFACE

What's new in this edition? The test bank has been completely revised: Additions and deletions reflect the new material in the textbook. We have increased both the quantity and quality of the questions. We have also increased possibilities for integration of other ancillary publications into an overall educational plan.

The manual continues to focus on the major goal of environmental education by emphasizing integration of concepts, values, participation, and skills needed by environmentally literate citizens. The format has been revised to more easily offer assistance for instructors working with their students in each of these areas.

ORGANIZATION

This manual is divided into two sections. Section One contains the following subsections for each chapter:

Thinking

- Concept Map. Overview concept maps are provided for each chapter. They are one way to show connections among concepts within the chapter. They can be used as introductory material when beginning a chapter and/or summarizing material when ending a chapter. They are most valuable when linked to the overview map for the whole textbook found in the front cover of the text and when students make the maps themselves.

- Goals. Goals are defined at the beginning of each chapter of the text. Page numbers are referenced in the manual.

- Objectives. These objectives will help students identify and master the most important content of the chapter. We have included objectives that require basic intellectual skills (define, describe, compare) and objectives that require more advanced intellectual skills (analyze, evaluate , predict).

- Key Terms. These terms are shown in boldface in the text. The list of terms also includes the text page on which it is found. It might be worthwhile to photocopy both the Objectives and Key Terms sections and assign them to your students.

- Multiple Choice Questions. This section refers to the page numbers of Section Two where multiple choice questions for the chapter will be found.

- More Depth: Conceptual Term Paper Topics. This section provides topics to deepen students' understanding of the concepts introduced in the chapter.

- More Breadth: Interdisciplinary Activities and Projects. This section provides ideas to broaden students' interdisciplinary understandings of environmental studies.

- Multisensory Learning: Audiovisuals. This section provides title, synopsis, source abbreviation, and other statistics for potentially useful films and videos. Full information to obtain each source is listed in Appendix A.

Attitudes/Values

- Assessment. Assessment offers some questions to evaluate students' awareness, attitudes, and values which are pertinent to the content of the chapter.

- More Depth: Discussion and Term Paper Topics. These topics raise questions which deepen students' understanding of how different worldviews, values and attitudes interact with environmental issues.

Participation

- Lifestyle and Campus Community. This section refers to appropriate chapters of the document *Green Lives/Green Campuses*, which is designed to help students explore how their lifestyles and campuses interact with the environment. These activities are designed to make the abstract environmental concepts of the textbook more concrete.

- More Depth: Action-oriented Term Paper Topics. These topics explore in more depth actions which can be taken from individual to global levels with respect to the chapter's environmental issues.

Skills

- Environmental Problem-Solving Skills: Projects. These projects offer students opportunities to develop research, problem-solving, and interpersonal skills needed to address environmental problems.

- Laboratory Skills. This section refers to appropriate chapters of the Laboratory Manual designed to go with the textbook.

- Computer Skills. This section refers to a variety of computer programs that enable students to explore interactions of many environmental variables appropriate for particular chapters.

Section Two contains a wide variety of multiple choice questions covering the material in each chapter. There are also two appendixes. Appendix A contains 25 Concept Maps that break down major concepts from the main text and show the interconnectedness of topics related to these concepts. Appendix B lists the sources and distributors of media referred to in the Multisensory Learning: Audiovisuals subsections in Section One.

ACKNOWLEDGMENTS

Thanks to David Cotter for his work on prior editions of the instructor's manual. Most of the questions, activities and projects, and suggestions for term paper topics are built from prior editions.

Thanks to Walter H. Corson of Global Tomorrow Coalition for advice on audiovisual aids.

Thanks to Gerald O. Barney of the Institute for 21st Century Studies for his advice on computer software programs.

Thanks to Gene Heinze-Fry for his computer skills and editorial comments on energy and global warming sections.

Thanks to Sybil, Bennie and Callie, Art and Lois, Gene, Jeffrey, and Ayla for perspectives over four generations.

Thanks to Cindy Gimbert and Joan Forman for their support and Ayla-care which helped create the time for this work.

SECTION ONE

TEACHING RESOURCES

PART I.
HUMANS AND NATURE: AN OVERVIEW

CHAPTER 1

ENVIRONMENTAL PROBLEMS AND THEIR CAUSES

THINKING

Concept Map

For an overview of the concepts of this chapter, see Appendix A, Map 1 in the back of this manual. The maps in Appendix A are also available in the set of black-and-white transparency masters accompanying the text. You might want to take this opportunity to look at the overview map for the whole textbook in the front cover of the text and see where this chapter fits in. Also note that Appendix 4 of the text gives directions on how to make concept maps. At the end of each chapter of the text, there is an optional question directing students to construct their own concept maps. The maps available in the Appendix are just one way of looking at the concepts. Instructors and students are encouraged to produce their own conceptual hierarchies and draw their own connections. I would be glad to receive copies of both instructor and student maps to use in future work in this area. Copies would be gratefully received by Jane Heinze-Fry, c/o Wadsworth Publishing Co., 10 Davis Drive, Belmont, CA 94002.

Goals

See short, bulleted list of questions on p. 5 of text.

Objectives

1. Define *Earth capital.* Distinguish between living off of principal and living off of interest. Analyze which of these behaviors humans are currently illustrating. Evaluate the possibility of continuing to live in our current style.

2. Briefly describe the *state of the world.* Indicate the basic condition and trends in land, water, air, biodiversity, and people resources.

3. Draw an exponential growth curve. Distinguish between exponential growth and linear growth. Describe what has happened to the length of the doubling time over the course of human history. Define *overpopulation.* Include the concept of carrying capacity. Distinguish between people overpopulation and consumption overpopulation.

4. Distinguish between the following terms: nonrenewable, renewable, and potentially renewable resources; reuse and recycle; point source of pollution and nonpoint source of pollution; degradable, slowly degradable, and nondegradable pollutants.

5. Define *biological diversity.* Briefly describe its three components. Analyze the relationship between biodiversity and human life. Based on current human behavior, project the change in biodiversity over time. Propose what could be done to ameliorate the current trend in biodiversity.

6. Define *sustainable yield.* Describe the relationship between sustainable yield and environmental degradation.

7. Describe the *tragedy of the commons*. List three approaches to lessen this problem.

8. Distinguish between pollution prevention and pollution cleanup. List the three R's of resource use. Evaluate the effectiveness of these two approaches in decreasing pollution.

9. Describe a simple model and a more complex model of relationships among population, resource use, technology, environmental degradation, and pollution. Evaluate which model is most useful to you. Assess which model would be most useful in explaining relationships to young children and which more closely resembles reality.

Key Terms

solar capital (p. 5)   biological diversity (p. 12)
Earth capital (p. 5)   biodiversity (p. 12)
atmosphere (p. 6)   genetic diversity (p. 12)
water resources (p. 6)   species diversity (p. 12)
oceans (p. 6)   ecological diversity (p. 12)
soil (p. 6)   sustainable yield (p. 12)
forests (p. 6)   environmental degradation (p. 12)
living species (p. 6)   pollution (p. 14)
population (p. 6)   common-property resources (p. 16)
sustainable society (p. 6)   tragedy of the commons (p. 16)
linear growth (p. 8)   point sources (p. 16)
exponential growth (p. 8)   nonpoint sources (p. 16)
doubling time (p. 8)   concentration (p. 16)
rule of 70 (p. 8)   degradable (nonpersistent) pollutants (p. 16)
economic growth (p. 9)   biodegradable pollutants (p. 16)
gross national product (GNP) (p. 9)   slowly degradable (persistent) pollutants (p. 16)
GNP per capita (p. 9)   nondegradable pollutants (p. 16)
more developed countries (MDCs) (p. 9)   pollution prevention (p. 16)
less developed countries (LDCs) (p. 9)   input pollution control (p. 16)
resource (p. 10)   pollution cleanup (p. 17)
nonrenewable (exhaustible) resources (p. 10)   output pollution control (p. 17)
recycling (p. 10)   overpopulation (p. 18)
reuse (p. 11)   carrying capacity (p. 18)
reserves (p. 11)   people overpopulation (p. 18)
renewable resource (p. 12)   consumption overpopulation (p. 18)
potentially renewable resource (p. 12)   overconsumption (p. 18)

Multiple Choice Questions

See Instructor's manual, Section Two, p. 103.

More Depth: Conceptual Term Paper Topics

1.   Population. UN population projections.

2.   Poverty. Definition; roots; worldwide distribution; possibilities to alter the current situation.

3.   Technology. Research, development, and distribution of new technologies in the United States.

4.   State of the world. Bibliography of current resources summarizing the state of the world; most important areas of concern.

More Breadth: Interdisciplinary Activities and Projects

1.   As a class project, "adopt" a less developed country. Assign teams of students to investigate various aspects of the nation's physical, population, economic, social, political, and other characteristics as well as lifestyle and life quality. Allocate class time for periodic brief reports and discussions of research results.

2.   Find and share with the class songs, essays, poems, paintings, and literary passages that are strongly pro- or anti-technology.

3.   As a class exercise, make lists of the beneficial and harmful consequences that have resulted from America's adoption of automobile technology.

Multisensory Learning: Audiovisuals

Conversations for a Sustainable Society; 1993; 45 min.; with Amory and Hunter Lovins, Dennis Meadows, Dana Jackson, David Orr, and others; GF.
The Living Earth; 1991; 25 min.; NG.
Pollution: World at Risk; 1989; 25 min.; NG.
In the Name of Progress, from *Race to Save the Planet;* 1990; 60 min.; development projects and environmental degradation; ACPB.
Our Planet Earth; a collection of astronauts' recollections as they viewed Earth from space; BFF.
The Island: Will We Save Our Planet Earth?; 14 min.; a National Geographic film that increases awareness of environmental problems; NG.
What Is the Limit?; 1987; 23 min.; a National Audubon Society film that probes linkages among human population growth, environmental degradation, resource depletion, habitat destruction, and ethical considerations for the future; NA.
Our Common Future; 1988; presents findings of the UN World Commission on Environment and Development; GTC.
Conserving Our Environment; 16 min.; pollution and ways to protect the environment; CMTI.
The Blue Planet; Earth, life, and space; 26 min.; CBS.
Gaia, the Living Planet; 1990; 42 min.; BFF.
Turning the Tide; seven videos; major environmental issues facing the planet; BFF.

Listings include year of release and length of film where available.

See Appendix B for suppliers.

ATTITUDES/VALUES

Assessment

1. Is the world overpopulated? Have you experienced countries which you felt to be overpopulated? Have you seen videos of countries which you felt to be overpopulated?

2. Is the country in which you live overpopulated? What factors contribute to your feelings?

3. Is your local community overpopulated? What do you feel are the costs and benefits of the population size of your community?

4. Do you favor reducing the population size of the world?

5. Do you favor reducing the population size of the country in which you live?

6. Do you favor growth management of the community in which you live?

7. Do you feel that the size of the human population is one of the top environmental issues?

8. How many children do you plan to have?

9. Do you believe that any person should be able to have as many children as they want?

10. Does your country consume too many resources?

11. Do you consume too many resources?

12. Do you think it's important to change your consumption patterns?

More Depth: Discussion and Term Paper Topics

1. What is quality of life?

2. What is the history of conflict among pollution control, environmental degradation, and employment in the United States? Possible cases of interest include the automobile industry and fuel efficiency standards, and the spotted owl and the logging industry.

3. Is the United States overpopulated? Explore people overpopulation and consumption overpopulation.

4. Is bigger better or is small beautiful? Explore Schumacher's philosophy.

(Also, see text, Critical Thinking, pp. 22–23)

PARTICIPATION

More Depth: Action-oriented Term Paper Topics

1. Computer modeling methods. *Limits to Growth* by Meadows and Meadows; today's global climate models.

2. National. National efforts to address environmental needs; *Blueprint for the Environment.*

3. Global. Global attempts to address the environment and the economy; the UN document *Our Common Future;* the Earth Summit at Rio de Janeiro in 1992.

SKILLS

Environmental Problem-Solving Skills: Projects

1. As a class exercise, compute the cost of a hamburger, a movie ticket, a single-family home, and/or other commodities 30 years in the future, assuming a steady inflation rate of 5% (or the current inflation rate).

2. As a class exercise, compile a list of resources that are considered important today but were not recognized as resources 100 years ago. What are some things that have ceased to be significant resources during the last 50 years? What resources of the present will probably be of little value 50 years from now?

3. Have the class make a list of changes in your community's environment that have occurred over the last 10 years. Have them vote on which changes they considered desirable and which undesirable. Discuss the changes on which there is least consensus about desirability. Clarify the differences in values that underlie differences in students' responses.

4. Have students assume roles as futurists. Have them describe life as they predict it will be in the year 2010.

Laboratory Skills

(none)

Computer Skills

Tragedy of the Commons
   -Illustrates Garrett Hardin's principle of the Tragedy of the Commons—that increased exploitation of a common resource is desirable in the short term for each individual in a community, but disastrous in the long term to the whole community.
   -National Collegiate Software, Duke University Press, 6697 College Station, Durham, NC 27708

International Futures Simulation (IFS)
   -Provides a framework for evaluating the widely different public statements made about the workings of the global development system and about probable global futures.
   -CONDUIT, The University of Iowa, Oakdale Campus, Iowa City, IA 52242

World Economic Model (WEM)
   -Offers a consistent quantitative framework for carrying out prospective analysis and policy design exercises in the area of international economic relations.
   -Dr. A.R. Gigengack, World Model Project, State University of Groningen, P.O. Box 800, 9700 AV Groningen, The Netherlands

*SimEarth
   -To modify, manage, and nurture a planet from creation, through the formation of life, to the development of technology; based on the Gaia theory.
   -Maxis, 2 Theatre Square, #230, Orinda, CA 94563 (510-254-9700)

Balance of the Planet
   -Role-play the "High Commissioner of the Environment" by levying taxes and granting subsidies to solve a variety of global environmental problems.
   -Accolade, 550 W. Winchester Blvd., Suite 200, San Jose, CA 95128 (800-245-7744)

CHAPTER 2

CULTURAL CHANGES, WORLDVIEWS, ETHICS, AND SUSTAINABILITY

THINKING

Concept Map

See Appendix A, Maps 2 and 3.

Goals

See short, bulleted list of questions on p. 25 of text.

Objectives

1. Briefly describe the Earth-Wisdom (Environmental) Revolution.

2. Briefly describe hunter-gatherer societies, focusing on division of labor and power, the relationship of humans to nature, and the impact of their societies on the environment.

3. Describe early forms of agriculture. Describe changes that occurred in human population distribution, employment, and relationships between societies as the Agricultural Revolution unfolded.

4. Briefly describe the Industrial Revolution, focusing on changes in energy consumption. Describe relationships between energy consumption and the production and consumption of material goods. List the benefits that are distributed to most citizens of industrial societies.

5. Describe the relationship between actions and worldview. Explain how cultural changes take place.

6. Compare the human-centered planetary-management worldviews with the life-centered Earth-wisdom worldviews. Focus on differences in beliefs about the relationship of humans to nature (including other species and use of resources), beliefs about human capability (especially through science and technology), and beliefs about human-to-human relationships through economics and politics.

7. Compare hunter-gatherer societies, agricultural societies, and industrial societies, focusing on division of labor and power, the relationship of population to food supply, the relationship of humans to nature, the use of resources (energy and materials) per person, and the environmental impacts. Project how a sustainable-Earth society would fit into this analysis.

8. Analyze the amount of time over the course of history that has been used to bring about cultural changes. Include a comparison of the length of time to bring about the Agricultural Revolution and the length of time to bring about the Industrial Revolution. Project an amount of time to bring about a cultural change to a sustainable-Earth society. Suggest modern capabilities that might enable this change to occur.

9. Summarize Earth ethics. Include ethical statements regarding ecosystems, species, cultures, and individual responsibilities.

10. List eight steps which would help emphasize Earth education within the educational system. Elaborate on the common traps that lead to inaction.

Key Terms

Earth-Wisdom (Environmental) Revolution (p. 26)
hunter-gatherers (p. 26)
Agricultural Revolution (p. 26)
slash-and-burn cultivation (p. 27)
shifting cultivation (p. 27)
subsistence farming (p. 27)
Industrial Revolution (p. 28)

worldviews (p. 29)
frontier worldview (p. 29)
planetary management worldview (p. 32)
Earth-wisdom worldview (p. 33)
mindquake (p. 34)
Earth ethics (p. 34)
Earth education (p. 34)

Multiple Choice Questions

See Instructor's manual, Section Two, p. 111.

More Depth: Conceptual Term Paper Topics

1. Time for cultural change. The change process and how the rate of cultural change has accelerated; the rate of change of Earth's processes (such as erosion and movement of tectonic plates; the rate of evolutionary change of organisms; the rate of human adaptation in the past and what it is likely to be in the future.

2. Cultural differences. Cultural views of the human-environment relationship; attitudes toward nature; distribution of labor, power, and wealth; relationships between the sexes; social structure; and political style; cultural views that offer a sustainable-Earth worldview.

3. Competition and cooperation. Costs and benefits of cooperation; costs and benefits of competition; the balance of cooperation and competition in different cultures; the roots of rugged American individualism; evaluation of the need for a "me generation" versus the need for a "we generation."

4. Environmental impacts of the cultural revolutions. Energy consumption and use of materials throughout history.

5. Worldview and reality. Thomas Jefferson's vision of an agrarian America, focusing on distribution of labor and power among humans; how the Industrial Revolution altered distribution of labor and power in the United States.

More Breadth: Interdisciplinary Activities and Projects

1. Find and share with the class songs, folklore, literary passages, and art works that reflect U.S. land-use values and ethics as they have evolved from the frontier era to the present. Be sure to include Native American works. What can be discerned about the relationship of humans to nature in different cultures through their expressions of art?

2. Invite an anthropologist to visit your class to discuss the similarities and contrasts between hunter-gatherer societies and industrial societies. What can the study of hunter-gatherer societies teach that will be useful in the future?

3. Ask a psychologist and/or sociologist to address your class on the subject of human aggression, with particular emphasis on methods that have been or could be used to reduce its injurious effects and channel competitive energies into productive areas. Discuss the balance of competition and cooperation in different societies. Have each student give an example of competition and cooperation in his or her life and in the school. Discuss the benefits and drawbacks of each style. Suggest what the relationship of competition and cooperation might be in a sustainable-Earth society.

4. For the benefit of your class, arrange a panel discussion among spokespersons for the religious community in your locale. Ask the panel to address the subject of religion as a driving force for sustainable-Earth behavior and to respond to critics who argue that Judeo-Christian beliefs are at the root of the environmental crisis.

Multisensory Learning: Audiovisuals

The Environmental Revolution, from *Race to Save the Planet;* 1990; 60 min.; relationship of humans to their environment over time; ACPB.
Voices of the Land; 1991; 20 min.; correlations between the destruction of sacred land and destruction of the human spirit; BFF.
Prophets and Loss; 1991; 49 min.; a range of informed and profound thinkers comment on the environmental crisis and a path to a viable future: Sagan, Ehrlich, Schneider; VP.
Spirit and Nature; 1991; 88 min.; hosted by Bill Moyers, representatives of five major religions explore the creation of an environmental ethic based on stewardship; VP.
The Global Brain; 1985; 35 min.; critical point of choice: a vision of humanity's potential; VP.
Aldo Leopold: A Sand County Almanac; 1983; 29 min.; Leopold's wonder on his Sand County farm in Southern Wisconsin; CP.
Aldo Leopold: His Life and Thought; 1983; 27 min.; thoughts from wildlife management to the land ethic; CP.
For Earth's Sake: The Life and Times of David Brower; 1990; 58 min.; BFF.
Small Is Beautiful; 28 min.; Schumacher's view of decentralized, small-scale technology; BFF.
Perspective: On the Margin; 1988; 28 min.; fish farming, deer farming, hydroponics in the wilds of Scotland; LTS.
New Alchemy: A Rediscovery of Promise; 28 min.; a pioneering appropriate technology group and its work on solar aquaculture, bioshelters, wind power, and organic agriculture; BFF.
The Sun Dagger; 1983; 29 min.; discovery of a celestial calendar used by the Anasazi Indians who thrived in Chaco Canyon, New Mexico, 1,000 years ago; BFF.

See Appendix B for suppliers.

ATTITUDES/VALUES

Assessment

1. How do you feel toward past cultural revolutions?

2. Do you think it's time for another cultural revolution? What factors contribute to your feelings?

3. Do you feel our current cultural stage can continue indefinitely?

4. What kinds of changes do you think would improve the quality of life on Earth?

5. What kinds of changes are you willing to make to improve your own quality of life?

6. What kinds of changes do you think industrialized countries might make to improve the quality of life?

7. What kinds of changes do you think developing countries might make to improve the quality of life?

8. Do you believe that we have an obligation to leave Earth for future generations of humans and other species in as good a shape as we found it, if not better? Did past generations do this for you?

9. Do you think your environment will be more livable, about the same, or less livable 10 years from now?

10. Do you think that humans can bring about a major change within your lifetime that involves helping sustain rather than degrade Earth?

11. Do you believe that most environmental issues are overblown by environmentalists and the media?

More Depth: Discussion and Term Paper Topics

1. What is the "good life"?

2. Do you feel a part of or apart from nature?

3. Do you think technology can solve our environmental problems?

4. Do you think human ingenuity and substitution for materials that are being used up quickly can create a good life for Earth's people?

(Also, see text, Critical Thinking, pp. 36–38)

PARTICIPATION

More Depth: Action-oriented Term Paper Topics

1. Individual. Changes in lifestyle; use of energy and materials; consumer choices; voting.

2. Group. Lobbying; environmental legislation; environmental regulations; environmental research; environmental communications.

3. History of the U.S. environmental movement. Muir; Marsh; Pinchot; Roosevelt; Leopold; Carson; Commoner.

4. International environmental movement. The Green Movement; environmental groups of the emerging eastern European countries; the Chipko movement; the Montreal Protocol; the Earth Summit at Rio de Janeiro in 1992.

SKILLS

<u>Environmental Problem-Solving Skills: Projects</u>

1. As a class project, poll the students at your school on the subject of environmental protection and its importance. Investigate their beliefs and opinions regarding a variety of environmental quality management issues and alternatives. Research the development of environmental law in the United States. Describe the relationship of humans to nature that has guided different legislation.

2. Have your students write scenarios describing what everyday life would be like in the United States after a transition is made to a sustainable-Earth society. Identify areas of consensus and disagreement.

3. As a class exercise, have each student write out and hand in anonymously a list of the essential components of "the good life." Read some or all of the lists aloud to the class. Write a composite list on the blackboard and discuss each component in terms of sustainable-Earth values and guidelines.

4. As a class project, analyze use of resources at your college. Consider all of the things that must be done to place the management of the institution and its physical plant on a sustainable-Earth footing—such as energy conservation, water conservation, frugal consumption of materials, reuse of materials and recycling, solid waste management, noise control, provision of wildlife habitat, integrated pest management, designing for people, toxic waste disposal, and environmental education. Study the decision-making structure of your institution. Try to convince the administration to implement your plan. Evaluate your success. Develop alternative strategies as needed.

<u>Laboratory Skills</u>

(none)

<u>Computer Skills</u>

(none)

PART II.
PRINCIPLES AND CONCEPTS

CHAPTER 3

MATTER AND ENERGY RESOURCES: TYPES AND CONCEPTS

THINKING

<u>Concept Map</u>

See Appendix A, Map 4.

<u>Goals</u>

See short, bulleted list of questions on p. 41 of text.

<u>Objectives</u>

1. Briefly describe how science works. State the questions science tries to answer. Summarize scientific methods.

2. Distinguish between science and technology.

3. Describe what distinguished environmental science from science in general. Describe the paralysis-by-analysis trap. Describe two other problems that arise when science is used to address environmental problems.

4   Define *matter*. Distinguish between forms of matter and quality of matter. State the law of conservation of matter.

5. Define *energy*. Distinguish between forms of energy and quality of energy.

6. Distinguish among physical, chemical and nuclear changes.

7. Distinguish between nuclear fission and nuclear fusion.

8. State the first and second laws of energy.

9. Describe the implications of the laws of matter and energy for a long-term sustainable-Earth society.

10. Distinguish among throwaway, matter-recycling, and sustainable-Earth societies.

<u>Key Terms</u>

science (p. 41)	scientific theory (p. 42)
scientific data (p. 41)	scientific methods (p. 42)
scientific law (p. 41)	technology (p. 42)
scientific hypothesis (p. 41)	environmental science (p. 42)
scientific models (p. 41)	matter (p. 42)

14   Section One/Chapter 3

elements (p. 42)

compounds (p. 43)

mixtures (p. 43)

atoms (p. 43)

ions (p. 43)

molecules (p. 43)

protons (p. 44)

neutrons (p. 44)

electrons (p. 44)

nucleus (p. 44)

atomic number (p. 44)

mass number (p. 44)

isotopes (p. 44)

chemical formula (p. 45)

matter quality (p. 45)

high-quality matter (p. 45)

low-quality matter (p. 45)

energy (p. 45)

kinetic energy (p. 46)

electromagnetic radiation (p. 46)

ionizing radiation (p. 46)

nonionizing radiation (p. 46)

potential energy (p. 46)

solar energy (p. 46)

energy quality (p. 48)

high-quality energy (p. 48)

low-quality energy (p. 48)

physical change (p. 49)

chemical change (p. 49)

chemical reaction (p. 49)

law of conservation of matter (p. 49)

nuclear change (p. 49)

natural radioactive decay (p. 49)

radioactive isotopes (p. 49)

radioisotopes (p. 49)

gamma rays (p. 50

alpha particles (p. 50)

beta particles (p. 50)

half-life (p. 50)

nuclear fission (p. 50)

critical mass (p. 51)

chain reaction (p. 51)

nuclear fusion (p. 51)

law of conservation of energy (p. 52)

first law of energy (p. 52)

first law of thermodynamics (p. 52)

energy quality (p. 52)

second law of energy (p. 52)

second law of thermodynamics (p. 52)

throwaway societies (p. 53)

matter-recycling society (p. 54)

sustainable-Earth (low-waste) society (p. 54)

Multiple Choice Questions

See Instructor's manual, Section Two, p. 117.

More Depth: Conceptual Term Paper Topics

1. The universe. Total amounts of matter and energy in the universe; the big bang theory of the origin of the universe; the role of entropy in the destiny of the universe.

2. Low-energy lifestyles. Individual case studies such as Amory Lovins and national case studies such as Sweden.

3. Nature's cycles and economics. Recycling attempts in the United States; bottlenecks that inhibit recycling; strategies that successfully enhance recycling efforts.

More Breadth: Interdisciplinary Activities and Projects

1. Ask a physics professor or physics lab instructor to visit your class and, by using simple experiments, demonstrate the matter and energy laws.

2. As a class exercise, try to inventory the types of environmental disorders that are created in order to maintain a classroom environment—the lighting, space heating and cooling, electricity for projectors, and other facilities, equipment, and services.

3. Invite a medical technician to speak to your class on the beneficial uses of ionizing radiation. What controls are employed to limit the risks associated with the use of radioisotopes for diagnostic and treatment procedures?

Multisensory Learning: Audiovisuals

Energy: What Energy Means; 1982; 15 min.; converting from one form of energy to another; NG.
The Forms of Energy; 30 min.; CBS.
The Flow of Heat Energy; 53 min.; CBS.
A Conversation with Stephen Jay Gould; 28 min.; how he became interested in science as a career; CBS.
Careers in Science; 60 min.; CBS.
Science and Human Values; 30 min.; CBS.
Radiation: Types and Effects; 22 min.; FHS.
Radiation: Origins and Controls; 27 min.; FHS.

See Appendix B for suppliers.

ATTITUDES/VALUES

Assessment

1. Do you feel a part of the flow of energy from the sun?

2. Do you feel you play a role in nature's cycles?

3. How do you feel when your home is air-conditioned? heated?

4. How do you feel when you turn on a light? the television? your CD player?

5. How do you feel on a sunny day? a cloudy day?

6. What right do you have to use of Earth's material resources? Are there any limits to your rights? What are they?

7. What rights do you have to Earth's energy resources? Are there any limits to your rights? What are they?

8. Do you believe that cycles of matter and energy flowing from the sun have anything to do with your lifestyle? with your country's policies?

More Depth: Discussion and Term Paper Topics

1. An evaluation of the positive and negative contributions of nuclear technologies. Nuclear weapons in World War II and the Cold War; radioisotopes in research and medical technology; nuclear power plants.

2. How much are you willing to pay in the short run to receive economic and environmental benefits in the long run? Explore costs and payback times of energy-efficient appliances, energy-saving lightbulbs, and weather stripping.

3. Can we get something for nothing? Explore the attempts of advertising to convince the public that we can indeed get something for nothing. Explore attempts to create perpetual motion machines. Explore the history of the *free lunch* concept.

4. Is convenience more important than sustainability? Explore the influence of U.S. frontier origins on the throwaway mentality.

(Also, see text, Critical Thinking, p. 55)

PARTICIPATION

More Depth: Action-oriented Term Paper Topics

1. Individual. Actions that improve energy efficiency and reduce consumption of materials.

2. Community. Enhance recycling efforts: curbside pick up versus recycling center dropoffs; high-tech versus low-tech sorting of materials; Osage, Iowa, a case study in community energy efficiency.

3. National energy policy. Evaluation of the current national energy policy proposals in light of the laws of energy and long-term economic, environmental, and national-security interests.

SKILLS

Environmental Problem-Solving Skills: Projects

1. A human body at rest yields heat at about the same rate as a 100-watt incandescent light bulb. As a class exercise, calculate the heat production of the student body of your school, the U.S. population, and the global population. Where does the heat come from? Where does it go?

2. As a class exercise, conduct a survey of the students at your school to determine their degree of awareness and understanding of the three basic matter and energy laws. Discuss the results in the context of the need for low-entropy lifestyles and sustainable-Earth societies.

Laboratory Skills

(none)

Computer Skills

(none)

CHAPTER 4

ECOSYSTEMS AND HOW THEY WORK

THINKING

Concept Map

See Appendix A, Map 5.

Goals

See short, bulleted list of questions on p. 57 of text.

Objectives

1. List six characteristics of life.

2. List and briefly describe four layers or spheres of the earth.

3. Compare the flow of matter and the flow of energy through the biosphere.

4. Define *biome*. Explain the major cause of the distribution of biomes on the Earth.

5. Define biological diversity and distinguish among three types of biodiversity. List the five kingdoms of life. Distinguish between evergreens and deciduous plants. Distinguish between vertebrates and invertebrates.

6. Distinguish between *abiotic* and *biotic* components of ecosystems. Describe how organisms are classified by how they get their nutrients. Be sure to distinguish between the following sets of terms: producer, consumer; photosynthesis, chemosynthesis, aerobic respiration; herbivore, carnivore, omnivore, detritivore.

7. Define *limiting factor*. Give one example of a limiting factor in an ecosystem.

8. Apply the second law of energy to food chains and pyramids of energy, which describe energy flow in ecosystems. Briefly describe two other ecological pyramids and indicate if their shape is always a pyramid.

9. Apply the law of conservation of matter to biogeochemical cycles, which describe the flow of matter through ecosystems. Briefly describe the following cycles: carbon, nitrogen, phosphorous, sulfur, hydrologic.

10. Define *niche*. Distinguish among the following sets of terms: specialist, generalist; fundamental niche, realized niche; interference competition, exploitation competition, resource partitioning; predation, parasitism, commensalism, mutualism.

Key Terms

cell (p. 57)	atmosphere (p. 57)
mutations (p. 57)	troposphere (p. 57)
natural selection (p. 57)	stratosphere (p. 57)
evolution (p. 57)	hydrosphere (p. 57)

lithosphere (p. 57)

ecosphere (p. 57)

biosphere (p. 57)

nutrient (p. 60)

nutrient cycles (p. 60)

biogeochemical cycles (p. 60)

ecology (p. 60)

organism (p. 60)

species (p. 60)

population (p. 61)

genetic diversity (p. 61)

habitat (p. 62)

community (p. 62)

biological community (p. 62)

ecosystem (p. 62)

climate (p. 62)

biomes (p. 62)

biological diversity (p. 62)

biodiversity (p. 62)

genetic diversity (p. 62)

species diversity (p. 62)

ecological diversity (p. 62)

eukaryotic (p. 65)

prokaryotic (p. 65)

Monera (p. 65)

bacteria (p. 65)

cyanobacteria (p. 65)

Protista (protists) (p. 65)

Fungi (p. 65)

Plantae (plants) (p. 65)

evergreens (p. 65)

deciduous plants (p. 65)

succulent plants (p. 65)

epiphytes (p. 65)

Animalia (animals) (p. 65)

invertebrates (p. 65)

vertebrates (p. 65)

biotic (p. 66)

abiotic (p. 66)

producers (p.66)

autotrophs (p. 66)

photosynthesis (p. 66)

chemosynthesis (p. 68)

consumers (p. 68)

heterotrophs (p. 68)

herbivores (p. 68)

primary consumers (p. 68)

carnivores (p. 68)

secondary consumers (p. 68)

tertiary (higher-level) consumers (p. 68)

omnivores (p. 68)

detritivores (p. 68)

detritus (p. 68)

decomposers (p. 68)

detritus feeders (p. 68)

aerobic respiration (p. 69)

range of tolerance (p. 70)

law of tolerance (p. 70)

acclimation (p. 70)

threshold effect (p. 70)

limiting factor principle (p. 71)

limiting factor (p. 71)

salinity (p. 71)

dissolved oxygen (DO) content (p. 71)

food chain (p. 71)

trophic level (p. 71)

food web (p. 73)

pyramid of numbers (p. 73)

biomass (p. 73)

pyramid of biomass (p. 73)

pyramid of energy flow (p. 73)

gross primary productivity (p. 74)

net primary productivity (p. 74)

carbon cycle (p. 76)

nitrogen cycle (p. 78)

nitrogen fixation (p. 78)

phosphorus cycle (p. 79)

sulfur cycle (p. 81)

hydrologic cycle (p. 81)

water cycle (p. 81)

native species (p. 82)

immigrant species (p. 82)

alien species (p. 82)

indicator species (p. 82)

keystone species (p. 82)

ecological niche (niche) (p. 83)

specialist species (p. 83)

generalist species (p. 85)

symbiotic relationships (p. 85)

fundamental niche (p. 85)

interspecific competition (p. 86)

interference competition (p. 86)

exploitation competition (p. 86)

resource partitioning (p. 86)

realized niche (p. 86)

competitive exclusion principle (p. 86)

predation (p. 86)

predator (p. 86)

prey (p. 86)

predator–prey relationship (p. 86)

parasite (p. 86)

host (p. 86)

mutualism (p. 87)

commensalism (p. 87)

Multiple Choice Questions

See Instructor's manual, Section Two, p. 127.

More Depth: Conceptual Term Paper Topics

1. Cycles of matter. Particular cycles of matter, clarifying chemical changes throughout the cycle; the processes of photosynthesis and respiration and how they connect autotrophic and heterotrophic organisms.

2. Energy flow. Energy flow in a particular ecosystem; relationships among species in a particular ecosystem; comparison of the life of a specialist with that of a generalist.

3. Humans trying to work with ecosystems. Composting; organic gardening; land reclamation; rebuilding degraded lands; tree-planting projects.

More Breadth: Interdisciplinary Activities and Projects

1. Organize a class trip to a natural area such as a forest, grassland, or estuary to observe the elements of ecosystem structure and function. Arrange for an ecologist or naturalist to provide interpretive services.

2. Bring a self-sustaining terrarium or aquarium to class and explain the structure and function of this conceptually tidy ecosystem. Discuss the various things that can upset the balance of the ecosystem and describe what would happen if light, food, oxygen, or space were manipulated experimentally.

3. Organize a class field trip to systematically investigate the ecological niches for plant and animal life existing in a landscape significantly modified by human activities. If possible, arrange to travel along a gradient that will take you from farmland to suburbs to city to central business district. (A simplified version of this exercise could be done by walking around campus.)

4. Have your class secure a large jar or glass container and equip it with a ventilated (finely perforated or meshed) top. Place a male and a female fruit fly in it, together with a plentiful supply of food. Set the jar aside and monitor the fly population as the days pass. Why does the population increase slowly at first and then very rapidly? What causes the inevitable population collapse? Why do all of the flies, rather than just the "surplus" population, die?

5. Find works of literature, art, and music that show human attachment to and destruction of natural ecosystems.

Community of Living Things, 4 modules: *Change, Diversity, Interrelationships, Energy;* 60 min.,
    45 min., 75 min., 45 min. respectively; PBS.
Life on Earth; four hours on two cassettes; CBS.
Photosynthesis: The Flow of Energy from Sun to Man; 15 min.; CBS.
The Water Cycle and Erosion; 15 min.; CBS.
Water in the Air; 81 min.; CBS.
Energy and Nutrients in Ecology, Parts 1 and 2; 12 min. each; CBS.
Enemies of the Oak; 54 min.; CBS.
Intimate Strangers: Symbiosis; 8 min.; CBS.
Voyage to the Enchanted Isles; 60 min.; the Galapagos; CBS.
The Sequoia Giants of Sequoia National Park; 28 min.; CBS.
Death Trap; 60 min.; carnivorous plants; CBS.
The Building Blocks of Life; 60 min.; two-part program examines the functioning structure of the
    cell; CBS.
Seasons in the Swamp; 1988; 27 min.; annual hydrologic cycle of a cypress swamp; FPL.
Creatures in the Great Lakes; 1985; 20 min.; underwater life of the Great Lakes; BP.
Blooming Secrets—The Ecology of Spring Wildflowers; 1986; 16 min.; wildflower survival
    strategies; MDC.
The Galapagos: My Fragile World; 1986; 48 min.; view of the islands through the eyes of a
    native; WETA.

See Appendix B for suppliers.

ATTITUDES/VALUES

Assessment

1. Do you feel you are part of an ecosystem? What niche do you fill?

2. Have you visited a variety of types of ecosystems? If so, where do you feel most at home?

3. Do you hold any particular feelings for producers? consumers? decomposers?

4. Do you have any particular feelings toward relationships demonstrated in ecosystems?
    Competition? Predation? Commensalism? Parasitism? Mutualism?

5. Do you feel there will always be enough matter and energy for the survival of all individuals
    of all species? Will the carrying capacity of the Earth be expanded by new technologies? Will
    nature be able to continually absorb "waste products" from human societies?

6. How do you feel when you think of a coyote eating a rabbit? How do you feel when you
    think of humans eating hamburgers?

7. Do humans have a right to domesticate and eat other species? Should there be any limits
    to the amount of meat products that an individual eats? If so, what should determine
    those limits?

8. Do you feel any responsibility to protect natural ecosystems? Would you support the
    preservation of representative ecosystems? If so, on what basis?

More Depth: Discussion and Term Paper Topics

1.   Should we eat lower on the food chain?

2.   Should we rely more on perpetual sources of energy?

3.   What do nature's cycles of matter suggest about landfills, incinerators, reducing consumption, and recycling?

(Also, see text, Critical Thinking, p. 89)

PARTICIPATION

More Depth: Action-oriented Term Paper Topics

1.   Field and laboratory methods used in ecological research. Measuring net primary productivity and respiration rates; analyzing for particular chemicals in the air, water, and soil; studying relationships among species; population studies; computer modeling of ecological interrelationships.

SKILLS

Environmental Problem-Solving Skills: Projects

1.   As a class exercise, have each student list the kinds and amounts of food he or she has consumed in the past 24 hours. Aggregate the results and compare them on a per capita basis with similar statistics derived from studies of dietary composition and adequacy in food-deficient nations. How many people with a vegetarian diet could subsist on the equivalent food value of the meat consumed by your class?

2.   Have the students debate the argument that eating lower on the food chain is socially and ecologically more responsible, cheaper, and healthier. (It is helpful to do this around a time when fasting is common.) Also, look at the long-term picture: will eating low on the food chain sustain an exponentially growing human population indefinitely?

3.   Define an ecosystem to study on campus. As a class project, analyze the abiotic and biotic components of the ecosystem. Draw webs and construct pyramids to show the relationships among species in the ecosystem. Project what might happen if pesticides were used in the ecosystem, if parts of the ecosystem were cleared for development, or if a coal-burning power plant were located upwind.

Laboratory Skills

Laboratory Manual for Miller's Living in the Environment and Environmental Science, Lab 1: Introduction to the Compound Microscope; Lab 2: Biological Classification; Lab 3: The Plankton Community.

Computer Skills

(none)

CHAPTER 5

ECOSYSTEMS: WHAT ARE THE MAJOR TYPES AND WHAT CAN HAPPEN TO THEM?

THINKING

Concept Map

See Appendix A, Maps 6 and 7.

Goals

See short, bulleted list of questions on p. 91 of text.

Objectives

1. Describe how climate affects the distribution of plant life on Earth. Compare the climate and adaptations of plants and animals in deserts, grasslands, and forests. Be sure to distinguish among the three major kinds of forests.

2. Evaluate the significance of the ecological contributions of the oceans. Distinguish between coastal zones and open sea. List and compare the four principal zones of an ocean.

3. Distinguish between coastal and inland wetlands. Describe the ecological functions performed by wetlands. Describe environmental problems associated with coastal and inland wetlands.

4. List and compare the four zones of a lake. Distinguish among oligotrophic, eutrophic, and mesotrophic lakes. Distinguish among the three zones of a freshwater stream.

5. Define *homeostasis*. Compare the abiotic and biotic factors that contribute to population growth (biotic potential) with the abiotic and biotic factors that limit population growth (environmental resistance). Distinguish among inertia, constancy, and resilience.

6. Define *carrying capacity*. Describe how carrying capacity affects exponential growth. Distinguish the concepts within each of the following sets of terms: *density-dependent* and *density-independent* population controls; *stable, irruptive,* and *cyclic* population change curves; *r-strategist* and *K-strategist; Type I, Type II,* and *Type III* survivorship curves.

6. Summarize chemical and biological evolution. Emphasize the oxygen revolution and its significance for humans.

7. Explain the importance of mutations and natural selection to adaptation and differential reproduction. Define coevolution. Summarize the roles played by mass extinctions and adaptive radiations in the evolutionary process.

8. Evaluate the statement "Diversity leads to stability." Distinguish among three kinds of stability and give one example of each.

9. Define *succession*. Describe how humans affect communities. State the first law of human ecology and describe its implications for human interactions with the environment.

10. List six key features of living systems. Describe implications these features carry for human lifestyles.

Key Terms

climate (p. 91)

biomes (p. 92)

desert (p. 92)

grasslands (p. 92)

forest (p. 93)

latitude (p. 93)

altitude (p. 93)

permafrost (p. 98)

coastal zone (p. 101)

coral reefs (p. 101)

estuaries (p. 101)

coastal wetlands (p. 101)

beaches (p. 102)

barrier islands (p. 102)

open sea (p. 107)

lakes (p. 107)

eutrophic lake (p. 107)

oligotrophic lake (p. 107)

mesotrophic lake (p. 107)

surface water (p. 108)

runoff (p. 108)

watershed (p. 108)

drainage basin (p. 108)

inland wetlands (p. 109)

homeostasis (p. 111)

inertia (p. 112)

persistence (p. 112)

constancy (p. 112)

resilience (p. 112)

population dynamics (p. 112)

biotic potential (p. 113)

environmental resistance (p. 113)

carrying capacity (K) (p. 113)

density-dependent check (p. 114)

density-independent check (p. 114)

r-strategists (p. 115)

K-strategists (p. 115)

survivorship curve (p. 115)

fossils (p. 115)

chemical evolution (p. 117)

oxygen revolution (p. 118)

biological evolution (p. 120)

evolution (p. 120)

theory of evolution (p. 120)

genes (p. 120)

gene pool (p. 120)

mutations (p. 120)

adaptation (p. 121)

differential reproduction (p. 121)

natural selection (p. 121)

coevolution (p. 122)

speciation (p. 122)

extinction (p. 122)

mass extinction (p. 123)

adaptive radiations (p. 123)

ecological succession (p. 124)

community development (p. 124)

primary succession (p. 124)

pioneer species (p. 124)

secondary succession (p. 124)

monocultures (p. 126)

first law of human ecology (p. 127)

principle of connectedness (p. 127)

Multiple Choice Questions

See Instructor's manual, Section Two, p. 142.

More Depth: Conceptual Term Paper Topics

1. Human activities interfering with ecosystems. Inorganic fertilizers; phosphate detergents and cultural eutrophication; sulfur compounds and acid deposition; CFCs and ozone depletion; combustion of fossil fuels and greenhouse gases; degradation and deforestation of tropical forests; wetlands development; intensively developed coastal areas.

2. Plant and animal adaptations to different biomes. Desert plants and animals; plants and animals of the tundra; mountain microclimates and vertically zoned vegetation; organisms of the Amazon.

3. Nature's response to human activities. Succession on deserted farmlands; succession after fire; dilution of pollution by streams; species migration; a closer look at homeostatic systems; population crashes; pioneer species; problems with crop monocultures.

4. Fragile ecosystems. Deserts; tropical forests; tundra.

More Breadth: Interdisciplinary Activities and Projects

1. Arrange a field trip providing opportunities to compare and contrast ecosystems of several different types, including some damaged or stressed by human activities. Invite an ecologist or biologist along to identify and discuss specific examples of species adaptation to environmental conditions. Do the boundaries between different kinds of ecosystems tend to be sharply delineated? Can you identify factors that limit the growth of certain species?

2. Ask students to bring to class and share examples of art, music, poetry, and other creative expressions of human thoughts and feelings about Earth's deserts, grasslands, forests, and oceans. Lead a class discussion on the subject of how human culture has been shaped to an important degree by the environmental conditions of each major biome.

3. As a class exercise, evaluate the diversity of your community using criteria such as ethnic, racial, religious, and socioeconomic groups; lifestyles; and industries, landscape features, and landscape forms. What elements of diversity have proved troublesome? What additional elements of diversity would improve your community?

4. Organize a local field trip for the class to examine recently disturbed areas for evidence of resilience.

Multisensory Learning: Audiovisuals

Biovideo: The Evidence for Evolution; 30 min.; CBS.
Selection in Action, Parts I, II, and III; 20 min. each, CBS.
Origins of Life; 60 min.; CBS.
Out of the Past; Parts I, II, and III; 20 min. each part; evolution of life through study of rocks and fossils; CBS.
Evolution I: Natural Selection; 31 min.; CBS.
Evolution II: Sources of Variety; 32 min.; CBS.
Evolution III: Speciation; 46 min.; CBS.
Land above the Trees; 1988; 19 min.; ecological connections and special adaptations to the alpine zone; NFB.
The Grassland; 20 min.; CBS.
The Chaparral; 21 min.; CBS.
The Desert; 20 min.; CBS.
Desert Animals and Plants; 30 min.; CBS.

Creatures of the Namib Desert; 60 min.; CBS.

Saguaro: Sentinel of the Desert; 54 min.; CBS.

The Environment Shapes the Forest; 12 min.; CBS.

The Deciduous Forest; 20 min.; CBS.

The Coniferous Forest; 22 min.; CBS.

The Moist Coniferous Forest; 18 min.; CBS.

The Rain Forest; 12 min.; CBS.

Antarctica; 11 min.; CBS.

Alaska; 12 min.; CBS.

Introduction to Marine Biomes; 15 min.; CBS.

The Marine Biome: The Physical Environment; 15 min.; CBS.

The Marine Biome: Shores and Estuaries; 15 min.; CBS.

The Marine Biome: The Open Ocean; 15 min.; CBS.

The Marine Biome: Coral Reefs; 15 min.; CBS.

The Marine Biome: Man and the Ocean; 15 min.; CBS.

The Enchanted Forest; 22 min.; CBS.

Animals of Africa: Impalas, Wildebeests, and the Gemsbok Reserve; 70 min.; CBS.

Animals of Africa: Wondrous Works of Nature; 70 min.; CBS.

Hummingbirds—Up Close; 55 min.; CBS.

Ocean Desert: The Sargasso Sea; 89 min.; CBS.

The Great Barrier Reef; 60 min.; CBS.

Within the Coral Wall; 60 min.; CBS.

Face of the Deep; 60 min.; CBS.

Creatures of the Mangrove; 60 min.; CBS.

The Tropical Kingdom of Belize; 60 min.; CBS.

Waves and Beaches; 20 min.; CBS.

Ocean Currents and Winds; 15 min.; CBS.

Estuary; 12 min.; use of underwater microphotography; BFF.

The Intertidal Zone; 17 min.; ecology of the intertidal zone and pollution effects on food chains; BFF.

The Living Ocean; 1988; 25 min.; interrelationships between the ocean and humans; NG.

The Salt Marsh; 1975; 22 min.; EBEC.

Wellsprings; 1976; 58 min.; MM.

Where the Bay Becomes the Sea; 30 min.; diversity of life in the Bay of Fundy; BFF.

Equatorial River: The Amazon; 1988; 22 min.; BFF.

See Appendix B for suppliers.

ATTITUDES/VALUES

Assessment

1. Have you ever suffered from environmental stress? If so, were you able to respond to the stress? Please describe.

2. Is there evidence of environmental stress in your community?

3. Is there evidence of ecosystem responses to stress in your community?

4. How do you feel when you see an ecosystem under stress?

5. Do you feel you have a right to create environmental stress? Are there limits to your rights? If so, what are they?

6. Do you feel you have a responsibility to protect natural ecosystems?

7. Do you feel nature's response mechanisms can continue to absorb whatever waste products are produced by human processes?

8. Do you feel that new technologies will be able to help natural ecosystems survive stresses under which they are placed?

More Depth: Discussion and Term Paper Topics

1. To what extent should we disrupt and simplify natural ecosystems for our food, clothing, shelter, and energy needs and wants?

2. Should we retreat from the beach?

3. Should our goal for human population be to reach Earth's carrying capacity?

(Also, see text, Critical Thinking, pp. 128–129)

PARTICIPATION

More Depth: Action-oriented Term Paper Topics

1. Scientific methods. Carrying capacity analysis; genetic engineering.

2. Regional. Restoration of degraded ecosystems such as Lake Erie; coastal zone management.

SKILLS

Environmental Problem-Solving Skills: Projects

1. What soil types and significantly different microclimates exist in your locale? As a class project, inventory these elements of diversity and relate them to observable differences in the distribution of vegetation, animal life, agricultural activities, and other phenomena.

2. Are inland wetlands being drained and filled in your locale? Is there a nearby stream or river being subjected to excessive levels of pollution? Are there ponds and lakes in your vicinity suffering from cultural eutrophication? Do you live where spartina marshes or estuaries are suffering from human-induced stresses? Is it feasible for you and your class to "adopt" one of these disturbed ecosystems and help restore it to health?

3. Arrange a debate on the problems and alternatives of coastal zone management. Debate the proposition that we should severely restrict engineering approaches to beach stabilization and adopt a "retreat from the beach" strategy, emphasizing the preservation of coastal ecosystems and the ecosystem services they provide.

4. As a class exercise, systematically study a modern freeway or interstate highway and trace its impact on the surrounding land in terms of succession, diversity, and stability.

Laboratory Skills

Laboratory Manual for Miller's Living in the Environment and Environmental Science. Lab 4: Ecological
    Succession.

Computer Skills

(none)

CHAPTER 6

THE HUMAN POPULATION: GROWTH, URBANIZATION, AND REGULATION

THINKING

Concept Map

See Appendix A, Map 8.

Goals

See short, bulleted list of questions on p. 131 of text.

Objectives

1. Define *birth rate, death rate, emigration rate,* and *immigration rate.* Write an equation to mathematically describe the relationship between these rates and the rate of population change. List at least five factors that affect birth rate and five factors that affect death rate.

2. Define *fertility rate.* Describe how fertility rate affects population growth.

3. Compare rates of population growth in MDCs and LDCs. Explain the differences you find.

4. Using population age structure diagrams, explain how the age structure of a country creates population growth momentum.

5. Describe the current worldwide trend in population distribution. Describe the three major shifts in population distribution in U.S. history.

6. Evaluate the costs and benefits of the automobile on U.S. society. List three alternative forms of transportation to the car, and evaluate the costs and benefits of each.

7. List seven resource and environmental problems faced by urban areas. Briefly describe the process of ecological land-use planning. Describe your conception of a sustainable urban environment.

8. Briefly describe the controversies which surround controlling population size through controlling migration and family planning.

9. List the four stages of the demographic transition. List social, biological, political, and economic issues that can be addressed to help LDCs undergo a demographic transition.

10. Compare and evaluate the population policies of India and China. List three actions that individuals who believe that exponential population growth is harmful can take to make things better.

Key Terms

birth rate (p. 131)	annual rate of natural population change (p. 131)
crude birth rate (p. 131)	
death rate (p. 131)	replacement-level fertility (p. 132)
crude death rate (p. 131)	total fertility rate (TFR) (p. 132)
zero population growth (ZPG) (p. 131)	life expectancy (p. 135)

The Human Population: Growth, Urbanization, and Regulation   29

infant mortality rate (p. 135)　　　　　　dust dome (p. 147)
migration (p. 136)　　　　　　　　　　　dust plume (p. 147)
age structure (p. 137)　　　　　　　　　land-use planning (p. 148)
population decline (p. 139)　　　　　　　zoning (p. 148)
urban area (p. 140)　　　　　　　　　　ecological land-use planning (p. 149)
degree of urbanization (p. 140)　　　　　demographic transition (p. 156)
urban growth (p. 140)　　　　　　　　　family planning (p. 157)
urban heat island (p. 147)

Multiple Choice Questions

See Instructor's manual, Section Two, p. 160.

More Depth: Conceptual Term Paper Topics

1.　Population growth. A case study of Mexico, China, India, Kenya; the geography of global population distribution; infant mortality trends and issues; illegal immigration into the United States; marriage age trends; fertility trends and the women's rights movement; factors influencing family size preferences; Earth's carrying capacity; LDCs trapped in phase two of the demographic transition; the World Bank and family planning; economics of fertility control technology in the United States; economic costs of childrearing in the United States.

2.　Environmental impacts of population. Air pollution in urban areas; land degradation from urban sprawl; deforestation and desertification in LDCs.

3.　Population control. Case studies: India, China, Japan.

4.　Urbanization, transportation, and land-use planning. Rural to urban migration patterns; the Sun Belt shift; central city lifestyles; mass transit systems: BART, METRO; case study in sustainable living: Davis, California; green spaces; building self-sufficient cities.

More Breadth: Interdisciplinary Activities and Projects

1.　Invite a public health official or nutritionist to your class to explain the factors involved in the decline in the global death rate over the past century and the decline in the infant mortality rate in the United States. Why is the latter rate higher in the United States than in many other developed nations?

2.　U.S. immigration policy had become a volatile political issue by the 1980s. Arrange a debate on this subject. Debate the proposition that the United States should enact and strictly enforce legislation that holds legal immigration to levels consistent with the achievement of ZPG within a few generations.

3.　Ask your students to share with the class poems, short stories, songs, paintings, collages, photographic displays, slide talks, or other works expressing their feelings about population issues and problems.

4.　Are there family planning clinics in your community that provide contraceptives and birth control counseling? Invite a family planning worker to visit your class and discuss the birth control aspects of family planning.

5.　As a class project, investigate the fate of agricultural land in your city's vicinity. Is anything being done to prevent prime agricultural land from being overtaken by urban sprawl?

The Population Reference Bureau rents through the mail over 50 video tapes, films, and
   slide/tape programs on population dynamics, the environment, and related topics. For a free
   list, send a self-addressed, stamped envelope to the Population Reference Bureau..
Silent Explosion; 1986; 20 min.; 1986 video; PI.
World Population; 1990; 7 min.; a depiction of human population growth from
   1 A.D. and projected to 2020; VP.
The Growth of Towns and Cities; 20 min.; FHS.
Urban Ecology; 24 min.; FHS.
Curitiba, Ecological Capital of Brazil; 20 min.; environmentally sensitive urban planning; LW.
World Population; 6 min.; graphic simulation of the history of human population growth; ZPG.
Seeds of Progress; 28 min.; Mexican rural development; WBP.
The Neighborhood of Coelhos; 28 min.; urban development program in Brazil; WBP.
The Miracle of Life; 60 min.; development from a cell to an embryo, to a fetus, then birth; CBS.
Dandora; 20 min.; urban development in Nairobi; WBP.

See Appendix B for suppliers.

ATTITUDES/VALUES

Assessment

1. Do you feel the size of the human population is an important environmental issue?

2. Do you feel consumption by the human population is an important environmental issue?

3. Do you feel that humans have the right to have as many children as they want? Are there any
    limits on this right? If so, what are they?

4. Do you feel that there should be a national population policy? What steps would
    you support?

5. Do you feel that teen pregnancy is a problem?

6. Do you feel that women's roles are important in addressing population size?

7. What are your feelings toward birth control? Population control?

8. Do you feel that the Earth will be able to sustain the projected increases in human
    population growth?

9. How do you feel when you see skyscrapers? clover-leaf highways? parks? greenspaces?

10. Are you familiar with mass transit possibilities in an urban center in your area?

11. Are you familiar with zoning procedures in an urban center in your area?

12. How do you feel toward urban sprawl? urban renewal?

13. Do you feel that some forms of urban growth are more desirable than others? Which ones?

14. Are there any limits to urban growth? What are they?

15. What efforts do you support to make cities more sustainable environments?

More Depth: Discussion and Term Paper Topics

1. Do you think the United States needs a population policy? Should the federal government stop subsidizing large families?

2. Evaluate U.S. immigration policy.

3. Do you think the United States should play a global leadership role in promoting stabilization of the world's human population?

4. Would you rather be a baby boomer or a baby buster?

(Also, see text, Critical Thinking, p. 162.)

PARTICIPATION

Lifestyle and Campus Community

See *Green Lives/Green Campuses*, Chapter 6, Growth Management: Population and Land Use, p. 39

More Depth: Action-oriented Term Paper Topics

1. Individual. Decisions individuals make about family size and urban conditions and ways individuals can influence government agencies and nongovernment institutions concerned with population.

2. Regional. Urban renewal programs; mass transit systems.

3. National. Zero Population Growth during the 1980s; ZPG analysis of the U.S. way of taxing.

4. Global. The UN International Conference on Population; UN Family Planning Association; International Planned Parenthood Federation.

SKILLS

Environmental Problem-Solving Skills: Projects

1. Survey the marriage and childbearing intentions of your female students. Find out at what age students' mothers married and the number of children each had. Tally the results and compare them with recent trends in marriage age and total fertility.

2. Survey your students to obtain age or lifespan information about their grandparents. Compare the results with the average life expectancy in the United States in the year 1900 (46 for men and 48 for women). Invite your students to discuss major implications of these findings.

3. As a class, research the environmental impact of the growing populations of the LDCs and MDCs. Find data comparing the impact of children from MDCs and LDCs. Collect data comparing population growth in MDCs and LDCs. Project the responsibilities for environmental degradation by future human populations from MDCs and LDCs . Collect data on the birth control policies of representative MDCs and LDCs. Hold a brainstorming session about strategies to control the human population. See if a consensus can be formed about appropriate strategies for limiting environmental damage of human populations.

4. As a class project, study print and broadcast advertising to determine whether small families are directly or indirectly encouraged as the ideal model for U.S. society. Compare and contrast the results with those obtained through a similar analysis of magazine advertising in the 1940s and 1950s.

5. Have your students analyze the political platforms of the major political parties in the United States. What positions do they take on the birth of American children and birth control? What positions do they take on the influence of the United States in global population growth patterns? To what extent does debate on population policy revolve around right to life, desired pregnancies, and quality of life for the children who are born?

6. Show your students some color slides or photographs of urban blight in your community. Ask them to identify the causes of these problems and to suggest a variety of approaches to solving them.

7. As a class project, analyze your local transportation network. Outline a socially and ecologically responsible plan for an improved transportation network featuring a mix of individual and mass transit methods. Investigate the extent to which bicycling has been or could be adopted as an integral part of your community's transportation system.

Laboratory Skills

Laboratory Manual for Miller's Living in the Environment and Environmental Science. Lab 5: Exponential Growth; Lab 6: Population Control.

Computer Skills

Microcomputer Programs for Demographic Analysis (MCPDA)
  - Performs a wide range of tests and analyses on demographic data.
  - Institute for Resource Development, Westinghouse, P.O. Box 866, Columbia, MD 21044

DYNPLAN
  -Calculates the effects that specific health care interventions and family-planning measures can be expected to have on the demography of a nation.
  -Stan Berstein, Department of Population and International Health, School of Public Health, University of Michigan, Ann Arbor, MI 48109

CHAPTER 7

ENVIRONMENTAL ECONOMICS AND POLITICS

THINKING

Concept Map

See Appendix A, Maps 9 and 10.

Goals

See short, bulleted list of questions on p. 164 of text.

Objectives

1. Distinguish among the concepts in the following sets of terms: *natural resources, manufactured capital, human capital; centrally planned economy, market economy, mixed economic system.*

1. Define *gross national product* (GNP). Evaluate the commonly held belief that GNP is an indicator of a country's well-being. Describe at least two alternative indicators that take social and environmental factors into account. Evaluate the accuracy of these indicators.

2. Define *externalities*. Give one example of an external cost and one example of an external benefit. Describe measures that can be taken to move toward full-cost pricing.

3. Evaluate the use of cost-benefit analysis. Define *discount rate*. Use your understanding of discount rate as you describe the pros and cons of using cost-benefit analysis.

4. Predict likely consequences for a society whose goal is zero pollution. Draw a figure illustrating how to find the optimal level of pollution. Distinguish between pollution control and pollution prevention strategies.

5. List seven strategies that can be used to modify a pure market system to improve environmental quality and reduce resource waste. List advantages and disadvantages of each tool.

6. Evaluate the potential impacts of global free trade.

7. Define *poverty*. Compare trickle-down and sustainable development approaches to poverty.

8. Describe a sustainable-Earth economy.

9. List five approaches of U.S. environmental legislation. List five principles used to set guidelines for environmental regulations. Describe at least five political strategies used by people to stay in power.

10. List three types of environmental leadership. Assess which style best fits you. List three particular applications of environmental leadership that might be of interest to you.

11. Distinguish between environmental groups and the wise-use movement. Distinguish between mainstream and grass-roots environmental groups.

12. Summarize the results of the 1992 Rio Earth Summit.

Key Terms

economy (p. 164)

economic decisions (p. 164)

economic resources (p. 164)

Earth capital (p. 164)

natural resources (p. 164)

manufactured capital (p. 164)

human capital (p. 164)

pure command economic system (p. 164)

centrally planned economy (p. 164)

pure market economic system (p. 164)

pure capitalism (p. 164)

market equilibrium (p. 164)

mixed economic systems (p. 164)

economic growth (p. 165)

gross national product (GNP) (p. 165)

real GNP (p. 166)

real GNP per capita (p. 166)

internal costs (p. 168)

externalities (p. 168)

external benefit (p. 168)

external costs (p. 168)

full-cost pricing (p. 169)

full cost (p. 169)

cost-benefit analysis (p. 170)

discount rate (p. 170)

poverty (p. 175)

politics (p. 179)

democracy (p. 179)

mainstream environmental groups (p. 184)

grass-roots environmental groups (p. 184)

anti-environmental movement (p. 185)

Multiple Choice Questions

See Instructor's manual, Section Two, p. 178.

More Depth: Conceptual Term Paper Topics

1. Traditional versus sustainable-Earth economics. Hazel Henderson's views of converting the economic pie into a layer cake (by adding social and environmental measures of well-being and worth); GNP, NEW, ISEW, and other economic indicators; pollution control as a growth industry; a U.S. energy policy.

2. Economic and political aspects of poverty. The sharing ethic and enlightened self-interest; land reform; the World Bank and development projects; debt-for-nature swaps; technology transfer.

3. Green groups. Germany's Green Party; Earth First!; Greenpeace; the Environmental Defense Fund; the Natural Resources Defense Council; Earth Day; Public Interest Research Groups (PIRGs); Sierra Club; the Nature Conservancy; the National Wildlife Association.

More Breadth: Interdisciplinary Activities and Projects

1. Will people refrain from polluting excessively if they understand that such behavior is socially and ecologically irresponsible? Discuss this question with your class and make a list of the various reasons people might have for ignoring moral persuasion and preaching. If possible, invite a social psychologist to address your class on the subject of attitude-behavior consistency and motivation.

2. Ask industrial and environmental lobbyists to visit the class and discuss their goals, methods, and problems.

Multisensory Learning: Audiovisuals

A Call to Action; 1994; 60 min.;what individuals have done to make a difference and alter human relationships with Earth; FHS.

Rocking the Boat: You Can Fight City Hall; 1994; 60 min.; how seemingly disenfranchised communities are working through community organizing, political and legal action, and public education; FHS.

It Needs Political Decisions, from *Race to Save the Planet;* 1990; 60 min; explores power and limitations of politics in protecting the environment; ACPB.

Now or Never, from *Race to Save the Planet;* 1990; 60 min; individual action; ACPB.

An Act of Congress; 58 min.; traces the progress of the Clean Air Act of 1977 through committee and the House of Representatives to enactment; CMTI.

Branches of Government: The Legislative Branch; 1982; 23 min.; shows the multiple interests that congressmen represent; NG.

Save the Earth: A How-To Video; 1990; 60 min.; individual actions matter; VP.

The Politicians; 19 min.; political pressures in Brazil's Amazonia; BFF.

Earth First: The Struggle for the Australian Rainforest; 1990; 58 min.; everyday people who stirred the conscience of a nation through their risk-taking actions; VP.

Environment Under Fire: Ecology and Politics in Central America; 1988; 28 min.; threats to Central America's environment and potential solutions; VP.

Stopping the Coming Ice Age; 1988; 45 min.; explores connections among human contributions to the greenhouse effect, global warming, and Ice Ages; proposes appropriate actions; VP.

Politics, People, and Pollution, Part A: Listening to America; PBS.

Introduction to Ecological Economics; 1991; 45 min.; GF.

Investing in Natural Capital; 1993; 45 min.; GF.

Mastering the Marketplace; 1994; 60 min.; real costs of pollution and conservation, "green" marketing, and life-cycle analysis; FHS.

The Barefoot Economist; 1989; 52 min.; the founder of Barefoot Economics synthesizes a vision of social justice, environmental health, and a sustainable future; LF.

River People: Behind the Case of David SoHappy; 1990; 50 min.; a clash of fishing rights and the right to religious freedom; FL.

Conservation of the Southern Rainforest; UF.

A Sense of Place: Tourism, Development, and the Environment; UF.

People Who Fight Pollution; 1971; 18 min.; MM.

The Power to Change; 1979; 28 min.; MM.

Energy and Morality; 33 min.; complex relationships among energy use, economics, and ethics; BFF.

Dialogue on International Development; 20 min.; design and evaluation; easing tension between LDCs and MDCs; BFF.

The Price of Progress; 54 min.; social, environmental, and economic costs of several World Bank programs; BFF.

See Appendix B for suppliers.

ATTITUDES/VALUES

Assessment

1. Do you believe that individuals and countries should have the right to consume as many resources as they can afford?

2. Do you believe that the most important nation is the one that can command and use the largest fraction of the world's resources to promote its own economic growth?

3. Do you believe that the more we produce and consume, the better off we are?

4. Do you believe that humans have a duty to subdue wild nature to provide food, shelter, and other resources for people and to provide jobs and income through increased economic growth?

5. Do you believe that resources are essentially unlimited because of our ability to develop technologies to make them available or to find substitutes?

6. Do you believe that environmental improvement will result in a net loss, a net gain, or no change in the total number of jobs in your country? In your community?

7. Do you believe that environmental improvement will result in a net loss, a net gain, or no change in the total number of jobs in your country? In your community?

8. Would you be in favor of improving the air or water quality in your community if this meant a net loss of local jobs?

9. Would you be in favor of improving the air or water quality in your community if this meant that you lost your job?

10. Would you favor requiring that the market cost of any product or service include all estimated present and future environmental costs?

11. Do you favor debt-for-nature swaps in which poor countries would be forgiven most of their debts to rich countries, in exchange for protecting specified wild areas of their country from harmful and unsustainable forms of development?

12. Have you ever had an opportunity to be a leader? What leadership style do you find most comfortable?

13. Have you met a legislator who worked on environmental policy? How did you feel about the experience?

14. Have you met a grass-roots environmental activist? How did you feel about the experience?

15. Do you feel the government can play a responsible role in establishing a sustainable relationship between humans and their environment?

16. Do you feel global environmental security is necessary for national security?

17. Would you support a 10% increase in income taxes if you knew this revenue would be used to improve environmental quality?

More Depth: Discussion and Term Paper Topics

1. What priorities should guide the design of a measure of sustainable economic welfare?

2. Are the best things in life not things? What are the basic material requirements for survival with dignity and security?

3. How useful is cost-benefit analysis?

4. Is global free trade a good thing?

5. What is the best way to address poverty?

(Also, see text, Critical Thinking, p. 191.)

PARTICIPATION

More Depth: Action-oriented Term Paper Topics

1. Individual. Exercising environmental ethics as a consumer and as a voter; strategies for being an agent of change; Mahatma Gandhi; Martin Luther King, Jr.

2. Environmental group strategies. Active interference with environmentally damaging activities: tree hugging and painting baby seals; lobbying; letters campaigning for and against particular legislation; buying land for conservation.

3. National government methods to control pollution. Pollution prevention vs. pollution control; pollution rights; taxes; payments and incentives for pollution control; environmental legislation (such as the National Environmental Policy Act of 1969); the Environmental Protection Agency.

4. Global. The UN Environment Program (UNEP).

SKILLS

Environmental Problem-Solving Skills: Projects

1. Have your class survey the economic growth that has taken place recently in your state or community. Make lists of the positive and negative consequences associated with this growth; then discuss the implications for human well-being and future life quality. Should growth in your state or community be redirected? If so, specifically how? Invite a professional planner to discuss this issue with your class.

2. As a class exercise, explore the agencies in your community or state that are responsible for recruiting new industries. What are their goals and methods? Compare these values and methods with sustainable-Earth values and methods.

3. Have your class design an indicator of sustainable economic welfare that can be applied to individual communities to monitor change.

4. Have your students make a list of the employers whose payrolls are very important to the economic health of your community. How would a transition to a sustainable-Earth economy affect the employment structure of your community?

5. As a class exercise, conduct a school or community poll to find out if people are willing to pay for pollution control. Have the entire class participate in the design of a brief opinion poll. The questions should be designed to find out what kinds of environmental qualities people want to see preserved and what they are willing to give up (in monetary or other terms) to ensure that these qualities are protected. Try to standardize the procedure and get as many respondents as practical. Analyze the results and discuss them in class.

6. As a class exercise, develop the basic elements of a federal budget for next year that includes realistic levels of spending for environmental quality management. Have students decide on a list of priorities for pollution control.

7. As a class exercise, use the *Congressional Record* (or equivalent state documents) to follow the progress of various pollution, land-use, energy, population, or other environmentally related bills. If possible, have your institution join an environmental network, such as Econet, that will allow students to access information about environmental legislation, the members of relevent congressional committees, and background material to understand different environmental issues.

8. Does the United States have a sustainable-Earth president ? Have students evaluate the current administration's performance from the point of view of sustainability. They should use specific references and examples. Have students locate resources (such as documents prepared by the League of Conservation Voters) that report the voting records of members of Congress on environmental legislation. Using those resources, have students evaluate representatives from the locations where they live. Have them evaluate how important their findings are in forming an opinion about their elected officials.

9. As a class project, identify a local environmental issue early in the semester or term and follow the actions of environmental groups addressing that issue. What strategies and tactics are used, and with what effects?

Laboratory Skills

(none)

Computer Skills

Dynamic Synthesis of Basic Macroeconomic Theory (DSBMT)
-Provides, in a single model, a synthesis of the major theoretical macroeconomic models used by economists to provide advice on the management of the economies of nations and thereby reduce disagreement among economists on economic policy.
-System Dynamics Group, Sloan School of Management, Massachusetts Institute of Technology, 50 Memorial Dr., Cambridge, MA 02139

Environmental Assessment System (EASY)
-Provides a flexible decision support system for political decisions involving multiple decision makers and complex issues, such as the environment.
-R. Janssen and W. Hafkamp, Institute for Environment and Energy, Free University, P.O. Box 7161, 1007 Amsterdam, The Netherlands

CHAPTER 8

RISK, TOXICOLOGY, AND HUMAN HEALTH

THINKING

Concept Map

See Appendix A, Map 11.

Goals

See short, bulleted list of questions on p. 193 of text.

Objectives

1. List four classes of common hazards and give two examples of each. List six cultural hazards in order of most to least hazardous.

2. List five principal types of chemical hazards and give two examples of each.

3. Define *epidemiology*. Draw a dose–response curve and explain how it can be used. Distinguish between acute and chronic effects.

4. Describe how earthquakes are caused. Describe how the severity of earthquakes is measured. Describe how prediction of earthquakes has improved.

5. Describe what causes a volcano. Describe how prediction of volcanic activity has improved.

6. Distinguish between transmissible and nontransmissible diseases. Explain which occurs most in LDCs and which occurs most in MDCs. Relate an epidemiologic transition to a demographic transition.

7. Describe how the hazards of smoking and sexually transmitted diseases could be reduced in the United States. List diet changes which can help prevent cancer.

8. Define *risk analysis*. Compare technology reliability to human reliability. Define *risk-benefit analysis* and *desirability quotient*. Give examples of technologies with high and low desirability quotients. List five problems with risk assessment.

9. List seven questions asked by risk managers. List seven cases in which the public generally perceives that a technology or product has a greater risk than the risk estimated by experts.

Key Terms

risk (p. 93)	carcinogens (p. 194)
probability (p. 193)	metastasize (p. 194)
risk assessment (p. 193)	mutagens (p. 194)
dose (p. 193)	teratogens (p. 194)
response (p. 193)	dose–response curve (p. 195)
toxic chemicals (p. 194)	lethal dose (p. 196)
hazardous chemicals (p. 194)	median lethal dose (LD_{50}) (p. 196)

epidemiology (p. 196)

earthquake (p. 196)

volcano (p. 198)

transmissible disease (p. 199)

nontransmissible diseases (p. 201)

risk analysis (p. 204)

risk-benefit analysis (p. 205)

desirability quotient (p. 205)

risk management (p. 208)

<u>Multiple Choice Questions</u>

See Instructor's manual, Section Two, p. 192.

<u>More Depth: Conceptual Term Paper Topics</u>

1.   Environmental risks. Brown lung disease and the textile industry; black lung disease and the coal mining industry; asbestos as a carcinogen.

2.   Lifestyle risks. Health effects from secondary smoke; the rising lung cancer rate for women.

3.   Transmissible disease risks. History of infectious disease control; vaccines and immunology; how smallpox was eradicated; waterborne diseases of the LDCs; the history of malaria; schistosomiasis; cholera; tuberculosis; AIDS.

<u>More Breadth: Interdisciplinary Activities and Projects</u>

1.   Have local public health officials discuss with your class the types and frequency of diseases in the local area and describe efforts for disease control.

2.   Invite a spokesperson for the American Cancer Society to address your class on the subject of "nonsmokers' rights." What specific things can a person do to minimize his or her passive exposure to cigarette smoke? What are the limits of smokers' rights to pollute air that nonsmokers cannot avoid breathing? Review the changes in attitudes and behaviors that people in the United States have shown toward smoking over the last 10 years.

3.   Ask a nutritionist to discuss problems with the typical U.S. diet and how to make changes recommended by the National Academy of Sciences and the American Heart Association.

4.   Assign several students to visit a store that specializes in organically grown and "natural" foods. Have them describe the advertising claims made on behalf of natural foods.

<u>Multisensory Learning: Audiovisuals</u>

Environmental Illness: Bad Chemistry; 60 min.; people who cannot tolerate common chemicals of modern life; FHS.

The American Cancer Society's "Freshstart": 21 Days to Stop Smoking; step by step advice; CBS.

The AIDS Epidemic: Is Anyone Safe?; 50 min.; CBS.

The Clinical Story of AIDS: An Interview with Dr. Paul Volberding; 28 min.; CBS.

AIDS: Can I Get It?; 48 min.; CBS.

AIDS Alert; 23 min.; CBS.

Causes of Cancer; 29 min.; CBS.

Cancer: The Disease; 35 min.; CBS.
Genetic Engineering; 16 min.; CBS.

See Appendix B for suppliers.

ATTITUDES/VALUES

Assessment

1. What kinds of risks do you take every day?

2. What kinds of risks from the natural environment occur in your area?

3. What kinds of risks are you exposed to each day over which you have no control?

4. Do you feel that societal risks should be distributed equally among all citizens?

5. Should people who choose unhealthy lifestyles be covered by national health insurance?

More Depth: Discussion and Term Paper Topics

1. When do smokers' rights infringe on nonsmokers' rights?

2. Should we continue government subsidies of the tobacco industry?

3. Should we build in earthquake zones?

(Also, see text, Critical Thinking, p. 210)

PARTICIPATION

More Depth: Action-oriented Term Paper Topics

1. Individual lifestyle changes that cut risks. Quitting smoking; changing to a low-risk diet; steps to prevent breast cancer; preventive medicine.

2. Groups spreading the word about risks. The American Cancer Society; the American Lung Association; the American Heart Association.

3. National. The National Centers for Disease Control; Occupational Safety and Health Administration (OSHA); U.S. efforts to export tobacco products; the tobacco lobby; the Food and Drug Administration; new ingredient labels for food.

4. International. Sweden's antismoking campaign.

SKILLS

Environmental Problem-Solving Skills: Projects

1. Have your students obtain mortality and morbidity data for people living in poor and affluent sections of your community to determine the frequency and types of illness. Compare results with national statistics and attempt to explain any significant local differences.

2. What occupational health hazards are prevalent in your community? What is being done to protect workers from these on-the-job hazards? Have some students investigate this subject and report the results to the class.

Laboratory Skills

(none)

Computer Skills

Waterborne Toxic Risk Assessment Model (WTRISK)
 -Estimates the risks of adverse human health effects from substances emitted into the air, surface water, soil, and groundwater from sources such as coal-fired power plants.
 -Manager, Software and Publications Distribution, Electric Power Research Institute, 3412 Hillview Ave., P.O. Box 10412, Palo Alto, CA 94303

PART III.
RESOURCES: AIR, WATER, SOIL, MINERAL, AND WASTES

CHAPTER 9

AIR

THINKING

<u>Concept Map</u>

See Appendix A, Map 12.

<u>Goals</u>

See short, bulleted list of questions on p. 213 of text.

<u>Objectives</u>

1. List and briefly describe the layers of the atmosphere. Compare the function of ozone in the troposphere with the function of ozone in the stratosphere.

2. List eight classes of outdoor air pollutants. Distinguish between a primary pollutant and a secondary pollutant. Distinguish between stationary and mobile sources of pollution.

4. Distinguish between photochemical smog and industrial smog. Describe a thermal inversion and conditions under which it is most likely to occur.

5. Define *acid deposition*. List seven effects of acid deposition.

6. Describe the significance of the problem of indoor air pollution. List the four most dangerous indoor air pollutants.

7. Using the diagram on p. 222, list three potential sources of indoor air pollution where you live. Describe the best ways to deal with asbestos and radon.

8. Summarize air pollution effects on human health, plants, aquatic life, and materials.

9. Analyze the graphs on p. 228. Determine which pollutant saw the greatest decline. Evaluate the effectiveness of pollution prevention vs. pollution cleanup strategies.

10. Evaluate alternatives to gasoline. Determine which alternatives look best to you.

11. List six legal steps that would reduce air pollution: three for indoor air pollution, three for outdoor air pollution.

12. List ten ways to protect the atmosphere. List three ways individuals can reduce air pollution.

Key Terms

atmosphere (p. 213)   secondary pollutants (p. 214)
troposphere (p. 213)   photochemical smog (p. 215)
stratosphere (p. 213)   industrial smog (p. 216)
mesosphere (p. 214)   temperature inversion (p. 217)
thermosphere (p. 214)   thermal inversion (p. 217)
primary pollutants (p. 214)   acid deposition (p. 219)

Multiple Choice Questions

See Instructor's manual, Section Two, p. 202.

More Depth: Conceptual Term Paper Topics

1.   Atmosphere. Air pollution meteorology as an applied science.

2.   Outdoor air pollution. The geographic distribution of air quality problems; air pollution and
     major ecosystem disruption; fine particulates as a health hazard; Donora, Pennsylvania air
     pollution disaster; reducing urban heat island effects.

3.   Indoor air pollution. Radon gas; sources, health effects, and control measures for indoor air
     pollution.

4.   Acid deposition. Tall smokestacks and acid deposition; acid deposition in the northeastern
     United States; Germany's waldsterben; liming lakes; acid deposition and freshwater
     ecosystems.

More Breadth: Interdisciplinary Activities and Projects

1.   Visit the chemistry department or invite a chemistry professor to visit your class to discuss
     measurement of air and water pollution. Have the professor show you instruments for
     measuring air and water pollutants in the parts per million or lower range and explain the
     difficulty of making accurate and reproducible measurements of such low concentrations.

2.   As a class exercise, interview farmers, foresters, and wildlife experts in your area to
     determine whether they know of any plant, fish, or animal damage from air pollution. Is acid
     deposition a problem in your locale? If so, what is the extent of the damage? Is anything
     being done about it?

3.   Have a doctor or health official visit your class to explain and illustrate the various types of
     lung disease and damage that can result from air pollution. If possible, have the expert show
     you specimens (or photographs) of lung tissue from a young child, an urban dweller, a rural
     dweller, a smoker, a nonsmoker, and from patients suffering from lung cancer, emphysema,
     and chronic bronchitis.

4.   Have an epidemiologist visit the class to present available evidence on relationships between
     air pollution and human health.

5.   Have someone knowledgable about automobile engine design visit your class and discuss
     some of the problems associated with designing engines that pollute less and deliver better
     fuel economy.

6. Have a meteorologist visit your class to discuss the weather and climate patterns of your locale. Find out if there are any atmospheric patterns in your area that aid or hinder air pollution effects.

Multisensory Learning: Audiovisuals

Do You Really Want to Live This Way? from *Race to Save the Planet;* 1990; 60 min.; air and water pollution caused by western lifestyles; ACPB.

Our Planet Earth; 23 min.; from space, shows how pollution travels through the atmosphere; BFF.

Breath Taken; 1990; 36 min.; comments from those who study and those who are victims of asbestos-related diseases; FP.

Eco: A Poisoned Future; 1987; 24 min.; lead levels in England; CTE.

Charlie Brown, Clear the Air; 7 min.; cartoon by the American Lung Association examining air pollution and ways to clean it up; FDC.

On the Road to Clean Air; 15 min.; produced by the American Lung Association, shows testing of tailpipe emissions ; FDC.

Pollution of the Upper and Lower Atmosphere; 17 min.; effects of technologies on Earth's atmosphere; CMTI.

Acid Rain; 20 min.; geological and meteorological interactions; sources and potential solutions; FHS.

Air Pollution: Outdoor; 16 min.; burning fuel and the status of research into new and cleaner fuels and combustion methods; FHS.

Lead Poisoning; 26 min.; health hazards of lead exposure and what is being done to prevent and treat it; FHS.

Radon; 26 min.; possible health effects of radon pollution, how to detect and minimize exposure; FHS.

Air Pollution: Indoor; 13 min.; the problem; FHS.

Smokers Are Hazardous to Your Health; 50 min.; FHS.

Noise Pollution; 26 min.; FHS.

See Appendix B for suppliers.

ATTITUDES/VALUES

Assessment

1. What is blowing in the wind in your community? How does the wind make you feel?

2. Do you feel that the development of your community is related to the climate of the area? Are there any connections you have experienced?

3. Are you aware of mountains or bodies of water in your area that affect local climate conditions?

4. Do humans have a right to use the atmosphere in any way they wish? Do you see limits to freedom of choice? If so, what determines those limits?

5. Do you feel that humans have the power to alter Earth's climate? Do you feel that humans can responsibly control their impact on the atmosphere?

6. Have you ever breathed pristine air? How did it feel?

7. Have you ever breathed highly polluted air? How did it feel?

8. Do humans have a right to breathe clean air?

9. Do you take steps to improve the indoor air quality of your living space?

10. Are you willing to drive your car less to create less air pollution?

11. Do you support strong emissions standards on automobiles and power plants?

More Depth: Discussion and Term Paper Topics

1. How much air pollution should we tolerate?

2. What are you willing to do to prevent acid rain from defacing statues and historical monuments?

3. How much responsibility for clean air should areas upwind carry for areas downwind?

4. How much do Americans value clean air in their homes?

(Also, see text, Critical Thinking, p. 233)

PARTICIPATION

Lifestyle and Campus Community

See *Green Lives/Green Campuses*, Chapter 3, Outdoor Air Quality, p. 16; Chapter 4, Indoor Air Quality, p. 24.

More Depth: Action-oriented Term Paper Topics

1. Methods of assessing and cleaning up air pollution. Scientific methods for measuring indoor and outdoor air pollutants; methods for measuring automobile emissions; stack scrubber technology; catalytic converters and their problems; electric automobiles.

2. Cities. The Pittsburgh air pollution cleanup; air pollution cleanup in Los Angeles.

3. National. The Clean Air Act amendments; the EPA's record on enforcement of the Clean Air Act.

4. International. The London air pollution cleanup; air pollution in eastern Europe; joint responsibility of Canada and the United States for acid deposition in the Great Lakes region.

SKILLS

Environmental Problem-Solving Skills: Projects

1. Have students survey corrosion and damage to buildings and statues that result from air pollution in your area. Try to estimate the total cost per year for replacement, repair, cleaning, and painting. Who pays for this?

2. Have your class make a communitywide survey of particulate fallout and plot the results on a map of your area. Obtain some small, open-top boxes—all the same size. Use masking tape to stick a clean piece of white typing paper in the bottom of each box. Place the boxes at various locations for a period of 24 hours and compare the relative darkness on the paper. Does the particulate fallout vary with height? (Try some rooftops.) Compare your map with any official air pollution monitor locations or test spots. Do the official monitors give a realistic picture? See if you can trace the major causes in heavy fallout areas. You can vary the experiment by using a strip of exposed masking tape to collect the solids and then observe or count them under a microscope.

3. Have your class use a local map to determine the distance from your classroom or school to mountains, hills, tall buildings, towers, and other recognizable landmarks at varying distances. Each day for several weeks estimate visibility by noting which of these landmarks you can see. Note whether the weather is rainy, cloudy, or hazy in order to determine if poor visibility is due primarily to weather conditions or to air pollution. Try to compare your results with official records (if available).

4. As a class project, test the vital capacity of the lungs of each member of the class. Your vital capacity is the total volume of air you can exhale in one breath. Low vital capacity indicates that only a small fraction of the volume of a person's lungs is being used for breathing and obtaining oxygen. It can also lead to an enlarged heart because the heart must work harder to pump blood through the lungs. Test lung capacity as a function of age and sex, and on track team members, smokers and nonsmokers, urban and rural dwellers, and people with asthma, bronchitis, and emphysema. A portable vital-capacity tester can easily be made by inserting a two-hole stopper in a gallon jug. Put a short glass tube with a 6-inch piece of rubber tubing attached through one hole and a long glass tube that almost reaches the bottom of the jug through the other hole. Attach a 15- to 24-inch rubber tube to this glass tube. Fill the jug half to two-thirds full of water. Fix the stopper tightly, take a deep breath, blow into the short rubber tube to force the water out of the jug and into another jug or container, and measure the water in the second jug. Run several trials and get an average for each person tested.

5. Ask your students to conduct a class or school survey to see what percentage of the respondents (a) can identify the major pollutants from automobiles, (b) know what air pollution devices are on their cars, (c) have their engines tuned on a regular basis, and (d) would be willing to pay extra for more effective air pollution control devices (including the maximum they would pay).

Laboratory Skills

Laboratory Manual for Miller's Living in the Environment and Environmental Science. Lab 12: The Human Respiratory System; Lab 13: Air Pollution; Lab 14: Thermal Inversion; Lab 17: Noise Pollution.

Computer Skills

Acid Deposition (ADEPT) Model
    -Analyzes alternative strategies for dealing with the problem of acid deposition.
    -Manager, Software and Publications Distribution, Electric Power Research Institute, 3412 Hillview Ave., P.O. Box 10412, Palo Alto, CA 94303

User's Network for Applied Modeling of Air Pollution (UNAMAP)
-Provides tools to analyze the implications for air quality of a wide variety of possible development projects and programs and for assessing alternative air-pollution control regulations
-Computer Products, National Technical Information Service, 5285 Port Royal Rd., Springfield, VA 22161

CHAPTER 10

CLIMATE, GLOBAL WARMING, AND OZONE LOSS

THINKING

<u>Concept Map</u>

See Appendix A, Maps 13 and 14.

<u>Goals</u>

See short, bulleted list of questions on p. 235 of text.

<u>Objectives</u>

1. Distinguish between weather and climate. List and briefly describe seven factors that determine patterns of global air circulation.

2. Describe patterns of timing of past glacial and interglacial periods. Indicate which type of period we are now enjoying.

3. List the four greenhouse gases that humans contribute to the atmosphere. Using Figure 10-9 on p. 240, construct a table which succinctly compares these gases for % contribution to human-caused greenhouse gases, main sources, length of time gas remains in the atmosphere, and relative strength of the gas per molecule compared to carbon dioxide.

4. Describe the change in global temperature since 1860.

5. Describe how future changes in climate are projected. List six projections made from major climate models. List eight uncertainties in the climate modeling process.

6. Briefly describe how global warming might affect agriculture, water supplies, forests, biodiversity, sea level, weather extremes, and human health.

7. Evaluate the need to respond to potential global warming. List three changes in energy use that could help prevent global warming. List four "technofixes" for global warming and consider the pros and cons of each. List eight ways to prepare for global warming.

8. List six measures individuals can take to reduce global warming.

9. Briefly describe what is happening to ozone in the stratosphere. Describe what is causing this change, what the likely effects of this change will be, and how this change can be prevented.

10. Using Table 10-1 on p. 254, evaluate the CFC substitutes. Name and defend your top three choices. List two technofixes for ozone loss.

11. List three ways individuals can help protect the ozone layer.

<u>Key Terms</u>

weather (p. 235)	greenhouse gases (p. 237)
climate (p. 235)	greenhouse effect (p. 237)

Multiple Choice Questions

See Instructor's manual, Section Two, p. 214.

More Depth: Conceptual Term Paper Topics

1. Climate patterns. El Niño–Southern Oscillation; weather control and modification; paleogeography; evidence from bubbles in Arctic ice; the ice ages; the Gulf Stream; upwellings; thermal inversions.

2. Greenhouse effect. Greenhouse gases; global warming and our coastlines; global warming and the incidence of severe storms—more Hurricane Andrews to come?; deforestation and global warming; climate and biodiversity; rates of global climate change and adaptation; global warming and agriculture.

3. Ozone loss. Health effects of increased ultraviolet radiation; distinguishing tropospheric and stratospheric ozone; CFCs: uses and control of production.

More Breadth: Interdisciplinary Activities and Projects

1. Invite an atmospheric scientist (preferably a climatologist) to address your class on the subject of regional and global climatic change and why it is so difficult to forecast climate several years or decades into the future. What progress has been made in the development and testing of very large computerized models of global climate?

2. Visit a first-class weather station.

3. Have local emergency officials explain to your class the precautions that have to be taken in adverse weather conditions.

4. Have experts in various fields present their views on the impact that global warming will have on their arenas: for example, how will it affect farming, forestry, coastal zone management, and town planning?

Multisensory Learning: Audiovisuals

Only One Atmosphere, from *Race to Save the Planet*; 1990; 60 min.; the environmental challenge of potential global warming; ACPB.
Greenhouse Crisis—The American Response; the Union of Concerned Scientists looks at the interactions of energy consumption, greenhouse effect, and global warming; VP.
Jessica Tuchman Mathews; 30 min.; How serious is the greenhouse effect on global climate and environmental issues? PBS.
The Heat Is On: The Effects of Global Warming; 26 min.; FHS.
The Greenhouse Effect; 26 min.; FHS.
Global Warming; 26 min.; FHS.
Danger Ahead: Is There No Way Out; 26 min.; some solutions to the threat of global warming; FHS.
Alterations in the Atmosphere; 18 min.; man-made pollutants implicated in climate changes; FHS.
Drought and Flood: Two Faces of One Coin; 18 min.; potential effects of global warming; FHS.
Prophets and Loss; 1990; 60 min.; environmental thinkers discuss climate change; VP.
The Hole in the Sky; 1987; 58 min.; NOVA's look at the causes and implications of the Antarctic ozone hole; CMTI.
Once and Future Planet; 23 min.; industry, lifestyle, and the Earth's atmosphere; BFF.

Hole in the Sky; 1994; 52 min.; evidence of continuing ozone depletion in the stratosphere; FHS.
Assault on the Ozone Layer; 18 min.; global connections in destruction of the ozone layer; FHS.

See Appendix B for suppliers.

ATTITUDES/VALUES

Assessment

1. How do you feel when you consider the possibility of global climate change?

2. Do you feel that the environment will be able to absorb any of the changes that human cultures bring about?

3. Do you feel that new technologies will be able to modify any harmful effects that humans bring about in the environment?

4. Do you feel that humans have a responsibility to ameliorate the causes of global climate change? If so, what steps are you willing to take to slow the rate of global climate change?

5. Have you ever experienced a severe sunburn?

6. How would you feel carrying a protective umbrella for life?

7. Do you feel that humans have a responsibility to ameliorate the causes of ozone depletion? If so, what steps are you willing to take to slow the rate of global climate change?

More Depth: Discussion and Term Paper Topics

1. Do humans have a commitment to future generations? If so, how much?

2. How much are humans willing to spend for energy efficiency in the short term to receive long-term economic payoffs and help slow potential global warming?

3. Would humans rather prevent global warming or adapt to global warming as it happens?

(Also, see text, Critical Thinking, p. 257.)

PARTICIPATION

More Depth: Action-oriented Term Paper Topics

1. Individual. Improving energy efficiency; recycling CFCs from appliances.

2. Global: Montreal Protocol; 1992 Earth Summit in Rio; International Atomic Energy Agency.

SKILLS

Environmental Problem-Solving Skills: Projects

1. Develop students' awareness of the effect of temperature by working out the following:
    •Heating degree days or cooling degree days (variation of daily mean in your locale from 65 degrees F.)

•Growing degree days for several crops (taking the base temperature for a given crop and comparing it to the daily mean in your locale)
•Temperature and humidity index using the formula

	THI	=	T – 0.53 (1 – RH) (T – 14)
where	THI	=	temperature and humidity index
	T	=	temperature
	RH	=	relative humidity

•Weather stress index (mean apparent temperatures averaged over 40 years compared to mean apparent temperature for a particular day)

2. As a class project, carefully prepare a questionnaire to investigate what people know about global warming (causes and potential effects), attitudes toward potential global warming, and actions (if any) they are willing to take to ameliorate rapid global climate change. Administer the questionnaire to a variety of citizens, deciding upon a sampling strategy in advance. Summarize your results in appropriate tables, graphs, charts, and written descriptions. What conclusions can be drawn from your results? Who might be interested in receiving a copy of your work?

Laboratory Skills

(none)

Computer Skills

Long-Term Global Energy–Carbon Dioxide Model
    -Makes long-term global projections concerning energy utilization and carbon dioxide emissions from the energy sector.
    -Thomas A. Boden, Oak Ridge National Laboratory, P.O. Box X, Carbon Dioxide Information Analysis Center, Building 2001, Oak Ridge, TN 37831

Atmospheric Greenhouse Model (AGM)
    -Analyzes the consequences for the global climate of various scenarios regarding the production of carbon dioxide from fossil fuel combustion.
    -Prof. L. D. D. Harvey, Department of Geography, University of Toronto, 100 St. George Street, Toronto, Ontario M5S 1A1, Canada

Nuclear Crash: The U.S. Economy After Small Nuclear Attacks
    -Studies nuclear bottleneck attacks on the United States and especially the time required for the U.S. economy to recover from such attacks.
    -Prof. Kosta Tsipis, Program in Science and Technology for International Security, Massachusetts Institute of Technology, 20A-011, Cambridge, MA 02139

CHAPTER 11

WATER

THINKING

<u>Concept Map</u>

See Appendix A, Maps 15 and 16.

<u>Goals</u>

See short, bulleted list of questions on p. 259 of text.

<u>Objectives</u>

1. List six physical properties that make water unique.

2. Briefly describe Earth's water supply. Compare amounts of salt water and fresh water. Compare amounts of frozen fresh water and water available for human use. Define *watershed* and *groundwater*. Distinguish between *water consumption* and *water withdrawal*.

3. Name and briefly describe two major water supply problems. Tell which of these problems is of most concern in the region where you live.

4. List four methods to increase water supply. Compare advantages and disadvantages of using large dams. State four ways to prevent unnecessary water waste.

5. Describe the history of the Everglades and the current steps being taken to correct its water supply problems.

6. List eight common types of water pollutants and give two examples of each. Distinguish between point and nonpoint sources of pollution.

7. Briefly explain the differences among streams, lakes, groundwater, and oceans which distinguish their vulnerability to pollution.

8. Draw an oxygen sag curve to illustrate what happens to dissolved oxygen levels in streams below points where degradable oxygen-demanding wastes are added.

9. Define *cultural eutrophication*. List three ways to reduce cultural eutrophication.

10. List two water pollution problems particular to oceans. Tell one way each might be prevented and one way each might be treated. List three ways to protect coastal waters.

11. Distinguish between nonpoint-source and point-source pollution. Describe at least three strategies to reduce nonpoint-source pollution. Briefly describe the Clean Water Act. State four ways it could be strengthened. Briefly describe and distinguish among primary, secondary, and advanced sewage treatment. List three ways to protect groundwater. List three things individuals can do to maintain water supply and quality.

Key Terms

surface water (p. 260)

watersheds (p. 260)

drainage basins (p. 260)

surface runoff (p. 260)

zone of saturation (p. 260)

groundwater (p. 260)

water table (p. 260)

aquifers (p. 261)

natural recharge (p. 261)

recharge area (p. 261)

water withdrawal (p. 261)

water consumption (p. 261)

droughts (p. 262)

floodplain (p. 263)

desalination (p. 272)

riparian rights (p. 275)

prior appropriation (p. 275)

biological oxygen demand (BOD) (p. 279)

point sources (p. 280)

nonpoint sources (p. 280)

cultural eutrophication (p. 284)

dredge spoils (p. 290)

sewage sludge (p. 290)

septic tank (p. 294)

primary sewage treatment (p. 295)

secondary sewage treatment (p. 295)

advanced sewage treatment (p. 296)

Multiple Choice Questions

See Instructor's manual, Section Two, p. 223.

More Depth: Conceptual Term Paper Topics

1. Water pollution. Animal feedlot wastes; electric power plants and thermal pollution; waterborne disease-causing agents and their control; waterborne disease problems in LDCs; pesticides; deep-well disposal and groundwater contamination; sanitary landfills and groundwater contamination; hazardous storage and disposal problems; *Exxon Valdez* oil spill; leaking underground gasoline tanks; salinity problems in irrigated areas.

2. Case studies in water pollution. The fight to save Lake Erie; Lake Baikal; James River kepone spill; Chesapeake Bay; ocean dumping in the New York Bight.

3. Water supply. Drought history of the African Sahel; trickle irrigation; urban construction and aquifer recharge problems; water diversion projects in China and India; groundwater hydrology: research needs; the Ogallala Aquifer; groundwater use and land subsidence in central Florida.

More Breadth: Interdisciplinary Activities and Projects

1. Ask your class to determine the local agricultural and industrial uses of water. Is irrigation used widely? What is the source of irrigation water? What water conservation practices are used by local government, industry, and agriculture?

2. Ask your students to bring to class and share paintings, photographs, poems, songs, or other expressions of intense human feelings about water as a life-sustaining and precious substance.

3. Invite a local, state, or federal water pollution control official to discuss water pollution control methods, progress, and problems with your class.

4. Visit a sewage treatment plant with your class. Find out what level of sewage treatment is used in your community. What is the volume of effluent discharged? If effluent is discharged into a river or stream, is the water subsequently used for drinking water supply? Are there bodies of water in your locale unfit for fishing or swimming because of inadequately treated sewage effluent? If so, is anything being done to correct the problem?

Multisensory Learning: Audiovisuals

How People Are Helping Coasts, item #58838, 1990; 60 min.; NWF.
Affluent Effluent; 31 min.; CBS.
Water: More Precious than Oil; 58 min.; PBS.
The Alternative Is Conservation; 28 min.; CBS.
Downwind/Downstream: Threats to the Mountains and Waters of the American West; 58 min.; water quality and mining operations; BFF.
Natural Waste Water Treatment; 1987; 29 min.; use of marsh plants in decentralized sewage treatment in Germany, Switzerland, and the Netherlands; BFF.
Not a Drop to Drink; 1987; 16 min.; five-part series produced by CNN Science and Technology unit; CNN.
Coastal Dunes; 20 min.; the precarious vegetation-soil balance; CBS.
The Construction of Hoover Dam; 35 min.; CBS.
No Dam Good; BFF.
The Navesink—Restoration of a River; 1987; 27 min.; a historic river meets nonpoint source pollution; NJN.
The Friends of the Clinch and Powell Rivers; 1987; 8 min.; natural and cultural heritage of these naturally diverse rivers; ST.
From Sea to Shining Sea; 1986; 20 min.; grass roots against ocean chemical discharge; BFF.
Pointless Pollution: America's Water Crisis; 28 min.; nonpoint source pollution; BFF.
Mad River: Hard Times in Humboldt County; 1982; 54 min.; collision of forest decline and economic pressures; FLP.
The Wasting of a Wetland; 23 min.; industry, agriculture, population growth, and the Everglades; BFF.
Run, River, Run; 28 min.; channel creation and rebuilding the wilderness of the Kissimee River; SFWMD.
Estuary; 12 min.; closeup of wetlands and waterways; BFF.
Are You Swimming in a Sewer?; 58 min.; NOVA film on wastewater management; CMTI.
Drowning Bay; 1970; 10 min.; MM.
Requiem or Recovery?; 29 min.; transnational threat of acid rain; NFBC.
Water: A Precious Resource; 1980; 23 min.; water cycle; water use and abuse; NG.
Land Use and Misuse; 20 min.; sources, hazards, and prevention of water pollution; CMTI.
Testing the Waters; 57 min.; BFF.
The Effects of Water Pollution; 19 min.; effects of water pollutants on the food chain with focus on the seals; FHS.
Through Fish Eyes: A Close-up Look at Water Pollution; FHS.
Seas under Siege; 56 min.; toxic contamination found off coastlines; FHS.
The Ocean Planet: The Death of the Mississippi; 23 min.; FHS.
Baikal: Blue Eye of Siberia; the damage to and political efforts to save Lake Baikal; FHS.
The Ocean Sink; 29 min.; consequences of using the sea as a dumping ground with case study on Minimata.

See Appendix B for suppliers.

ATTITUDES/VALUES

<u>Assessment</u>

1. Do you consume too much water? Does your community consume too much water?

2. Do you feel your community is doing enough to provide water? Do you feel other strategies should be tried?

3. Are you confident that your community has an adequate water supply for the needs of the community for the next decade? Do you feel other strategies should be tried?

4. What are your feelings toward increasing water supply through building dams?

5. What are your feelings toward increasing water supply through more withdrawal of groundwater?

6. Do you favor metering water use and charging water consumers the full cost of providing fresh water?

7. Do you favor increasing the price of irrigation water to reflect its true cost and encourage conservation among farmers?

8. Do you favor local ordinances which conserve water (such as the low-use toilet requirement for new housing)?

9. Would you support sharp increases in monthly water bills for all homes, buildings and industries to discourage water waste?

<u>More Depth: Discussion and Term Paper Topics</u>

1. How responsible are upstream communities for ensuring that high-quality water is delivered to downstream communities?

2. What role has water supply played in wars between countries in the last decade?

3. What are human attitudes toward cleanliness of water resources?

4. Is the public ready for water recycling?

(Also, see text, Critical Thinking, p. 300.)

PARTICIPATION

<u>Lifestyle and Campus Community</u>

See *Green Lives/Green Campuses*, Chapter 1, Water Supply,
p. 1; Chapter 2, Water Quality, p. 8.

<u>More Depth: Action-oriented Term Paper Topics</u>

1. Scientific methods of analyzing water. Groundwater pollution testing and monitoring; water quality testing, such as with Hach kits.

2. Individuals acting to conserve water and prevent water pollution. New uses for residential "gray" water; water-saving showers, faucets, and toilets.

3. State. The California Water Plan.

4. National. The Safe Drinking Water Act of 1974; protection of groundwater; the Clean Water Act; problems in enforcing water quality standards; the role of the Environmental Protection Agency in water quality management; the Toxic Substances Control Act; the National Eutrophication Survey; the Coastal Barrier Resources Act; water-rights battles in the West.

5. Global. UN Conference on the Law of the Sea.

SKILLS

Environmental Problem-Solving Skills: Projects

1. Have your students explore community water resources. Where does your town get its water? What is the average daily use in summer? In winter? What are the major uses in your area? (List the 10 biggest users.) How much does your class use?

2. Have students pick out finished products in your classroom or home and, beginning with the raw materials, trace the use of water in giving you the final product.

3. If a dam has been constructed or is being built in your area, visit the site with your class. Ask students to find pictures of the area and its water control problems before the dam was built, evaluate whether the dam should have been built, and substantiate their claims. Were there alternatives to dam construction? What is the expected lifetime of the dam?

4. Have your class identify the present or potential sources of contaminants in your community's drinking water supply. If surface water from a river or lake is used, how and to what degree is it being polluted before it is withdrawn for your use? If groundwater is used, is the aquifer subject to contamination by leaking sanitary landfills, improperly functioning septic tanks, uncontained hazardous wastes, or other sources of pollution? (Ask a public health and/or environmental official to discuss these problems with your class.)

5. Ask your students to explore the principal sources of industrial water pollution in your community. What specific types of chemicals are removed in the treatment of these industrial wastes? How is this accomplished?

6. With your class, take a field trip to the nearest lake, river, or stream, preferably with a biologist. Note its smell, appearance, taste (if safe), flow, and ecological characteristics. How have these changed over the past 20 years? What plants and animals do you find living in or near the water? What are their functions? If possible, measure oxygen content, temperature, and pH. Use the library to determine what types of fish and other plant and aquatic life might exist under these conditions. What shifts would happen if acidity, dissolved oxygen, or temperature were individually or collectively increased? decreased? What might happen if acidity decreased and temperature increased? Try to visit other sites upstream and downstream from your town to compare water quality. Which sites did you prefer? Why?

7. With your class, visit the nearest reservoir, pond, or lake and try to find evidence of natural eutrophication and human-induced eutrophication. How deep is the body of water? How does depth and water quality vary throughout the year? How old is the body of water? What factors appear to limit growth of organisms in the body of water? What might the normal life span of the body of water be? Its actual life span? If possible, get a chemist or biologist to help you gather physical data (pH, salinity, turbidity, algae counts, species diversity, depth of bottom sediments) to establish the stage of succession. Try to find people who have lived near the body of water for a number of years and ask them to describe changes they have observed.

Laboratory Skills

Laboratory Manual for Miller's Living in the Environment and Environmental Science. Lab 7: The Bacteria: Representatives of Kingdom Monera; Lab 8: Water Quality Testing I: The Coliform Test; Lab 9: Water Quality Testing II: Dissolved Oxygen and Biochemical Oxygen Demand; Lab 10: Water Quality Testing III: pH; Lab 11: Wastewater Treatment.

Computer Skills

Water Supply Simulation Model (WSSM)
    -Evaluates the physical and economic characteristics of a water supply system.
    -U.S. Environmental Protection Agency, Office of Research and Development, Water Engineering Research Laboratory, ATTN: Dr. James A. Goodrich, Environmental Scientist, Systems and Cost Evaluation Staff, Drinking Water Research Division, Cincinnati, OH 45268

CHAPTER 12

MINERALS AND SOIL

THINKING

Concept Map

See Appendix A, Maps 17 and 18.

Goals

See short, bulleted list of questions on p. 302 of text.

Objectives

1. Briefly describe the layers of the Earth's interior. Describe the internal and external Earth processes responsible for forming Earth's landscape. Be sure to distinguish three different tectonic plate boundaries and the geologic features often found at each. Explain how this knowledge is significant for the formation and discovery of mineral deposits.

2. List and define three broad classes of rock. Briefly describe the rock cycle and indicate interrelationships among these classes.

3. Briefly describe methods of surface and subsurface mining.

4. List environmental impacts of extracting, processing, and using mineral resources.

5. Distinguish between mineral resources and mineral reserves. Draw a hypothetical depletion curve. Project how this curve would be affected by the following changes in assumptions: (a) recycling of the resource is increased, (b) discoveries of new deposits of the resource are made, (c) prices rise sharply, (d) a substitute for the resource is found.

6. Distinguish between critical minerals and strategic minerals. Based on Figure 19-8 (on p. 521 in the text), analyze which foreign sources are most critical to U.S. needs.

7. Assess the possibility of increasing mineral resource supplies through finding new deposits, improving technology of mining low-grade ore, getting minerals from the ocean, and finding substitutes.

8. Define soil horizon. Briefly describe six soil layers. Using Figure 12-4 on p. 304 in the text, compare soil profiles of five important soil types.

9. Describe a *fertile soil.* In doing so, be sure to refer to soil texture, porosity, loam, and acidity.

10. Describe the problem of soil erosion. Be sure to state two major causes of soil erosion, describe both world and U.S. situations, and explain why most people are unaware of this problem.

11. List and briefly describe five soil conservation strategies. Describe how land-use planning can help prevent soil erosion.

12. Describe the problem of salinization and waterlogging of soils and how they can be controlled.

Key Terms

core (p. 302)

mantle (p. 302)

crust (p. 302)

plates (p. 302)

lithosphere (p. 304)

plate tectonics (p. 304)

divergent plate boundary (p. 304)

convergent plate boundary (p. 304)

subduction zone (p. 304)

transform fault (p. 305)

erosion (p. 305)

weathering (p. 305)

mineral (p. 306)

rock (p. 306)

ore (p. 306)

igneous rock (p. 306)

sedimentary rock (p. 306)

metamorphic rock (p. 306)

rock cycle (p. 307)

subsurface mining (p. 308)

surface mining (p. 308)

overburden (p. 308)

spoils (p. 308)

open-pit mining (p. 308)

dredging (p. 308)

strip mining (p. 308)

tailings (p. 311)

smelting (p. 311)

identified resources (p. 312)

reserves (p. 312)

undiscovered resources (p. 312)

other resources (p. 312)

depletion time (p. 312)

critical minerals (p. 312)

strategic minerals (p. 312)

soil (p. 318)

soil horizons (p. 319)

soil profile (p. 319)

humus (p. 319)

infiltration (p. 321)

leaching (p. 321)

humification (p. 321)

mineralization (p. 321)

soil texture (p. 321)

loams (p. 321)

soil porosity (p. 321)

soil permeability (p. 321)

soil structure (p. 321)

soil acidity (pH)(p. 321)

soil erosion (p. 322)

desertification (p. 325)

soil conservation (p. 328)

conventional-tillage farming (p. 328)

conservation-tillage farming (p. 328)

terracing (p. 328)

contour farming (p. 329)

strip cropping (p. 329)

alley cropping (p. 329)

agroforestry (p. 329)

gully reclamation (p. 329)

windbreaks (p. 330)

shelterbelts (p. 330)

organic fertilizer (p. 330)

commercial inorganic fertilizer (p. 330)

animal manure (p. 330)

green manure (p. 331)

compost (p. 331)

crop rotation (p. 331)

salinization (p. 331)

waterlogging (p. 332)

Multiple Choice Questions

See Instructor's manual, Section Two, p. 243.

More Depth: Conceptual Term Paper Topics

1. Geologic processes. Plate tectonics; placer formation.

2. Resource demands. The commodities market; aluminum industry in the United States; steel alloys; industrial use of chromium and other strategic minerals.

3. Mining processes. Manganese-rich nodule mining methods; Frasch process sulfur mining; improved technology for mining low-grade ore.

4. Mining and the environment. Modern surface mining reclamation methods; surface mining and the acid runoff problem; smelting and air pollution problems.

5. Increasing resource supply. History of resource substitution; LANDSAT technology and mineral resources; extracting minerals from seawater.

6. Soil. The web of life in the soil; soil formation and pioneer ecological succession; soils of your locale.

7. Human impact on the soil. Overgrazing and desertification; acid deposition as a threat to soil quality; sediment: how serious a water pollutant?

More Breadth: Interdisciplinary Activities and Projects

1. Invite an economic geologist to your class to discuss state-of-the-art technology for exploration and development of metal and nonfuel minerals.

2. Are any surface mining operations taking place in your vicinity? If so, arrange for a spokesperson to explain how the mining company complies with the specifications of the Surface Mining Control and Reclamation Act.

3. Invite a Soil Conservation Service representative to your class to discuss local soil conservation problems and erosion-control methods.

4. Take a class field trip to several farms or ranches in your locale that offer you the opportunity to contrast excellent soil management practices with poor ones.

5. As a class exercise, discuss the economic, political, social, and environmental consequences that might ensue if the fertile soils of the Great Plains and the Corn Belt were ruined by human-accelerated soil erosion.

6. With your class, visit several construction sites in your locale. Look for evidence of human-accelerated soil erosion and methods or practices employed to minimize it.

7. Have your students find poems, songs, or paintings that express intense human feelings about the land and soil of working farms or ranches. Discuss these feelings in the context of modern large-scale commercial agriculture or agribusiness.

8. Ask an experienced practitioner of organic farming or gardening to visit the class and describe methods used to preserve the soil and maintain its fertility without using inorganic fertilizers and chemical poisons.

Multisensory Learning: Audiovisuals

Geological Processes

Born of Fire; 60 min.; movement of Earth's plates; CBS.
The Eruption of Mt. St. Helens; 30 min.; CBS.
Plate Tectonics; 14 min.; CBS.
Plate Tectonics: A Revolution in the Earth Sciences; 28 min.; PBS.
Volcanoes; 14 min.; CBS.
Valley Glaciers; 20 min.; CBS.
Glacial Deposits; 20 min.; CBS.
Earth: The Geology of an Ever-Changing Planet; CBS.
The Shape of the Land; 15 min.; CBS.
Erosion: Carving the Landscape; 15 min.; CBS.
Our Dynamic Earth; 1979; 23 min.; plate tectonics, earthquakes, volcanoes; NG.

Minerals

The Age of Metals: Can It Last?; 60 min.; PBS.
The Minerals Challenge; 27 min.; technological advances meeting nonrenewable mineral resource
    needs; USBM.
Mining: Who Needs It?; 15 min.; mining's contribution to our way of life; 15 min.; AMC.
Broken Arrow; 1986; 69 min.; 12,000 Navajos relocated from their sacred lands to facilitate uranium
    development at Big Mountain, Arizona; DC.
Resource in Crisis; 1975; 60 min.; mineral industry, taxation, profits, environmental impacts, and
    occupational health issues; NI.

Soils

Soils: Profiles and Processes; 20 min.; CBS.
On American Soil; 1983; 28 min.; soil erosion; economics and politics of farming; BFF.

See Appendix B for suppliers.

ATTITUDES/VALUES

Assessment

1. What types of geological formations are found in your area?

2. Where is the nearest tectonic plate junction to your area? What type of junction? What are the
    likely geological processes to occur at that junction?

3. How do you feel when you see the results of a volcano, flood, or earthquake on the news?

4. Do you feel that human behavior should be modified by knowledge of geology? If so, how?

5. Would you support policies which discourage human development in areas where natural
    hazards are most likely to occur? If so, please specify.

6. Do you think new technologies will enable us to reduce the effects of natural disasters
    on humans?

7. How do you feel as a member of the human species when you consider the full span of geological time?

8. What soil type is most common in your area?

9. What are the most common soil problems in your area?

10. What feelings do you have toward the soil?

11. Do you feel humans have a right to use the soil in any way they choose? If not, what are the limits?

12. Do you feel nature can take care of any harm humans bring the soil?

13. Do you feel new technologies will solve any problems humans create with the soil?

14. Do you feel humans have a responsibility to protect the quality and fertility of the soil? If so, what steps do you think should be taken to protect the soil?

More Depth: Discussion and Term Paper Topics

1. Should mineral-rich but underexplored Antarctica be opened to mineral exploration and development without delay?

2. Should seabed mineral deposits such as manganese-rich nodules be declared an ecosphere resource to which landlocked less-developed nations may rightfully stake a fair-share claim?

3. Do you have an attachment to any particular piece of land? Explore the roots of your attachment. Is the land protected from erosion and other forms of land degradation?

4. Do you think land-use management and zoning are good practices? Why or why not?

5. Is the rapid deterioration of agricultural soils in the United States a sufficiently serious problem to warrant strict federal laws with heavy fines for farmers or ranchers failing to employ wise soil conservation methods? Arrange a class debate on this issue.

6. What impact have the Dust Bowl, Hugh Bennett, and the Soil Conservation Service had on this country?

(Also, see text, Critical Thinking, p. 334)

PARTICIPATION

Lifestyle and Campus Community

See *Green Lives/Green Campuses*, Chapter 5, Soil and Food, p.31.

More Depth: Action-oriented Term Paper Topics

1. Individuals. Soil testing methods and procedures; what individuals can do to prevent soil erosion and nutrient depletion on their own property; agricultural practices that restore nutrients and prevent erosion; composting; no-tillage farming; crop rotation; windbreaks; forestry practices that minimize erosion; ranching management that minimizes erosion.

2. City/Regional. Land-use planning and zoning; urban soil management problems; the Soil Conservation Service.

SKILLS

Environmental Problem-Solving Skills: Projects

1. As a class exercise, identify the minimum metal and mineral resources required to support human life at an acceptable level of quality. (Decide for yourself what acceptable life quality is.) What is the present and anticipated supply of these resources? Which essential element of the mix is least dependable in future supply? Write comprehensive and internally consistent scenarios describing what life would be like without this resource (or an affordable substitute).

2. Have your students consult recent issues of the *Wall Street Journal, Forbes, Fortune,* and other reputable sources of commodities information and forecasts. Note the metal and nonfuel mineral resources that are in the news because of supply and demand problems. Are the concerns and opinions expressed in market publications reflected adequately in the print and broadcast media reporting that is oriented to the public?

3. As a class, conduct a school and community survey to assess people's beliefs about and attitudes toward America's reliance on foreign sources for various strategic materials.

4. As a class exercise, assign appropriate roles and research assignments to students (or teams of students) and stage mock negotiations among several less-developed nations seeking to establish a cartel for controlling the supply and price of the vitally important industrial commodity X. Assume that the cartel is successful and stage additional mock negotiations between the cartel and more-developed nations whose economies are suffering grievously for want of affordable supplies of commodity X.

5. As a class exercise, have your students create a soil management plan (illustrated by sketches, drawings, or photographs) for a hypothetical badly eroded farm.

6. Visit the town planning office with your class. See if there is a land-use plan for the town. Are there zoning laws that prevent the development of certain areas, such as prime farmland and wetlands? Why or why not?

7. Take a field trip around the community with your class. See if you can identify any sloped areas that are eroding significantly. Try to discern if the land erosion is resulting in sediment pollution in surface waters. Investigate if anything is being done about it. Draw up a plan that would prevent further erosion. Share it with people who might be interested in implementing the plan.

<u>Laboratory Skills</u>

Laboratory Manual for Miller's Living in the Environment and Environmental Science. Lab 15:
   Introduction to the Arthropods; Lab 16: Soil Diversity.

<u>Computer Skills</u>

(none)

CHAPTER 13

WASTES: REDUCTION AND PREVENTION

THINKING

<u>Concept Map</u>

See Appendix A, Map 19.

<u>Goals</u>

See short, bulleted list of questions on p. 336 of text.

<u>Objectives</u>

1. List the three largest sources of solid waste in the United States. List EPA's three strategies for dealing with solid waste in priority order. Evaluate how well those priorities are being met.

2. List seven ways to reduce production of solid waste. Summarize the strategy of reuse of solid wastes: beverage containers, grocery bags, diapers, tires.

3. Summarize the strategy of recycling of solid wastes: compost, wastepaper, aluminum, plastics. Distinguish between closed-loop recycling and open-loop recycling. Evaluate which is better. Compare centralized materials-recovery facilities and source-separation strategies.

4. List three obstacles to reuse and recycling strategies in the United States.

5. Summarize the strategy of incineration of solid wastes. Compare U.S. and Japanese incineration strategies.

6. Summarize the strategy of sanitary landfills, including pros and cons.

7. Define *hazardous waste*. List nine substances which are not included in the narrow definition of hazardous waste. Using Table 13-1 on p. 353, list the hazardous wastes you use.

8. Summarize the state of our knowledge about lead and dioxins.

9. List in priority order our options for dealing with hazardous waste. Using Table 13-2 on p. 359, list household chemicals you use and less harmful alternatives you could try.

10. Summarize the Resource Conservation and Recovery Act. Evaluate the effectiveness of Superfund. Briefly describe the environmental justice movement.

11. Distinguish among a high-waste throwaway system, a moderate-waste resource recovery and recycling system, and a low-waste sustainable-Earth system. Indicate how each system would deal with aluminum cans. List three ways individuals can move toward a low-waste sustainable-Earth system.

Key Terms

solid waste (p. 336)   sanitary landfill (p. 350)
municipal solid waste (p. 336)   hazardous waste (p. 353)
compost (p. 343)   dioxins (p. 355)
incineration (p. 349)   Superfund (p. 362)

Multiple Choice Questions

See Instructor's manual, Section Two, p. 263.

More Depth: Conceptual Term Paper Topics

1.   Solid waste. Sanitary landfill space shortages; Virginia Beach's Mt. Trashmore; *Mobro*, the garbage barge; plastics in the environment.

2.   Hazardous waste. Definitions: what is included and what is excluded; household hazardous wastes and alternatives; lead; dioxin.

More Breadth: Interdisciplinary Activities and Projects

1.   Visit a community recycling center and observe its operations. If no recycling program for household waste is available in your vicinity, why not lobby for one or set one up as a class project?

2.   Does your state require refundable deposits on all beer and soft-drink containers? If so, investigate the extent to which the program is living up to expectations. If not, invite spokespersons for both sides of the issue to debate the matter for the benefit of your class.

3.   Invite a city or county official responsible for solid waste disposal to discuss related economic, political, and logistical problems. Ask about possible plans for future improvements in collection and resource recovery plants.

4.   Invite public health officials to address your class on the subject of hazardous waste risks to public health in your community.

5.   Invite a toxicologist to visit your class and discuss the problem of metal toxicity.

Multisensory Learning: Audiovisuals

Solid Waste

Waste Not, Want Not, from *Race to Save the Planet*; 1990; 60 min; potential solutions to waste problems; ACPB.
The W.O.W. Series of Video Tapes on the Solid Waste Crisis including *Waste Management as if the Future Mattered*; 1989; VAP.
Fueling the Future: No Deposit, No Return; 1988; 58 min.; investigates how wastefulness carries energy costs, depletes resources, and increases pollution; VP.
The Garbage Crisis: What Do We Do with It All? 1986; 23 min.; the solid-waste crisis in New Jersey; HMDC.
What About Waste? MDNR.
The Trash Monster and the Wizard of Waste; 12 min. film strips; CSWMB.

Riches from the Earth; 1982; 20 min.; discovery, depletion, technology, conservation, and recycling of minerals; NG.

The Disposable Society; 26 min.; solid waste; FHS.

Down in the Dumps; 26 min.; running out of landfill space and a look at alternatives; FHS.

Recycling

Recycling: The Endless Circle; 1992; 25 min.; NG.

There's More to Mining than Meets the Eye; 25 min.; mining processes and reclamation; AMC.

Re-use It or Lose It; 1991; 20 min.; Sierra Club film explaining how individuals, communities, and businesses can recycle to decrease the amount of waste they produce; CMS.

Recycling in the USA: Don't Take No for an Answer; 1987; VAP.

Recycling: Waste into Wealth; 1984; 29 min.; state-of-the-art techniques; BFF.

Recycling; 21 min.; EPA film with a variety of methods for treating municipal solid waste; FDC.

Waste Not: Reducing Hazardous Waste; UF.

The Village Green; 1975; 16 min.; EPA film about a self-sustaining recycling center in Greenwich Village, New York; FDC.

Incineration

Europeans Mobilizing Against Trash Incineration; 1990; VAP.

The Rush to Burn; 1989; 35 min.; examines incineration; VP.

Restoring the Environment; 26 min.; incineration of hazardous wastes; FHS.

Hazardous Waste

Global Dumping Ground; 60 min.; investigative reporting by Bill Moyers; SLP.

Waste Not: Reducing Hazardous Waste; 1988; 35 min.; industrial efforts at waste reduction; UF.

Toxic Waste; 60 min.; development, effects, and control of organic chemicals; BFF.

Toxic Chemicals: Information Is the Best Defense; 1985; two parts, each 26 min.; business leaders, local officials, and concerned citizens working together to create a model hazardous materials-disclosure ordinance; BFF.

The Toxic Trials; 1988; 60 min.; a NOVA case study; CMTI.

In Our Own Backyard: The First Love Canal; 1982; 59 min.; BFF.

Toxic Racism; 1994; 60 min.; grass-roots movement to prevent toxic waste dumping and industrial pollution of poor and minority neighborhoods.

Sowing the Seeds of Disaster; 26 min.; biotechnology to produce pollutant-eating organisms and frost-resistant plant strains as well as a consideration of dangers of introduction of nonnatural substances into our ecosystem; FHS.

The Toxic Goldrush; 26 min.; growth of the waste cleanup industry; FHS.

PCBs in the Food Chain; 18 min.; how marine pollution moves through the food chain from plankton through dolphins; FHS.

Poisonous Currents of Air and Sea; 18 min.; focus on PCBs moving globally; FHS.

See Appendix B for suppliers.

ATTITUDES/VALUES

Assessment

1. Have you visited a landfill? How did you feel during your visit?

2. Have you visited an incinerator? How did you feel during your visit?

3. Have you visited a recycling center? How did you feel during your visit?

4. Do you feel that natural ecosystems will be able to continue to absorb the wastes from human activities?

5. Do you feel that new technologies will be able to eliminate our current solid waste problems?

6. Do you feel that solid waste issues are one of the top three environmental concerns?

7. Are you willing to separate your trash, carry reusable shopping bags, and purchase products with reduced packaging?

8. Are you willing to make purchases based on lifetime cost of items rather than just the initial cost?

9. Would you favor a nationwide law requiring a 25¢ refundable deposit on all bottles and cans to encourage their recycling or reuse?

10. Would you support a law requiring everyone to separate their trash into paper, bottles, aluminum cans, steel cans, and glass for recycling and to separate all food and yard wastes for composting?

11. Would you support a law that bans all throwaway bottles, cans, and plastic containers and requires that all beverage and food containers be reusable (refillable)?

12. Would you support a law requiring that at least 60% of all municipal solid waste be recycled, reused, or composted?

13. Would you support a law banning the construction of any incinerators or landfills for disposal of hazardous or solid waste until at least 60% of all municipal solid waste is recycled, reused, or composted and the production of industrial hazardous waste has been reduced by 60%?

14. Have you ever visited a landfill or incinerator that handles hazardous waste? How did you feel?

15. Should industries and other producers of hazardous waste be allowed to inject such waste into deep underground wells?

16. Would you support a law banning the emission of any hazardous chemicals into the environment, with the understanding that many products you use now would cost more and some would no longer be made?

17. Would you support a law banning the export of any hazardous wastes and pesticides, medicines, or other chemicals banned in your country to any other country? Would you also support a law banning export of such wastes from one part of a country to another so that each community is responsible for the waste it produces?

More Depth: Discussion and Term Paper Topics

1. Trace the roots of the throwaway mentality.

2. Should disposable goods and built-in obsolescence be discouraged by legislation and economic means (such as taxes)?

3. Should urban incinerators be encouraged as an alternative to sanitary landfills?

4. Should we redefine hazardous wastes?

(Also, see text, Critical Thinking, pp. 369–370.)

PARTICIPATION

Lifestyle and Campus Community

See *Green Lives/Green Campuses*, Chapter 7, Solid Wastes, p. 47; Chapter 8,
    Hazardous Wastes, p. 57.

More Depth: Action-oriented Term Paper Topics

1. Recycle, reuse, reduce, rethink. Garage sales; source separation of household wastes; appliances built to last; what consumers can do about excessive packaging; recycling centers; resource recovery plants: the Saugus model.

2. City/County. Municipal resource recovery plants; recycling industrial wastes; scrapyards; anti-litter campaigns; incinerators; sanitary landfills.

3. State. Bottle bills.

4. National. The 1976 Resource Conservation and Recovery Act; EPA's Superfund program.

5. International. Recycling programs in Sweden and Switzerland; Germany's tough packaging law.

SKILLS

Environmental Problem-Solving Skills: Projects

1. If possible, take a class field trip to an open dump, a sanitary landfill, a secured landfill, and an incinerator. Observe problems associated with each approach to waste management.

2. Encourage your students to find out how your school and community dispose of wastes. Are recycling centers available? Are they conveniently located? What materials do they accept? Do any local factories or other industries accept wastes for recycling? How much is recycled? Would a source separation program be feasible?

3. Have students who live at home maintain a record of solid wastes discarded by their families in the course of one week. What percentage of this material could actually be recycled?

4. As a class, survey excess packaging in various products at local supermarkets. (Ask permission first; many supermarket managers are cooperative, but some are not.) Make up ecological ratings for each category based on the concept that packages inside of packages are very undesirable. Write manufacturers about the results of your findings. See if store managers would make your results available to customers at an environmental education stand or bulletin board.

5. Is classic lead poisoning a serious health problem in your community? Have your students consult with public health officials and find out what has been done to make people aware of potential hazards and to reduce the likelihood of dangerous exposure. Are there houses and other buildings in your community with flaking lead-based paint?

Laboratory Skills

Laboratory Manual for Miller's Living in the Environment and Environmental Science. Lab 20: Solid Waste Prevention and Management.

Computer Skills

(none)

PART IV.
BIODIVERSITY: LIVING RESOURCES

CHAPTER 14

FOOD RESOURCES

THINKING

Concept Map

See Appendix A, Map 20.

Goals

See short, bulleted list of questions on p. 373 of text.

Objectives

1. Using Figure 14-3 on p. 375, list four types of agriculture. Compare the inputs of land, labor, capital, and fossil-fuel energy of these systems. Evaluate the green revolution. What were its successes? Its failures? Describe four interplanting strategies.

2. Describe the trends in world food production since 1950. Summarize food distribution problems. Define *malnutrition* and *undernutrition*. Indicate how many people on Earth suffer from these problems. List six steps proposed by UNICEF which would help address these problems.

3. From what nutritional malady do people in MDCs suffer? Describe the health implications and what steps can be taken to alleviate the problem.

4. List 12 environmental effects of agriculture.

5. Describe the possibilities of increasing world food production by increasing crop yields, cultivating more land, and using unconventional foods and perennial crops. Briefly describe the role of location, soil, insects, and water as limiting factors of food production.

6. Describe trends in the world fish catch since 1950. Assess the potential for increasing the annual fish catch. Evaluate the potential of fish farming and fish ranching for increasing fish production.

7. Assess the pros and cons of agricultural subsidies and international food relief. Describe strategies that you feel would be most sustainable.

8. Describe a sustainable-Earth agricultural system. List three steps that could be taken to move the United States toward more sustainable agriculture.

9. List three things individuals can do to move society toward more sustainable agriculture.

Key Terms

industrialized agriculture (p. 373)	overnutrition (p. 378)
plantation agriculture (p. 373)	food additives (p. 378)
traditional subsistence agriculture (p. 373)	marasmus (p. 378)
traditional intensive agriculture (p. 373)	kwashiorkor (p. 378)
green revolution (p. 373)	fisheries (p. 384)
interplanting (p. 375)	sustainable yield (p. 386)
polyvarietal cultivation (p. 375)	optimum yield (p. 386)
intercropping (p. 375)	overfishing (p. 386)
agroforestry (p. 375)	commercial extinction (p. 386)
alley cropping (p. 375)	aquaculture (p. 387)
polyculture (p. 375)	fish farming (p. 387)
undernutrition (p. 378)	fish ranching (p. 387)
malnutrition (p. 378)	sustainable-Earth agricultural systems (p. 389)

Multiple Choice Questions

See Instructor's manual, Section Two, p. 277.

More Depth: Conceptual Term Paper Topics

1. Agricultural systems. Inorganic fertilizers; history of development of one crop or livestock species; green revolution; crops with designer genes; politics of American agriculture; feedlot beef cattle production in the Corn Belt; range livestock production in the American West; urban growth and the loss of prime cropland; modern food storage and transportation; comparisons of environmental impacts of traditional and industrial agricultural practices.

2. Hunger and food distribution. History of great famines; malnutrition and learning; the geography of malnutrition.

3. Fishing. Overfishing; aquaculture; the Peruvian anchovy story.

More Breadth: Interdisciplinary Activities and Projects

1. Have your students locate and bring to class photographs, paintings, or history passages describing the effects of hunger and starvation.

2. Ask students to find and bring to class photographs, songs, paintings, and literature reflecting human feelings for fishermen, whalers, and farmers.

3. Invite an agricultural economist to your class to discuss shifts in the United States from farming to agribusiness and the historical role of subsidies in agriculture.

4. Invite a representative from the United States Department of Agriculture (or some other informed source) to your class to discuss how U.S. political decisions such as emergency foreign aid and global trade affect U.S. farmers.

5. Invite a county agricultural agent to your class to discuss local agricultural problems and opportunities. What major changes in agricultural practices are likely to occur in the coming decades? with what consequences? What types of farming activities are carried on in your locale? What is the balance between large and small farms? What are the major products? How much of the produce is used in local areas? How much is shipped out and where does it go?

6. Invite an organic farmer or experienced organic gardener to address your class on the subject of alternatives to energy-intensive agriculture. If possible, arrange a field trip to investigate organic farming practices.

Multisensory Learning: Audiovisuals

Quantity of Food

Save the Earth—Feed the World, from *Race to Save the Planet*; 1990; 60 min; intensive and traditional farming techniques; ACPB.
Circle of Plenty; 1987; 27 min.; biointensive agriculture; BFF.
Will the World Starve?; 58 min.; farmers, government economic policies, land degradation, poverty, and soil erosion in Nepal, China, and Ethiopia; CMTI or CFS.

Quality of Food/Natural Foods

Diet for a New America; 1991; 60 min.; food choices, personal health, and the planet's health; VP.
Nutrition: Eating Well; 25 min.; balanced eating; NG.
The Victory Garden; 60 min.; CBS.
Edible Wild Plants: Video Field Guide to 100 Useful Wild Herbs; 60 min.; CBS.

Agricultural Practices

The Miracle of Guinope; 16 min.; Honduran village successfully ends hunger and slash-and-burn agriculture; WN.
Fueling the Future: Hot Wiring America's Farms; 1988; 58 min.; examines impacts of energy-intensive farming and explores efficient alternatives; VP.
Wheat Today, What Tomorrow?; 1990; 32 min.; desertification and agriculture; BFF.
Sowing for Need or Sowing for Greed?; 56 min.; impacts of modern agricultural methods exported to developing countries; BFF.
The Desert Doesn't Doesn't Bloom Here Anymore; 58 min.; a NOVA film examining water and irrigation policies and how they affect soil quality; CMTI.
The End of Eden; cattle ranching, development, and the environment; TBS.
Land Use Today for Tomorrow; 36 min.; soil erosion and competition for land use; CFS.
Fragile Harvest; 1986; 49 min.; global agriculture and threatened genetic diversity; BFF.
The Great Gene Robbery; 1986; 26 min.; BFF.
Eco: The Gene Drain; 1987; 24 min.; lack of genetic diversity in crops; CTE.

Fish

Riches from the Sea; 1984; 23 min.; fish, petroleum, minerals; NG.
An Incidental Kill; 1986; 28 min.; environmental hazards from super-efficient ocean fishing nets; DWP.
Where Have All the Dolphins Gone?; 1989; 48 min.; failure of U.S. government laws to protect marine mammals; VP.

See Appendix B for suppliers.

ATTITUDES/VALUES

Assessment

1. Have you ever fasted? If so, how did it feel?

2. Do you feel everyone has a right to a healthy diet?

3. Do you favor greatly increased foreign aid to poor countries to help them reduce poverty, to improve environmental quality, and to develop sustainable use of their own resources?

4. Do you favor a more equitable distribution of the world's resources and wealth to greatly reduce the current wide gap between the rich and the poor, even if this means less for you?

More Depth: Discussion and Term Paper Topics

1. What is the best way to manage food distribution for foreign aid?

2. Is using lifeboat ethics the best way to decide who gets to eat?

3. Aldo Leopold's land ethic.

(Also, see text, Critical Thinking, p. 392.)

PARTICIPATION

Lifestyle and Campus Community

See *Green Lives/Green Campuses*, Chapter 5, Soil and Food, p. 31.

More Depth: Action-oriented Term Paper Topics

1. Individuals: sustainable agriculture. Organic home gardening; neglected edible plants; composting; crop rotation; organic fertilizers; windbreaks.

2. Cities. Land-use planning and zoning.

3. Global. UN food conferences; 1982 UN Conference on the Law of the Sea; agricultural training and research centers in the LDCs.

SKILLS

Environmental Problem-Solving Skills: Projects

1. As a class, plan a daily menu for a family of four receiving minimum welfare payments (consult local welfare agencies for current payment levels and use current food prices). Ask your students how they would like subsisting solely on this diet.

2. As a class exercise, determine what percentage of your diet—as individuals and as a group—consists of meat. What are some ecological implications of this amount of meat in the diet? What are the health implications? What are the alternatives?

3. Arrange a class debate on the proposition that food-exporting nations should use population control and resource development as criteria to determine which of the food-importing nations will receive top priority. Conduct a mock triage and follow it with mock appeals for hearings for denied nations.

<u>Laboratory Skills</u>

(none)

<u>Computer Skills</u>

Computerized System for Agricultural and Population Planning Assistance and Training (CAPPA)
  -Facilitates use of a multisectoral scenario approach to agricultural planning.
  -Chief, Development Policy Training and Research Service, Policy Analysis Division, Food and Agriculture Organization of the United Nations, Via delle Terme di Caracalla, 00100, Rome, Italy

Standard National (Agricultural) Model (SNM)
  -Analyzes the consequences of domestic or international policy changes for a nation's domestic food situation.
  -Director, Food and Agriculture Program, International Institute for Applied Systems Analysis, Schloss Laxenburg, A-2361, Laxenburg, Austria

CHAPTER 15

PROTECTING FOOD RESOURCES: PESTICIDES AND PEST CONTROL

THINKING

Concept Map

See Appendix A, Map 21.

Goals

See short, bulleted list of questions on p. 394 of text.

Objectives

1. Compare first-generation and second-generation pesticides. Distinguish between broad-spectrum and narrow-spectrum agents.

2. Using information about persistence and biological amplification in Table 15-1 on p. 396, evaluate which pesticide is most hazardous.

3. Make a case for using pesticides. List five characteristics of the ideal pesticide.

4. Describe the consequences of relying heavily on pesticides. Summarize threats to wildlife and the human population.

5. Name the U.S. law that controls pesticide regulation. Evaluate the effectiveness of this law. State four steps recommended by the National Academy of Sciences regarding this law.

6. List and briefly describe nine alternative pest management strategies.

7. Define *integrated pest management.* Analyze the pros and cons of using IPM.

8. Name three steps individuals can take to control pests in an environmentally sound way.

Key Terms

pest (p. 394)
pesticides (p. 394)
insecticides (p. 394)
herbicides (p. 394)
fungicides (p. 394)

nematocides (p. 394)
rodenticides (p. 394)
pesticide treadmill (p. 397)
integrated pest management (IPM) (p. 405)

Multiple Choice Questions

See Instructor's manual, Section Two, p. 291.

More Depth: Conceptual Term Paper Topics

1. Pesticides. Pesticides as hazardous waste; pesticide hazards to agricultural workers; chlorinated hydrocarbons; organophosphates and carbamates; pyrethroids and rotenoids; biological amplification of persistent pesticides; DDT and malaria control; Agent Orange; the Bhopal accident; pesticide residues in foods; pesticide runoff as a threat to agriculture.

2. Pesticide alternatives. Integrated pest management; food irradiation; genetic control by sterilization: the screwworm fly; pheromones.

More Breadth: Interdisciplinary Activities and Projects

1. Invite a county agricultural agent to address your class on the subject of pesticide use and abuse in your locale. Try to determine what factors, including government programs, combine to keep farmers on the pesticide treadmill.

Multisensory Learning: Audiovisuals

Inert Alert: Secret Poisons in Pesticides; 17 min.; documentary; NCAP.
Love, Women, and Flowers; 1988; 56 min.; women in the Colombian flower industry exposed to highly toxic pesticides that have been banned in industrialized countries; WMM.
The Wrath of Grapes; 1987; 15 min.; health dangers posed to U.S. farm workers; UFW.
Common Ground: Farming and Wildlife; 1987; 57 min.; chemicals dividing farmers and environmentalists; WETA.
The Secret Agent; 1983; 55 min.; dioxin, the world's most toxic man-made chemical; GMPF.
Integrated Pest Management in Agriculture; 29 min.; how and why IPM was developed; SLVP.
Lights Breaking: Ethical Questions about Genetic Engineering; 1986; 59 min.; ethical and practical concerns raised by genetic engineering; BFF.
Seeds; 1987; 28 min.; importance of genetic diversity to the world's food supply; BFF.
Spiders; 30 min.; CBS.
Cleaning up Toxics at Home; 1990; 25 min.; produced by the League of Women Voters; VP.
Cleaning up Toxics in Business; 1990; 25 min.; produced by the League of Women Voters; VP.
The Military and the Environment; 1990; 29 min.; the military's handling of toxics; VP.
Putting Aside Pesticides; 26 min.; long-term effects of pesticides and a look at alternatives; FHS.

See Appendix B for suppliers.

ATTITUDES/VALUES

Assessment

1. Have you ever eaten food grown with fertilizers and pesticides? How did it taste?

2. Have you ever eaten organically grown food? How did it taste?

3. Are you aware of places to obtain organically grown food in your area?

4. Do you prefer perfect looking fruits and vegetables grown with pesticides to slightly blemished fruits and vegetables grown without pesticides?

5. Do you favor regulation of pesticides exported from the United States?

More Depth: Discussion and Term Paper Topics

1. Evaluate pesticide advertising. What does it tell? What doesn't it tell?
2. Which do you prefer: unblemished fruits and vegetables that may contain pesticide residues or blemished fruits and vegetables without pesticide residues?

3. Rachel Carson's *Silent Spring.*

4. Which is better: a broad-spectrum or a narrow-spectrum pesticide?

(Also, see text, Critical Thinking, pp. 406–407.)

PARTICIPATION

More Depth: Action-oriented Term Paper Topics

1. Individual. Safe disposal of household pesticides; homeowner strategies and tactics to reduce pesticide use.

2. National regulation. Federal Insecticide, Fungicide, and Rodenticide Act.

3. Global. International sales of U.S.–produced pesticides whose use is banned in the United States; General Agreement on Tariffs and Trade (GATT) and its implications for U.S. regulations regarding pesticide levels in American foods.

SKILLS

Environmental Problem-Solving Skills: Projects

1. With the help of a chemist or other appropriate consultant, have your students evaluate the ingredients, uses, and warning labels of a representative sample of pesticides sold for home and garden applications. Are the instructions for use, storage, and disposal adequate? How much additional information should be supplied to further reduce the likelihood of harm to people and wildlife?

2. Are people generally aware of and concerned about the hazards of using pesticides on a large-scale, long-term basis? As a class project, conduct a survey of students or consumers to address these and related questions. What do the results imply for the role that education should play in dealing with pesticide-related problems?

3. Have your students interview the college landscaping staff about which pesticides, if any, they use on campus. What tradeoffs did they consider when deciding to use those pesticides?

Laboratory Skills

(none)

Computer Skills

(none)

CHAPTER 16

BIODIVERSITY: SUSTAINING ECOSYSTEMS

THINKING

Concept Map

See Appendix A, Map 22.

Goals

See short, bulleted list of questions on p. 409 of text.

Objectives

1. Describe the importance of biological diversity. List five types of public lands in the United States. Explain the mission and principles of management of each.

2. Distinguish between old-growth and second-growth forests. Briefly describe the commercial and ecological significance of forests.

3. Distinguish between the goals of even-aged management and uneven-aged management. List five types of tree harvesting, indicating which type of management they are most likely to be used for.

4. Summarize the best current strategies for protecting forests from pathogens and insects, fires, air pollution, and climate change.

5. List seven steps to move toward sustainable forestry.

6. Analyze the national forest situation in the Northwest. List 10 steps which would help reform federal forest management.

7. Describe the current state of tropical forests. Describe the significance of tropical forests. List three underlying causes and seven direct causes of tropical forest destruction. List economic, biological, and political steps that can be taken to preserve the tropical rain forests.

8. Create as many linkages among tropical forests, energy needs, biodiversity, and economics as you can. Evaluate Costa Rica's ecological protection and restoration plan.

9. Briefly describe U.S. rangelands, national parks, and wilderness areas. Give one problem of each area and explain how that problem might be managed better.

Key Terms

old-growth forests (p. 412)   seed-tree cutting (p. 416)
second-growth forests (p. 413)   clear-cutting (p. 416)
even-aged management (p. 413)   stripcutting (p. 416)
uneven-aged management (p. 414)   debt-for-nature swaps (p. 431)
selective cutting (p. 416)   rangeland (p. 433)
shelterwood cutting (p. 416)   carrying capacity (p. 435)

Biodiversity: Sustaining Ecosystems

overgrazing (p. 435)   wilderness (p. 444)

riparian zones (p. 436)

Multiple Choice Questions

See Instructor's manual, Section Two, p. 303.

More Depth: Conceptual Term Paper Topics

1. Tropical rain forests. U.S. imports and tropical deforestation.

2. Multiple-use and moderate-use public lands in the United States. Bureau of Land
    Management policies and programs in the arid West; the U.S. Forest Service; Gifford Pinchot
    and the forest conservation movement; sustainable forestry's answer to clear-cutting; the role
    of fire in forestry management; the Sagebrush Rebellion.

3. U.S. restricted-use lands. The National Parks System; the National Wilderness System.

4. Wilderness preservation in LDCs.

More Breadth: Interdisciplinary Activities and Projects

1. Invite a National Park Service or state official to your class to discuss park problems and
    future management plans.

2. As a class field trip, visit a forest managed for pulp and paper production or industrial
    timbering. What specific methods are used to maximize economic returns and to curb
    ecosystem damage? Contrast the appearance of commercial forestland and relatively
    undisturbed forestland. Which do you like best? Why?

Multisensory Learning: Audiovisuals

Ecosystems

Amazonia: A Celebration of Life; 1984; 20 min.; a World Wildlife Fund production presenting a
    diversity of species of the rain forest; SUNY.
Creatures of the Mangrove; 1986; 59 min.; a National Geographic film portraying the biodiversity of
    a tidal forest; NG.
Rivers to the Sea; 46 min.; abundant life and stresses of coastal river systems; BFF.
Keepers of the Forest; 1988; 28 min.; destruction of the tropical rain forest and how to stop it; UF.
The Fragile Forest; 1987; 28 min.; working with the wood-energy crisis in the Himalayas of
    Nepal; LTS.
Replanting the Tree of Life; 1987; 20 min.; an inspiring reminder of the part trees play in our lives
    and in the life of the planet; BFF.
Man of the Trees; 1982; 26 min.; one man's life dedicated to saving the world's trees; VP.
Rain Forests: Proving Their Worth; 1990; 30 min.; new movement to market sustainably collected
    forest products; VP.
The Forest Through the Trees; 1990; 58 min.; different perspectives on the redwoods and the spotted
    owl and a look toward alternative futures; VP.
Ancient Forests; 1992; 25 min.; NG.
Rain Forest; 1983; 60 min.; NG.
Korup: An African Rain Forest; 1984; interrelationships of organisms in a rain forest; PF.

The Tropical Rain Forest; 28 min.; ecosystem focus on rich biodiversity adapted to heavy rainfall; FHS.

Preserving the Rain Forest; 24 min.; from indigenous cultures to forest depletion through shifting agriculture and technological efficiency; case study of controlled agriculture and industrial activity in a natural reserve in the Ivory Coast; FHS.

National Parks

Yellowstone in Winter; 60 min.; animals struggle for survival; CBS.

Denali Wilderness; 30 min.; pristine Alaskan wilderness; CBS.

Yosemite: Seasons and Splendor; 40 min.; CBS.

Glacier National Park, Montana; 30 min.; CBS.

Grand Teton National Park; 30 min.; CBS.

Shenandoah: The Gift; 30 min.; CBS.

Acadia National Park and Cape Cod National Seashore; 30 min.; CBS.

Rocky Mountain National Park; 30 min.; CBS.

The Complete Yellowstone; 60 min.; CBS.

Bryce, Zion, and the North Rim of the Grand Canyon; 25 min.; CBS.

Grand Canyon, Petrified Forest, and Painted Desert; 45 min.; CBS.

Death Valley National Monument; 26 min.; CBS.

Carlsbad Caverns and Guadalupe Mountains National Park; 53 min.; CBS.

Mesa Verde National Park; 23 min.; CBS.

Mt. Rushmore and the Black Hills; 30 min.; CBS.

Mt. Rainier National Park; 28 min.; CBS.

Voices from the Ice: Alaska; 20 min.; CBS.

Mammoth Cave National Park; 30 min.; CBS.

Halting the Fires; 1990; 52 min.; the United Kingdom, West Germany, and Brazil examine the economic, environmental, and social consequences of the huge number of fires set in the Amazon; FL.

Olympic; 1990; 30 min.; mood film, multi-media portrait of Olympic National Park; AV.

To Protect Mother Earth (Broken Treaty II); 1989; 60 min.; the Shoshones struggle to save their ancestral lands from strip-mining, nuclear tests, and oil drilling; CPr.

Antarctica; 1987; 26 min.; Antarctica: wilderness, garbage dump, ozone hole; FHS.

Windy Bay: Wilderness under Siege; 1987; 49 min.; a clash between indigent people and clear-cut logging; CBC.

The Law of Nature: Park Rangers in Yosemite Valley; 1987; 28 min.; how rangers' work has shifted from resource management to people management; UF.

The Development Road; 12 min.; inefficient development in Brazil's Rondonia province; BFF.

The Living Planet, Part 4, The Jungle; 1984; 55 min.; explores stratification in a rain forest from the top of a kapok tree to the forest floor; PSU.

See Appendix B for suppliers.

ATTITUDES/VALUES

Assessment

1. What wildlife is most common in your area?

2. Where are the nearest locations in your area to go to observe wildlife?

3. What are your feelings toward wildlife species? What relationship between humans and wildlife do you find most desirable?

4. Do you feel that humans have the right to relate to other species in any way they wish? If not, what limits do you see on human behavior toward other species?

5. Do you use products that come from the tropical forest? Do the products you use result in destruction of forest or continued sustainable use of the forest?

6. How do you feel when you see pictures of the destruction of ancient forests?

7. Do you feel we can continue to find substitutes for losses we suffer when ancient forests are destroyed?

8. Do you feel nature can continue to replenish forests at any rate humans choose to harvest the forests?

9. How do you feel when you see pictures of unemployed loggers unable to support their families?

10. Do you feel it is right to destroy cultures that exist sustainably in the tropical rain forests? If not, what steps do you support to protect these cultures?

11. What steps do you feel should be taken to support human cultures and wildlife species in ways that create sustainable societies?

12. Have you ever visited a mine? How did you feel about the mine? What benefits do you enjoy as a result of mining activity?

13. Have you ever visited rangeland? How did you feel about the land? What benefits do you enjoy as a result of cattle grazing?

14. Have you ever visited a national forest? How did you feel about the forest? What benefits do you enjoy as a result of lumbering activity?

15. Have you ever visited a wilderness area? How did you feel about the wilderness? What benefits do you enjoy as a result of protection of wilderness areas?

16. Would you support classifying a much larger proportion of the public lands (such as parks, forests, and rangeland) in your country as wilderness and making such land unavailable for timber cutting, livestock grazing, mining, hunting, fishing, motorized vehicles, or any type of human structure?

More Depth: Discussion and Term Paper Topics

1. Alaska's value: wilderness or oil supply?

2. Should fires be allowed to burn in forests on public lands?

3. Should products that result in destruction of tropical forests be banned in the United States?

4. Should mining be allowed in national wildlife refuges?

5. Should parts of the wilderness areas be set aside for wildlife only?

(Also, see text, Critical Thinking, pp. 448–449.)

PARTICIPATION

More Depth: Action-oriented Term Paper Topics

1. Individuals. Recycling wastepaper: obstacles and overcoming them.

2. Groups. The Nature Conservancy; the Wilderness Society; reducing the pressure of people on the national parks.

3. National laws. The Alaskan Land-Use Bill; the Endangered American Wilderness Act of 1978; the Wild and Scenic Rivers Act of 1968; the 1974 and 1976 Forest Reserves Management Acts.

4. Global. The UN World Heritage Trust; debt-for-nature swaps; solar cookers and the world firewood crisis.

SKILLS

Environmental Problem-Solving Skills: Projects

1. As a class project, compile a list of commodities for sale in your community whose production or harvesting contributes to the destructive exploitation of tropical forests. Are vendors and consumers aware of the consequences? Do they care about the consequences?

2. If there are rangelands in your locale, try to schedule a class visit to examples of poorly and well-managed grazing lands. Compare the quantity and quality of vegetation present.

3. If possible, visit a national park or wilderness area. Assess its current problems and analyze plans to address those problems.

Laboratory Skills

(none)

Computer Skills

Range, Livestock, and Wildlife
>   -Helps decision makers understand and evaluate policy alternatives for rangeland management.
>   -Paul Faeth, World Resources Institute/International Institute for Environment and Development, 1709 New York Avenue, N.W., 7th Floor, Washington, D.C. 20006

BIOCUT
>   -Assesses the economic viability of alternative designs and management strategies for wood energy plantations.
>   -National Technical Information Service, U.S. Dept. of Commerce, 5285 Port Royal Rd., Springfield, VA 22161

CHAPTER 17

BIODIVERSITY: SUSTAINING WILD SPECIES

THINKING

Concept Map

See Appendix A, Map 23.

Goals

See short, bulleted list of questions on p. 451 of text.

Objectives

1. Describe the economic, medical, aesthetic, ecological, and ethical significance of wild species.

2. Describe how species become extinct. List and describe six ways that humans accelerate the extinction rate.

3. Distinguish between endangered species and threatened species. Give three examples of each.

4. List nine characteristics that make species extinction prone.

5. List 10 goals of the World Conservation Strategy.

6. List and briefly describe three approaches to protect wild species from extinction. State one advantage particular to the ecosystem approach.

7. List and briefly describe three additional strategies to help protect species.

8. Describe how wildlife populations can be managed by manipulating successional stages of the habitat.

9. Describe how fish and game populations are managed in order to sustain the population. Analyze the lessons to be learned from the decline of the whaling industry.

10. List three ways individuals can help maintain wild species.

Key Terms

wildlife resources (p. 451)   threatened species (p. 455)
game species (p. 452)   wildlife management (p. 470)
endangered species (p. 454)   flyways (p. 471)

Multiple Choice Questions

See Instructor's manual, Section Two, p. 320.

More Depth: Conceptual Term Paper Topics

1. Significance of wildlife. Medicines derived from plants and animals; commercial products from wildlife; aesthetic and recreational significance of wildlife; ecological significance of wildlife.

2. Endangered and threatened wildlife. Tropical deforestation and species extinction; the international trade in endangered species and exotic pets; lead poisoning in waterfowl and the American bald eagle; Florida's alien species problem; the California condor; the Florida manatee; the blue whale.

3. Protecting wildlife. Gene banks; zoos and captive breeding programs; habitat management; artificial reef-building materials and methods.

More Breadth: Interdisciplinary Activities and Projects

1. Are there zoos, aquariums, botanical gardens, or arboretums in your locale operating programs designed to increase the populations of endangered species? If so, invite a spokesperson to explain one or more of these programs to your class.

2. Ask your students to bring to class and share paintings, sketches, poetry, songs, and other artistic creations depicting the beauty and wonders of wildlife.

3. Have a game warden address your class about management of populations of fish and animals that are hunted for sport.

Multisensory Learning: Audiovisuals

Biodiversity

The Biodiversity Crisis: Gone Before You Know It; 1994; 15 min.; how diversification happens and is threatened through case studies in Hawaii; FHS.

Plants in Peril; 26 min.; need to preserve plants for food and medicine and genetic diversity; FHS.

Biodiversity: The Videotape; 45 min.; adapted from the 1986 National Forum on Biodiversity; NAP.

Diversity Endangered; 1987; 10 min.; introduction to the meaning and importance of biological diversity; SITES.

Remnants of Eden, from *Race to Save the Planet;* 1990; 60 min.; growing human populations and biodiversity; ACPB.

Our Threatened Heritage; 1988; 19 min.; loss of biodiversity, deforestation, and climate change; VP.

Our Threatened Heritage; 1987; 50 min.; political aspects and environmental consequences of tropical deforestation; NWF.

Wildlife Conservation Strategies

Saving a Species; 26 min.; species depletion through habitat loss and how science is trying to reverse the trend; FHS.

Garden of Eden; 1983; 28 min.; need for corporate and conservation interests to work together to protect natural diversity; DC.

Amazonia: A Burning Question; 1987; 58 min.; Dr. Thomas Lovejoy describes efforts to understand and conserve the rain forest; SUNY.

Conservation of the Southern Rain Forest; 1988; 60 min.; Scientists, conservationists, native healers work together exploring alternative solutions for preserving the forest; BF.

Landscape Linkages; 1988; 25 min.; wildlife photography and computer graphics explain the corridor approach to wildlife conservation; FF.

Serving Time (The Nature of Things); 1987; 47 min.; zoos maintaining genetic variation in captive species; FI.

Return from Forever; 1987; 30 min.; reestablishment of a breeding population of osprey in Pennsylvania; WRCF.

Mooselift: A Reintroduction of Moose to Michigan's Upper Peninsula; 1987; 15 min.; documentary of the international cooperative effort to restore moose to Michigan; MDNR.

In Celebration of America's Wildlife; 1987; 57 min.; how U.S. wildlife is being restored; CF.

Individual Wildlife Species

African Wildlife; 60 min.; CBS.

Animals of Africa: Africa in Flight; 70 min.; CBS.

World of Audubon: Messages from the Birds; 1988; 48 min.; shorebirds as indicator species; TBS.

Shorebirds—A Rite of Spring; 1986; 28 min.; Delaware Bay and shorebird migration; NJN.

Save the Panda; 60 min.; CBS.

Polar Bear Alert; 60 min.; CBS.

Spirit of the Eagle; 1991; 30 min.; VP.

Where Have All the Dolphins Gone?; 1991; 58 min.; VP.

Cane Toads, An Unnatural History; 1988; 46 min.; wildly entertaining film about the importation of 102 cane toads into Australia; FRF.

World of Audubon: Grizzlies & Man: An Uneasy Truce; 1988; 48 min.; from invincible creature to threatened species; TBS.

Farewell, Ancient Mariner; 1988; 25 min.; plight of endangered sea turtles; OFP.

Cry of the Muriqui; 1982; 28 min.; endangered monkeys in Brazil; SUNY.

Bats: Myth and Reality; CBS.

Lions of the African Night; 60 min.; CBS.

California Condor; 60 min.; CBS.

The Great Whales; 60 min.; CBS.

White Wolf; 60 min.; CBS.

The Grizzlies; 60 min.; CBS.

Rhino on the Run; 54 min.; CBS.

Frozen Eden; 22 min.; CBS.

Gorilla; 60 min.; CBS.

Among the Wild Chimpanzees; 60 min.; CBS.

Song Dog; 27 min.; coyotes as predators; CBS.

Kindness Kills; 22 min.; human handouts and black bears in Banff National Park; CBS.

Close Encounters of the Deep Kind; 22 min.; CBS.

Tiger, Tiger; CBS.

Ducks under Siege; 1987; 48 min.; ducks facing drought, agricultural expansion, and human activities; WETA.

On the Edge of Extinction: Panthers and Cheetahs; 1987; 48 min.; attempts to save these breeds' habitats; deals with genetic variation, human overpopulation and interference, and poor natural resource management; WETA.

The Bats of Carlsbad; 1986; 5 min.; bat flights; NAVC.

The Elephant Seal: Living on the Edge of Extinction; 14 min.; genetic diversity; BFF.

Realm of the Alligator; 1986; 60 min.; the Okefenokee swamp, home of the alligator; NG.

Woodstork: Barometer of the Everglades; 58 min.; a National Audubon Society film looking at the woodstork as an indicator species; FA.

See Appendix B for suppliers.

ATTITUDES/VALUES

Assessment

1. Do you believe that humans have a duty to subdue wild nature to provide food, shelter, and other resources for people and to provide jobs and income through increased economic growth?

2. Do you believe that every living species has a right to exist, or at least struggle to exist, simply because it exists?

3. Do you believe that we have an obligation to leave Earth for future generations of humans and other species in as good a shape as we found it, if not better? Did past generations do this for you?

More Depth: Discussion and Term Paper Topics

1. Should animals be used for medical research? as sources of organs for surgical implants in humans? as sources of food, fur, fat, oils, and other commercially valuable products?

2. Are extremist tactics by Greenpeace and Earth First! necessary or justifiable?

3. Should sport hunting be used as a wildlife management tool?

4. Should limits be placed on genetic engineering for economic, aesthetic, ecosystem services, or other purposes?

5. Ancient forests: Is it only the spotted owl that is at stake?

(Also, see text, Critical Thinking, pp. 477–478)

PARTICIPATION

Lifestyle and Campus Community

See *Green Lives/Green Campuses*, Chapter 9, Wildlife: Plants and Animals, p. 67.

More Depth: Action-oriented Term Paper Topics

1. Scientific methods for estimating wildlife populations and successional stages of ecosystems.

2. Groups. Ducks Unlimited; the National Wildlife Federation; the Audubon Society; Greenpeace; Earth First!

3. National. America's National Wildlife Refuge System.

4. Global. The World Wildlife Fund.

SKILLS

Environmental Problem-Solving Skills: Projects

1. Compile a list of the wildlife species in your locale that have been officially designated as threatened or endangered. As a class project, find out what specific actions are being taken to assist these species.

2. As a class, examine and evaluate the goals of the World Conservation Strategy. Develop objectives that could help implement the goals that are agreed upon by the class.

Laboratory Skills

(none)

Computer Skills

Biological and Conservation Data (BCD) System
   -Provides an inexpensive, effective tool to inventory, rank, protect, and maintain endangered species.
   -The Nature Conservancy, Data Systems Divisions, 1815 North Lynn Street, Rosslyn, VA 22209

*Audubon Wildlife Adventures
   -Explore wildlife conservation issues of grizzlies and whales using scientific information, state-of-the-art graphics, computerized data bases, and on-line guidebooks.
   -Advanced Ideas, Inc., 591 Redwood Highway, #2325, Mill Valley, CA 94941 (415-388-2430)

*Wildways: Understanding Wildlife Conservation
   -Emphasizes the need for wildlife conservation through integration of biology, geology, and sociology in sections on Earth and life, basic necessities of life, importance of wildlife, population ecology, community ecology, extinction, wildlife management, and citizen action.
   -Opportunities in Science, Inc., P.O. Box 1176, Bemidji, MI 56601 (218-751-1110)

CHAPTER 18

SOLUTIONS: ENERGY EFFICIENCY AND RENEWABLE ENERGY

THINKING

Concept Map

See Appendix A, Map 24.

Goals

See short, bulleted list of questions on p. 481 of text.

Objectives

1. List the five key questions that must be asked about each energy alternative to evaluate energy resources.

2. List the advantages and disadvantages of improving energy efficiency so that we do more with less. Define *net energy*.

3. List the advantages and disadvantages of using direct solar energy to heat buildings and water and to produce electricity.

4. List the advantages and disadvantages of using hydropower, tidal power, wave power, ocean thermal currents, and solar ponds to produce electricity.

5. List the advantages and disadvantages of using wind to produce electricity.

6. List the advantages and disadvantages of using biomass to heat space and water, produce electricity, and propel vehicles.

7. List the advantages and disadvantages of using geothermal energy to produce electricity.

8. List the advantages and disadvantages of using hydrogen gas to heat space and water, produce electricity, and propel vehicles. Name the energy source that is needed to produce hydrogen to create a truly sustainable future.

9. List three ways that individuals can move toward use of perpetual and renewable energy resources.

Key Terms

net energy (p. 483)
energy efficiency (p. 484)
life-cycle cost (p. 484)
cogeneration (p. 486)

passive solar heating system (p. 492)
active solar heating system (p. 496)
photovoltaic cells (p. 500)
solar cells (p. 500)

hydroelectric power (p. 502)   biomass plantations (p. 506)

solar ponds (p. 504)   solar-hydrogen revolution (p. 508)

biomass (p. 505)   geothermal energy (p. 509)

<u>Multiple Choice Questions</u>

See Instructor's manual, Section Two, p. 334.

<u>More Depth: Conceptual Term Paper Topics</u>

1. Improving energy efficiency. Energy-efficient office buildings; earth-sheltered houses; retrofitting energy-wasting houses; superinsulation; earth tubes; evaporative coolers; energy-efficient appliances; compact fluorescent light bulbs; "smart" windows; superinsulated windows; roof-attachable solar cell rolls; the Albers Technologies air conditioner.

2. Solar technologies. The solar power tower; the Odeillo furnace; solar power satellites; photovoltaics; active solar systems; passive solar heating; microprocessors to control house temperatures.

3. Biomass. Modern wood stoves; bagasse as a biomass fuel; biogas digesters in the LDCs; gasohol; methanol.

4. Wind. Wind farming in California; wind turbine designs.

5. Water power. Large-scale hydropower projects in the LDCs; rehabilitating small-scale hydroelectric plants in New England; wave power devices—a comparison of various approaches; ocean thermal energy conversion; solar ponds; the Bay of Fundy tidal power project.

6. Hydrogen gas: a versatile fuel of the future.

<u>More Breadth: Interdisciplinary Activities and Projects</u>

1. Ask an architect or contractor with experience in renewable energy resource systems to visit the class and discuss the practical aspects of designing, financing, and installing small-scale solar, wind, and biogas systems for individual residences, farms, businesses, or factories.

2. Find out if representatives from your local electrical utility offer customers energy audits of their homes. If so, ask them to come to your class and tell what they look for in homes and what seem to be the most frequent ways that customers can increase their energy efficiency.

3. Have your students find out if your institution's electrical utility has a conservation program. Does it have policies that encourage customers to purchase energy-efficient appliances and use energy-efficient light bulbs?

4. Organize a class field trip featuring guided tours of homes and/or other buildings that have solar heating systems. If possible, include examples of both passive and active systems and an earth-sheltered house.

<u>Multisensory Learning: Audiovisuals</u>

More for Less, from *Race to Save the Planet;* 1990; 60 min; ways to reduce fossil fuel consumption and improve energy efficiency; ACPB.

Green Energy; 26 min.; renewable energy alternatives with focus on biomass; FHS.

Energy Alternatives: Solar; 26 min.; FHS.

Tomorrow's Energy Today; 1994; Free while supplies last; NREL.

Tomorrow's Energy Today: The Energy-Efficiency Option; 1994; Free while supplies last; NREL.

Building a Solar Culture; 28 min.; CBS.

Fueling the Future: No Place Like Home; 1988; 58 min.; examines how communities evolved in an era of cheap energy and highlights how some are working to develop a more energy-efficient approach; VP.

Lovins on the Soft Path: An Energy Future with a Future; 36 min.; BFF.

A Visit with Amory and Hunter Lovins; 1986; 14 min.; a tour of the Lovins' home/office as a model of energy efficiency; BFF.

Dawn of the Hydrogen Age; 5 min.; renewable energy alternatives; EP.

Energy Efficiency; 23 min.; meeting U.S. energy needs and energy standards; BFF.

Old House, New House; 1982; 27 min.; making a Victorian house more energy efficient; FL.

Harness the Wind; 12 min.; history and potential of wind power; BFF.

Kilowatts from Cowpies: The Methane Option; 25 min.; BFF.

Fields of Fuel: The Ethanol Debate; 1982; 28 min.; ethanol in the energy future; IFB.

The Solar Hydrogen Economy; 5 min.; a practical alternative; AHA.

Solar Energy Now?; 30 min.; PBS.

Energy: The Problems and the Future; 1978; 23 min.; renewable energy resources; NG.

If You Can See a Shadow; 1979; 28 min.; passive solar power projects; BFF.

Solar Energy: The Great Adventure; 1979; 18 min.; MM.

Solar Promise; 1979; 28 min.; MM.

Geothermal Energy; 28 min.; PBS.

See Appendix B for suppliers.

ATTITUDES/VALUES

<u>Assessment</u>

1. What is the major energy source for heating your living space?

2. What is the major energy source for providing electricity to your living space?

3. How do you feel toward different energy sources?

4. What would be the best alternative energy source in your area?

5. Do you feel a responsibility to use energy wisely? What steps are you willing to take to reduce your energy consumption?

6. How do you feel toward decentralization of the power grid?

7. Do you favor policies which encourage energy conservation and more development of renewable and perpetual energy sources?

More Depth: Discussion and Term Paper Topics

1. Choose: free-flowing streams or a network of small-scale hydropower facilities?

2. Should millions of homeowners erect small wind turbines for electrical production?

3. Should building codes be required to include passive solar concepts?

(Also, see text, Critical Thinking, p. 514.)

PARTICIPATION

Lifestyle and Campus Community

See *Green Lives/Green Campuses*, Chapter 10, Energy: Generating heat, p. 73; Chapter 11, Energy: Generating Electricity; Chapter 12, Energy: A Generation on the Go, p. 88.

More Depth: Action-oriented Term Paper Topics

1. Individual. Household energy savings.

2. Industry. The horizontal integration of large energy companies.

3. Policy. Government taxing and subsidizing policies and energy conservation; the National Audubon Society intermediate national energy strategy; an evaluation of the current administration's energy plan.

SKILLS

Environmental Problem-Solving Skills: Projects

1. As a class project, identify the major energy-related economic, political, environmental, and social problems in your community and state. What specific actions are being taken to alleviate these problems? Are these piecemeal efforts, or are they components of comprehensive and internally consistent energy plans?

2. Bring to class recent issues of periodicals devoted exclusively to renewable energy industry reporting. As a class exercise, scan the contents of these periodicals to see what research and development is on the cutting edge of progress.

3. Have your students audit energy use and waste on your campus and in activities (such as commuting) associated with the operation of your campus. Are opportunities to conserve significant amounts of energy going unrecognized or ignored?

4. As a class project, conduct a survey of students at your school to determine what beliefs and attitudes they have regarding sustainable-Earth energy alternatives that entail a loss of convenience or additional expenditures of time and money on the part of energy users. Are young people today willing to significantly alter their lifestyles to use and waste less energy?

5. As a class project, develop a simple questionnaire or test that can be used to measure a person's knowledge about the efficiency of various energy conversion devices, appliances, and systems in common use. What is the "energy IQ" of the average student on your campus? To obtain a crude measure, administer the test to a random sample of students and analyze the results.

6. Have your students examine recent issues of weekly news magazines, local newspapers, and nationally recognized newspapers. How much space is devoted to energy-related reporting and analysis? What is the relative degree of emphasis placed on developing alternative energy sources and systems? Curbing energy waste? Lifestyle adjustments that reduce energy needs? How are energy-related topics and issues handled in the *Wall Street Journal* and *Fortune?* How does corporate advertising address energy topics and issues? In these and other ways, try to determine how thoroughly and accurately the public is being informed about matters of critical importance to the nation's energy future.

Laboratory Skills

Laboratory Manual for Miller's Living in the Environment and Environmental Science. Lab 19: Energy Conservation; Lab 18: Energy Alternatives.

Computer Skills

(none)

CHAPTER 19

NONRENEWABLE ENERGY RESOURCES

THINKING

Concept Map

See Appendix A, Map 25.

Goals

See short, bulleted list of questions on p. 517 of text.

Objectives

1. Distinguish among primary, secondary, and tertiary oil recovery. List the advantages and disadvantages of using conventional oil, oil from oil shale, and oil from tar sands to heat space and water, produce electricity, and propel vehicles.

2. Distinguish among natural gas, liquefied petroleum gas, and liquefied natural gas. List the advantages and disadvantages of using natural gas as an energy source.

3. List and describe three types of coal. Indicate which is preferred for burning and which is most available. List advantages and disadvantages of using coal as a fuel source.

4. List two processes used to convert coal to synfuels. List the advantages and disadvantages of using synfuels.

5. Briefly describe the components of a conventional nuclear reactor. List advantages and disadvantages of using conventional nuclear fission to create electricity. Be sure to include aspects of the whole nuclear fuel cycle, including disposal of radioactive wastes, safety and decommissioning of nuclear power plants, and the potential for proliferation of nuclear weapons. List three ways to decommission a nuclear power plant. Assess which method is most likely to be used.

6. Describe the potential use of breeder nuclear fission and nuclear fusion as energy sources.

7. Using Table 19-1 (on pp. 536–537 of the text), design an energy strategy for the United States for the short term, intermediate term, and long term.

8. List three ways that individuals can contribute to a more sustainable-energy future for the United States.

Key Terms

petroleum (p. 517)   petrochemicals (p. 517)
crude oil (p. 517)   oil shale (p. 520)
primary oil recovery (p. 517)   kerogen (p. 520)
secondary oil recovery (p. 517)   shale oil (p. 520)
tertiary oil recovery (p. 517)   tar sand (p. 520)
enhanced oil recovery (p. 517)   bitumen (p. 520)

natural gas (p. 521)

liquefied petroleum gas (LPG) (p. 522)

liquefied natural gas (LNG) (p. 522)

coal (p. 523)

synfuels (p. 525)

coal gasification (p. 525)

synthetic natural gas (SNG) (p. 525)

coal liquefaction (p. 525)

radioactive wastes (p. 531)

breeder nuclear fission reactors (p. 535)

nuclear fusion (p. 535)

Multiple Choice Questions

See Instructor's manual, Section Two, p. 349.

More Depth: Conceptual Term Paper Topics

1. Oil and natural gas. Oil prices and economic development in the LDCs; enhanced oil-recovery techniques; shale oil extraction; petrochemicals; heavy oils from Athabascan tar sands; Alaska's Prudhoe Bay gas deposits.

2. Coal. Low-sulfur coal reserves in the United States; geographic distribution of coal-burning power plants in the United States; fluidized-bed combustion; the U.S. Synthetic Fuels Corporation.

3. Nuclear fission. Centralized energy planning in France; genetic damage to A-bomb survivors; Three Mile Island; Chernobyl; how nuclear fuel assemblies are made; radioactive tailings as a health hazard; geologic repositories for high-level radioactive wastes; commercial low-level nuclear waste dump sites; storing high-level liquid wastes; geographic distribution of nuclear power plants in the United States; keeping weapons-grade nuclear materials "out of the wrong hands"; nuclear reprocessing plants.

4. Breeder nuclear fission and nuclear fusion.

More Breadth: Interdisciplinary Activities and Projects

1. Arrange a class excursion to a coal-burning power plant in your vicinity. Have a company spokesperson explain the electricity generating process and the design and operating features of equipment and systems that control air pollution emissions and reduce thermal water pollution.

2. If there is a nuclear power plant operating in your vicinity, schedule a guided tour for your class.

3. If there is a nuclear power plant operating in your vicinity, invite a spokesperson from your local emergency disaster preparedness agency to present a guest lecture explaining the emergency evacuation plan for this facility.

Multisensory Learning: Audiovisuals

General

Energy: The Fuels and Man; 1978; 23 min.; what energy is; past and present uses; renewable and nonrenewable; NG.

The Power Struggle; 1986; 58 min.; problems of relying on nonrenewable energy sources and strategies to create a more reliable energy future; BFF.

Energy Supply; 36 min.; nonrenewables and renewables; BFF.
Fueling the Future: Running on Empty; 1988; 58 min.; the U.S. automobile; VP.
Toast; 12 min.; documents fossil-fuel inputs to make toast; BFF.

Nuclear

Deafsmith: A Nuclear Folktale; 1990; 43 min.; citizens' attempts to defend the land from nuclear
    pollution; VP.
Tjernobyl Efteraret (Chernobyl Autumn); 1987; 29 min.; Chernobyl's effect on Lapland; SF.
Chernobyl: The Taste of Wormwood; 1987; 52 min.; on-site photography of the Chernobyl blast site
    with interviews; FHS.
Radioactive Waste Disposal: The 10,000–Year Test; 50 min.; FHS.
The Transportation of Nuclear Materials; 40 min.; FHS.
Chernobyl: Chronicle of Difficult Weeks; 1986; 54 min.; directed by Vladimir Shevchenko, two weeks
    to three months after the incident; VP.
The Four Corners—A National Sacrifice Area?; 1983; 59 min.; cultural and ecological impact of
    energy development in the American Southwest; BFF.
Nuclear Energy: The Question Before Us; 1981; 26 min.; how a nuclear reactor works; pros and cons
    of nuclear power; NG.
Bound by the Wind; 1991; 40 min.; international impact of nuclear testing on people who live
    downwind; VP.
A Question of Power; 1986; 58 min.; perspectives on Diablo Canyon in California and applications
    to the present; VP.
The River That Harms; 1987; 45 min.; documentary of the largest radioactive waste spill in U.S.
    history and its impact on the Navajo Indians of New Mexico; VP.

Other Nonrenewable Energy Sources

Introduction to Petroleum in Michigan; 1987; 12 min.; geology and drilling of oil and gas formations
    in Michigan; BP.
Fusion: Work in Progress; 1989; 25 min.; NG.

See Appendix B for suppliers.

ATTITUDES/VALUES

Assessment

1.   Do you favor requiring all cars to get at least 21 kilometers per liter (50 miles per gallon) and
     vans and light trucks to get at least 15 kilometers per liter (35 miles per gallon) of gasoline
     within the next 10 years?

2.   Would you favor much stricter, twice-a-year inspections of air pollution control equipment
     on motor vehicles and tough fines for not keeping these systems in good working order?

3.   Would you favor a $2 tax on a gallon of gasoline and heating oil to help reduce wasteful
     consumption, extend oil supplies, reduce air pollution, delay projected global warming, and
     stimulate improvements in energy efficiency and the use of less harmful energy sources?

4.   Would you vote for anyone proposing a program to add a $2 per gallon tax on gasoline and
     heating oil, assuming all other factors are the same?

5. Would you support laws requiring that all new homes and buildings meet high energy efficiency standards for insulation, air infiltration, and heating and cooling systems?

6. Would you favor such a law for existing homes and buildings?

More Depth: Discussion and Term Paper Topics

1. Do you feel that the application process for nuclear power plants should be streamlined to limit citizen input?

2. Do you think emphasizing nuclear power and oil as the primary U.S. energy sources shows adequate concern for future generations?

3. Whistle-blowers. The Karen Silkwood story.

(Also, see text, Critical Thinking, p. 539.)

PARTICIPATION

Lifestyle and Campus Community

See *Green Lives/Green Campuses*, Chapter 10, Energy: Generating heat, p. 73; Chapter 11, Energy: Generating Electricity; Chapter 12, Energy: A Generation on the Go, p. 88.

More Depth: Action-oriented Term Paper Topics

1. National. The Atomic Energy Commission (1946–1975); the Nuclear Regulatory Commission; the Energy Research and Development Administration (ERDA); the Nuclear Safety Analysis Center; the Institute of Nuclear Power Operations; the Price-Anderson Act.

SKILLS

Environmental Problem-Solving Skills: Projects

1. As a class exercise, (a) obtain cost estimates for the construction of a single large-scale synfuels plant and (b) calculate how many soft path energy facilities or systems of various types (such as solar water heaters, biogas digesters, and photovoltaic devices) could be installed with an equivalent amount of money.

2. As a class exercise, make a list of the geopolitical responsibilities and costs incurred by the United States in association with maintaining an uninterrupted supply of affordably priced oil from foreign sources.

3. A new power plant must be built in your community, but it remains to be decided whether it will be a fossil fuel or a nuclear plant. As a class exercise, set up a mock public hearing to present the arguments for both sides. Make specific role assignments so that prepared statements will accurately reflect varying points of view such as those of contractors, environmentalists, project engineers, state energy officials, and concerned citizens.

Laboratory Skills

(none)

<u>Computer Skills</u>

ENERPLAN
    -Performs basic energy analysis for a nation, province, or community.
    -Mr. Nicky Beredjick, Director, National Resources and Energy Division, Department of Technical Cooperation for Development, United Nations, New York, NY 10017

Estimating Fossil Fuel Resources (EFFR)
    -Simulates the global exploitation of oil resources and evaluates alternative resource-estimation techniques.
    -Prof. John D. Sterman, Sloan School of Management, Massachusetts Institute of Technology, 50 Memorial Dr., Cambridge, MA 02139

SECTION TWO

MULTIPLE CHOICE QUESTIONS

CHAPTER 1
ENVIRONMENTAL PROBLEMS AND THEIR CAUSES

1.   All of the following illustrate exponential growth *except*
      a.   the king who promised to double the number of grains of wheat he put on each successive square of a checkerboard.
      b.   human population growth.
   * c.   driving 10 mph for 1 minute; then 20 mph for one minute; then 30 mph for one minute; then 40 mph for one minute.
      d.   money in a savings account.
      e.   folding paper in half 50 times so that the thickness would reach the sun.

2.   Most of the environmental problems we face are
      a.   increasing linearly.
      b.   decreasing linearly.
   * c.   increasing exponentially.
      d.   decreasing exponentially.
      e.   increasing logarithmically.

1-1   LIVING SUSTAINABLY

3.   Earth's capital includes all of the following *except*
      a.   wildlife.
      b.   air.
      c.   water.
      d.   soil.
   * e.   stocks and bonds.

4.   Which of the following categories are increasing in land surface each year?
      a.   tundra
      b.   forests
      c.   wetlands
      d.   grasslands
   * e.   deserts

5.   We are protected from excess ultraviolet radiation by the _____ in our atmosphere.
      a.   carbon dioxide
      b.   carbon monoxide
   * c.   ozone
      d.   nitrogen
      e.   oxides of nitrogen and sulfur

6.   The Union of Concerned Scientists points out all of the following observations *except*
      a.   stratospheric ozone is being depleted.
      b.   the total marine catch is now at or above the estimated maximum sustainable yield.
      c.   irreversible loss of species is particularly serious.
   * d.   the Earth is capable of detoxifying and balancing human impacts.
      e.   depletion of groundwater and pollution limits water supply.

7.   In order to move toward more sustainable societies, we must do all of the following *except*
      a.   bring environmentally damaging activities under control.
      b.   stabilize human population growth.
   * c.   define sex roles so that each has its own niche.
      d.   reduce current poverty levels.
      e.   manage resources crucial to human welfare more effectively.

8.   A sustainable society
      a.   manages its economy and population size without doing irreparable
            environmental harm.
      b.   satisfies the needs of its people without depleting Earth capital.
      c.   protects the prospects of future generations of humans and other species.
      d.   works with other countries.
  *   e.   all of the above.

1-2   POPULATION GROWTH AND THE WEALTH GAP

9.   Exponential growth
      a.   remains constant.
      b.   starts out slowly and remains slow.
  *   c.   starts out slowly then becomes very rapid.
      d.   starts rapidly and remains rapid.
      e.   starts out rapidly and becomes very slow.

10.   Linear growth
      a.   is characterized by a rapidly growing population.
  *   b.   is demonstrated by the sequence 12, 13, 14, 15, 16.
      c.   is illustrated by the numbers 2, 4, 8, 16, 32.
      d.   is characterized by resource use and consumption.
      e.   results in much more rapid growth than exponential growth.

11.   If the world's population grew by 2% in 1994 and continued at that rate, how long
      would it take Earth's population to double?
      a.   20 years
      b.   25 years
      c.   30 years
  *   d.   35 years
      e.   40 years

12.   Which of the following choices would make the statement *false*? Economic growth
      a.   provides goods and services for final use.
  *   b.   is accomplished by minimizing the flow of matter and energy through an economy.
      c.   is encouraged by population growth.
      d.   is encouraged by increased consumption per capita.
      e.   is generally considered to be a good thing.

13.   More developed countries (MDCs)
  *   a.   are highly industrialized.
      b.   have low gross national products per person.
      c.   are generally located in Asia.
      d.   make up about 80% of the world's population.
      e.   use about one-fourth of the world's energy resources.

14.   Which of the following statements about LDCs is *true?*
      a.   They are highly industrialized.
      b.   They have high average GNPs per person.
      c.   They include the United States, Canada, Japan, the Commonwealth of Independent
            States, and European countries.
      d.   They make up about one-fifth of the world's population.
  *   e.   They use about one-fifth of the world's resources.

15. About ___% of the world's human population lives in the less developed countries (LDCs).
    a. 49
    b. 58
    c. 67
  * d. 78
    e. 86

16. Since 1960, the gap between rich and poor, as measured by GNP per capita, has
    a. decreased, then increased since 1980.
    b. increased, then substantially decreased since 1980.
  * c. increased, then substantially increased since 1980.
    d. decreased, then substantially decreased since 1980.
    e. remained constant.

1-3   RESOURCES

17. For something to be classified as a natural resource, it must
  * a. satisfy a human need.
    b. be steadily renewed or replenished.
    c. be a form of matter.
    d. exist in great abundance.
    e. be used at sustainable-yield levels.

18. An example of a resource would be
    a. fresh air.
    b. fresh water.
    c. fertile soil.
    d. solitude.
  * e. all of the above.

19. We can extend use of nonrenewable resources by
    a. reducing direct consumption of the resource.
    b. reusing the same form of a particular resource many times.
    c. recycling a resource into new products.
    d. finding substitutes for a resource.
  * e. all of the above.

20. Resources that are called nonrenewable
    a. are also called perpetual resources.
    b. are only resources that are alive.
  * c. are capable of economic depletion.
    d. none of the above.
    e. b and c.

21. Nonrenewable resources include
    a. energy resources.
    b. nonmetallic resources.
    c. metallic resources.
  * d. all of the above.
    e. none of the above.

22. Nonrenewable resources include
    a. oil.
    b. salt.
    c. copper.
  * d. all of the above.
    e. none of the above.

23. All nonrenewable resources can be
    a. converted to nonmetallic minerals.
    b. converted to renewable ones.
  * c. exhausted or depleted.
    d. recycled or reused.
    e. used perpetually.

24. Which of the following is an example of recycling?
  * a. collecting and remelting aluminum beer cans
    b. cleaning and refilling soft-drink bottles
    c. selling used clothing at a garage sale
    d. saving leftovers in a peanut butter jar
    e. rescuing and repainting a toy you find at the dump

25. Reserves
    a. indicate limitless supplies of a resource.
    b. indicate established limits of a resource.
  * c. can be increased when new deposits are found.
    d. can be increased when price falls.
    e. a and c.
    f. b and d.

26. All of the following are renewable resources *except*
    a. groundwater.
    b. trees in a forest.
    c. fertile soil.
    d. air.
  * e. oil.

27. Use of a natural resource based on sustainable yield applies to
  * a. renewable resources.
    b. nonrenewable resources.
    c. perpetual resources.
    d. amenity resources.
    e. a and b.

28. Which of the following statements *best* illustrates the tragedy of the commons?
  * a. A factory pollutes a river as much as the law allows.
    b. Some levels of pollution are life-threatening.
    c. Some activities harm the environment, but others do not.
    d. Irrigated cropland can be ruined by salinization.
    e. People who walk on the commons ruin the grass.

29. Which of the following *best* describes the concept of environmental degradation?
    a. using solar power at a rapid rate.
    b. using oil.
    c. cutting trees for wood products.
  * d. letting agricultural runoff cause oxygen depletion and fish kills downstream.
    e. fertilizing crops.

30. Pollution includes
    a. dumping detergents into streams, causing fish kills.
    b. spraying with DDT, lowering the eagle population.
    c. releasing gases from coal combustion, causing acid rain.
    d. erupting volcanoes, destroying a forest ecosystem.
    e. a, b, and c.
 * f. all of the above.

31. Compared to pollutants from natural sources, pollutants resulting from human activities tend to be ___ concentrated in a particular area and ___ easily decomposed by natural processes.
 * a. more . . . less
    b. more . . . more
    c. less . . . less
    d. less . . . more
    e. similarly . . . similarly

32. Point sources of pollution
    a. enter ecosystems from dispersed and often hard-to-identify sources.
    b. include runoff of fertilizers and pesticides from farmlands and suburban lawns.
 * c. are cheaper and easier to identify than nonpoint sources.
    d. are more difficult to control than nonpoint sources.
    e. are described by all of the above.

33. Nonpoint sources of pollution
    a. enter ecosystems from single identifiable sources.
 * b. are more difficult to control than point sources.
    c. include smokestacks and automobile exhaust pipes.
    d. are cheaper and easier to identify than point sources.
    e. are described by all of the above.

34. Effects of pollution might include
    a. being unable to see the top of skyscrapers because of the smog.
    b. acid rain-induced destruction of a statue in your city park.
    c. spread of disease from an open dump.
    d. less diversity of stream life because of road salt runoff.
    e. b, c, and d.
 * f. all of the above.

35. Which of the following is *not* important in determining the damage produced by a pollutant?
    a. concentration
    b. persistence
 * c. origin
    d. chemical nature
    e. interaction with other chemicals

36. Degradable pollutants include
 * a. human sewage.
    b. DDT.
    c. aluminum cans.
    d. lead.
    e. mercury.

37. Persistent pollutants include
   a. DDT.
   b. aluminum cans.
   c. lead.
   d. mercury.
   * e. a and b.
   f. c and d.

38. Nondegradable pollutants include
   a. DDT.
   b. aluminum cans.
   c. lead.
   d. mercury.
   e. a and b.
   * f. c and d.

39. Pollution prevention strategies include
   a. reduce.
   b. reuse.
   c. recycle.
   * d. all of the above.
   e. none of the above.

40. Pollution cleanup strategies include
   a. reduce.
   b. reuse.
   c. recycle.
   d. all of the above.
   * e. none of the above.

41. You generally buy and eat microwave dinners. After dinner, cardboard tops and plastic trays remain. The *least* effective way to deal with this type of solid waste problem would be to
   a. buy microwave dinners with only cardboard components.
   b. prepare large quantities of food and divide it into your own reusable microwave containers.
   c. donate the plastic containers to the local nursery schools to use with preschoolers.
   d. recycle the components.
   * e. collect the components and incinerate them so they don't take up landfill space.

42. Pollution cleanup efforts can be overwhelmed by
   a. population growth.
   b. growth in consumption levels.
   c. poverty.
   * d. all of the above.
   e. none of the above.

43. Root causes of unsustainability include all of the following *except*
   a. resource waste.
   b. a utilitarian attitude toward the environment.
   c. loss of biodiversity.
   * d. inclusion of environmental and social costs in market prices.
   e. widespread use of fossil fuels.

44. Environmental impact of a human culture is determined by the population's
    a. size.
    b. affluence.
    c. consumption patterns.
    d. technological capability.
  * e. all of the above.

45. Efforts to improve environmental quality in the United States are predominantly based upon _____ wastes and pollutants.
    a. reducing
    b. reusing
    c. recycling
  * d. cleaning up
    e. none of the above.

1-5   THE CRISIS OF UNSUSTAINABILITY: PROBLEMS AND CAUSES

46. A very simple model of environmental degradation and pollution would include all of the following *except*
    a. number of people.
  * b. the climate in which the people live.
    c. average number of units of resources each person uses.
    d. amount of environmental degradation/pollution generated when each unit of resource is produced.

47. Which statement best illustrates people overpopulation?
    a. Air pollution is serious in the CIS.
  * b. Malnutrition is widespread in the LDCs.
    c. Per capita resource use is high in the United States.
    d. Japan's population is still growing.
    e. Many European countries are about at zero population growth.

48. Two children in the United States have as much environmental impact as _____ children in an LDC.
    a. 5–10
    b. 10–20
    c. 50–100
  * d. 100–200
    e. 400–500

49. All of the following are components of a multiple-factor model to account for environmental degradation and pollution *except*
    a. poverty.
    b. inappropriate application of technology.
  * c. realistic market prices.
    d. overconsumption and waste.
    e. none of the above; all are part of the model.

50. Paul Ehrlich is *least* likely to say:
    a. In absolute numbers, more people are hungry today than ever before.
    b. Land availability is very likely to constrain world agriculture.
  * c. Trends in world forests are little cause for concern since most biodiversity is found in marine environments.
    d. Water depletion is a problem in many areas.
    e. The environmental damage caused by burning oil is more of a problem than oil supply or demand.

51. Julian Simon is *least* likely to say:
    a. Although many people are still hungry, the food supply has been improving since World War II.
    b. Land availability won't constrain world agriculture in the coming decades.
    c. Trends in world forests are not worrisome.
    d. Water does not pose a problem of physical scarcity.
    * e. Renewable energy resources are the best choice for energy policy for the coming decades.

CHAPTER 2
CULTURAL CHANGES, WORLDVIEWS, ETHICS, AND SUSTAINABILITY

2-1   CULTURAL CHANGES

1.   The species *Homo sapiens sapiens* has lived on Earth about _____ years.
   a. 4,000
   b. 12,000
 * c . 40,000
   d. 75,000
   e. 100,000

2.   Exponential growth over time is characteristic of
   a.   human population growth.
   b.   resource consumption.
   c.   energy use per capita.
   d.   pollution and environmental degradation.
 * e.   all of the above.

3.   Which of the following statements does *not* characterize human skills and relationships within hunter-gatherer societies?
   a.   They lived in small groups of 50 or less.
   b.   They depended on sun, fire, and muscle power for energy.
 * c.   They had little knowledge about their natural surroundings.
   d.   They gradually developed tools and hunting weapons.
   e.   They learned to hunt large game cooperatively.

4.   Which of the following statements does *not* characterize relationships between hunter-gatherers and the environment?
   a.   They were nomadic.
   b.   They exploited their environment for food and other resources.
   c.   They were experts in survival and had a great understanding of nature.
   d.   Population size reflected food availability.
 * e.   They caused major environmental impacts.

5.   Advanced hunter-gatherer societies did all of the following *except*
   a.   make many kinds of tools and weapons.
   b.   use fire to flush out animals and to stampede herds.
   c.   use fire to convert forests into grasslands.
 * d.   subdue and dominate most other forms of life.
   e.   probably contribute to extinction of some large game animals.

6.   The Agricultural Revolution is characterized by
   a.   breeding and raising wild animals.
   b.   cultivating wild plants near settled communities.
   c.   fertilizing to improve soil fertility.
 * d.   a and b.
   e.   all of the above.

7.   Domestication of wild plants and animals occurred about ___ years ago.
   a.   5,000
 * b.   10,000
   c.   15,000
   d.   20,000
   e.   25,000

8. All of the following are characteristic of the first agricultural communities *except*
   a. slash-and-burn cultivation.
   * b. specialized farming of one crop.
   c. shifting cultivation.
   d. subsistence agriculture.
   e. placing roots and tubers in holes.

9. Slash-and-burn cultivation
   * a. leaves ashes from burned vegetation, which add plant nutrients to the soil.
   b. contours and terraces the land.
   c. ultimately leads to desertification.
   d. rotates crops yearly.
   e. a and b.

10. Shifting cultivation
    * a. alternates planting periods with fallow periods.
    b. permanently depletes the soil nutrients.
    c. ultimately leads to desertification.
    d. can be eliminated in stable societies.
    e. all of the above.

11. Subsistence farmers
    a. use draft animals to pull plows.
    b. require large, flat fields in grassland areas.
    * c. grow only enough food to feed their families.
    d. tend to cause severe deforestation.
    e. rapidly degrade the soil.

12. Which of the following human-resource relationships does *not* characterize a shift from hunter-gatherer to agricultural societies?
    a. Use of domesticated animals increased average energy use per person.
    b. Population increased with the increased food supply.
    c. More land was cleared and irrigated.
    d. People began accumulating material goods.
    * e. People used muscle, sun, and coal as energy sources.

13. The Agricultural Revolution resulted in all of the following *except*
    * a. protection of wild plants and animals.
    b. increased soil erosion.
    c. increased deforestation.
    d. salt buildup from irrigation.
    e. increased desertification.

14. The Industrial Revolution began in
    a. the United States.
    b. Japan.
    * c. England.
    d. France.
    e. Germany.

15. The Industrial Revolution started in the United States in the
    a. 1500s.
    b. 1600s.
    c. 1700s.
    * d. 1800s.
    e. 1900s.

16. The Industrial Revolution is characterized by all of the following *except*
    a. increased average per capita energy consumption.
    b. increased ability to utilize Earth's resources.
    c. increased economic growth.
    d. increased trade and distribution of goods.
    * e. increased social concern for workers.

17. A major stimulus for the Industrial Revolution was
    a. the bubonic plague.
    b. European wars.
    * c. shortage of wood for fuel and construction.
    d. poverty.
    e. food shortage.

18. Energy use during the Industrial Revolution
    a. was based primarily on wood.
    b. was based primarily on solar power.
    c. was based primarily on labor by human muscle.
    d. was based primarily on labor by domesticated animals.
    * e. shifted from renewable to nonrenewable sources.

19. The early Industrial Revolution brought
    a. a movement of workers to cities.
    b. an increased number of assembly-line jobs.
    c. an increased number of coal-mining jobs.
    d. accelerated exponential human population growth.
    * e. all of the above.

20. Benefits bestowed on most citizens of industrialized countries include all of the following *except*
    a. more affordable material goods.
    b. increase in average agricultural production per person.
    c. higher average life expectancy.
    * d. continued exponential growth of human population.
    e. better nutrition, medicine, and sanitation.

21. An important change to create a sustainable-Earth society would be
    a. a shift from habitat protection to species protection.
    b. a more efficient economic system.
    c. a streamlined political system.
    * d. a shift from pollution cleanup to pollution prevention.
    e. a strong national government to impose pollution control standards.

2-2 WORLDVIEWS: A CLASH OF VALUES AND CULTURES

22. Basic beliefs of planetary management worldviews include all of the following *except*
    a. We are the planet's most important species and nature requires our management.
    b. All economic growth is good.
    c. Our success depends on how well we control the planet for our benefit.
    * d. A healthy economy depends on a healthy environment.
    e. There is an unlimited supply of resources for our use.

23. Planetary management worldviews include all of the following variations *except*
    a. enlightened self-interest.
    b. free-market school.
    c. "no problem" school of thought.
    d. stewardship.
    * e. ecocentrism.

24. The "no problem" variation of the planetary management worldview is based on the general belief that
    a. pure capitalism should be used to make our economic decisions.
  * b. better science and technology can fix our problems.
    c. an ethical responsibility to "tend our garden" would improve most technological–economic growth worldviews.
    d. a mixture of market-based competition, improved technology, and government intervention can solve our problems.
    e. a shift to a female-oriented society will solve our problems.

25. People calling for more stewardship generally believe that
    a. pure capitalism should be used to make our economic decisions.
    b. better science and technology can fix our problems.
  * c. an ethical responsibility to "tend our garden" would improve most technological–economic growth worldviews.
    d. a mixture of market-based competition, improved technology, and government intervention can solve our problems.
    e. a shift to a female-oriented society will solve our problems.

26. A spaceship-Earth strategy is often used by which of the following worldviews?
    a. "no problem" school of thought
    b. free-market school
  * c. enlightened self-interest
    d. stewardship
    e. ecocentrism

27. Which of the following worldviews is *most* likely to support the view that most public property should be turned over to private ownership?
    a. "no problem" school of thought
  * b. free-market school
    c. enlightened self-interest
    d. stewardship
    e. ecocentrism

28. Which worldview is *most* likely to be based on a belief that better Earth care is better self-care?
    a. "no problem" school
    b. free-market school
  * c. responsible planetary management
    d. stewardship
    e. ecocentrism

29. Which of the following worldviews is based on a belief in the least government interference?
    a. "no problem" school
  * b. free-market school
    c. responsible planetary management
    d. stewardship
    e. ecocentrism

30. Which of the following would *not* characterize an Earth-wisdom worldview?
    a. Nature exists for all of Earth's species; not just us.
    b. Earth's resources are limited and should be conserved.
    c. Some forms of economic growth are beneficial and some are harmful.
    d. A healthy economy depends on a healthy environment.
  * e. Our success depends on exploiting natural resources in the most efficient ways.

31.   In describing the relationship between humans and ecosystems, biocentrists are *least* likely to assert that
      a.   we should sustain Earth's physical, chemical and biological capital.
      b.   we should try to understand and cooperate with nature.
      c.   when we use nature's capital, we should do the least harm.
      d.   our plans should include environmental impact analyses.
   * e.   our plans should focus on the needs of the present and remember that future generations will focus on their own problems.

32.   In describing human relationships with species and human cultures, biocentrists are *least* likely to say
      a.   each species has an innate right to struggle to live.
      b.   we should protect species from premature extinction.
   * c.   the best way to protect species is through passing legislation which protects endangered species.
      d.   no human culture should become extinct because of our actions.
      e.   we have a right to defend ourselves against species that harm us.

33.   In describing individual human responsibility, biocentrists are least likely to say we should
   * a.   use other species to meet all our needs and wants.
      b.   prevent excessive human births.
      c.   leave Earth in as good shape as we find it.
      d.   live lightly on the earth.
      e.   develop a sense of place.

34.   Based on his philosophy of Earth education, Tyler Miller would *least* appreciate a student who
      a.   drew connections among all the chapters of the book.
      b.   showed a deep respect for all life.
      c.   tried to apply ecological understandings to a sustainable lifestyle.
   * d.   memorized the entire textbook.
      e.   made a commitment to lifelong learning.

35.   Ways to improve Earth education would include all of the following *except*
      a.   exposing all teachers, media, corporate and government leaders to Earth education.
   * b.   offering more in-depth courses at the elementary level.
      c.   setting up a dual-track specialist-holistic system.
      d.   imbedding examples of holistic thinking into all teaching materials.
      e.   developing better measures of ecological literacy.

36.   Earth education calls for formal educational systems to become more
      a.   reductionistic.
      b.   discipline-oriented.
      c.   back to basics.
   * d.   holistic.
      e.   all of the above.

37.   It is easier to live sustainably when you
      a.   have a sense of place.
      b.   listen to children.
      c.   focus on the bioregion in which you live.
      d.   live simply.
   * e.   all of the above.

38. All of the following are traps that block living sustainably *except*
    a. a faith in simple, easy answers.
    b. gloom-and-doom pessimism.
    * c. critical evaluation of experts and leaders.
    d. blind technological optimism.
    e. paralysis-by-analysis.

39. According to Peter Montague, the environmentalism of the 1990s will bring clean production involving minimization of damage to natural ecosystems when
    a. raw materials are selected, extracted, and processed.
    b. products are designed.
    c. products are transported.
    d. products are used in and discarded from industries or homes.
    * e. all of the above.

40. Which question is Lester Brown *least* likely to ask in evaluating sustainable development policies?
    * a. Does it give investors the best quarterly returns?
    b. Does it slow population growth?
    c. Does it increase tree cover?
    d. Does it reduce generation of carbon emissions and toxic wastes?
    e. Does it protect the planet's soil and biodiversity?

PART II. PRINCIPLES AND CONCEPTS

CHAPTER 3
MATTER AND ENERGY RESOURCES: TYPES AND CONCEPTS

1.   Osage, Iowa made national news by
     a.   having a nuclear power plant disaster.
 *   b.   becoming the energy-efficiency capital of the United States.
     c.   finding the largest new shale oil deposit in the United States.
     d.   producing vast amounts of energy through burning garbage.
     e.   showing its appreciation of other towns by sending $1.2 million in energy expenses out of the local community.

2.   In his campaign to improve energy efficiency in Osage, Iowa, Wes Birdsall advised all of the following *except*
 *   a.   replacing fluorescent bulbs with energy-incandescent bulbs.
     b.   turning down temperatures on water heaters.
     c.   wrapping water heaters with insulation.
     d.   thermograms of buildings.
     e.   low-flow showerheads.

3-1   SCIENCE AND ENVIRONMENTAL SCIENCE

3.   Science
     a.   studies the past to predict the future.
     b.   attempts to discover order in nature to interpret the past.
     c.   is best described as a collection of facts found through using scientific methods.
 *   d.   uses data to formulate scientific laws.
     e.   is Absolute Truth.

4.   Discovering and formulating scientific laws requires
     a.   logic.
     b.   imagination.
     c.   use of scientific methods.
     d.   intuition.
 *   e.   all of the above.

5.   Which of the choices makes the following statement *false?*
     Technology _____
     a.   is the creation of new products and processes.
     b.   is supposed to improve our quality of life.
 *   c.   is the development of scientific laws and theories.
     d.   resulted in lasers and pollution control devices.
     e.   advances are often kept secret until new processes and products are patented.

6.   Which of the following statements does *not* describe the scientific enterprise?
     a.   Science is the acceptance of what works and the rejection of what does not.
 *   b.   Once established, scientific theories are rarely challenged and continue to hold true into the future.
     c.   Advances in scientific knowledge are often based on vigorous disagreement, speculation, and controversy.
     d.   Scientific laws and theories are based on statistical probabilities, not certainties.
     e.   Science has been advanced primarily through the use of reductionism.

7. Environmental science integrates knowledge from the disciplines of
    a. chemistry and physics.
    b. ecology and demography.
    c. resource technology and engineering.
    d. economics and politics.
    *   e. all of the above.

3-2 MATTER: FORMS, STRUCTURE, AND QUALITY

8. Matter is anything that
    *   a. has mass and occupies space.
    b. has the capacity to do work.
    c. can be changed in form.
    d. can produce change.
    e. can be conserved.

9. All of the following are elements *except*
    *   a. water.
    b. oxygen.
    c. nitrogen.
    d. hydrogen.
    e. carbon.

10. Liquid, solid, and gas are
    *   a. physical forms of matter.
    b. chemical forms of matter.
    c. mixtures.
    d. compounds.
    e. elements.

11. Protons, neutrons, and electrons are all
    a. forms of energy.
    b. equal in mass.
    *   c. subatomic particles.
    d. negative ions.
    e. positively charged.

12. N_2 and O_2 are examples of
    a. compounds consisting of two different elements.
    b. elements consisting of a compound and an ion.
    c. molecules consisting of two elements of the same compound.
    *   d. molecules consisting of two atoms of the same element.
    e. molecules consisting of two ions of the same element.

13. The atomic number is the number of
    a. atoms in a molecule.
    *   b. protons in an atom.
    c. nuclei in a molecule.
    d. electrons in an atom.
    e. neutrons in an atom.

14. The atomic mass is equal to the sum of the
    a. neutrons and isotopes.
    b. neutrons and electrons.
    *   c. neutrons and protons.
    d. protons, neutrons, and electrons.
    e. protons and electrons.

15. Isotopes differ from each other by their number of
    a. ions.
    b. protons.
    c. atoms.
 * d. neutrons.
    e. electrons.

16. All organic compounds are characterized by the presence of
 * a. carbon.
    b. hydrogen.
    c. oxygen.
    d. nitrogen.
    e. sulfur.

17. Which of the following sources of iron would be of the highest quality?
    a. iron deposits on the ocean floor
    b. a shelf of iron supplements in the pharmacy
 * c. a large, scrap metal junkyard
    d. a one-half-mile deep deposit of iron ore
    e. a field of spinach

3-3   ENERGY: TYPES, FORMS, AND QUALITY

18. Energy can be formally defined as
    a. the random motion of molecules.
 * b. the ability to do work or produce heat transfer.
    c. a force that is exerted over some distance.
    d. the movement of molecules.
    e. anything that occupies space and has mass.

19. Most forms of energy can be classified as either
    a. chemical or physical.
    b. kinetic or mechanical.
    c. potential or mechanical.
 * d. potential or kinetic.
    e. electrical or physical.

20. All of the following are examples of kinetic energy *except*
    a. a speeding bullet.
 * b. a stick of dynamite.
    c. a flow of electric current.
    d. a falling rock.
    e. a waterfall.

21. An example of potential energy is
    a. electricity flowing through a wire.
 * b. the chemical energy in a candy bar.
    c. a bullet fired at high velocity.
    d. a leaf falling from a tree.
    e. a perpetual motion machine.

22. All of the following are examples of ionizing radiation *except*
    a. cosmic rays.
    b. gamma rays.
 * c. microwaves.
    d. X rays.
    e. ultraviolet rays.

23. Ionizing radiation can
    a. interfere with body processes.
    b. cause cancer.
    c. change neutral atoms to positively charged ions.
    d. disrupt living cells.
    * e. all of the above.

24. Nonionizing radiation is emitted from all of the following *except*
    a. TV sets.
    b. video display terminals.
    c. electric blankets.
    d. overhead electric power lines.
    * e. X-ray machines.

25. All of the following are *indirect* forms of solar energy *except*
    a. waterfalls.
    b. biomass.
    * c. sunlight.
    d. wind.
    e. streams.

26. Which of the following is an example of direct solar energy that has been converted to an indirect form of solar energy and then stored as chemical energy?
    a. gravity
    b. wind
    * c. coal
    d. glaciers
    e. a waterfall

27. What percentage of the energy used to make the temperature of Earth's surface livable comes from commercial energy?
    * a. 1%
    b. 5%
    c. 10%
    d. 25%
    e. 50%

28. The most important supplement to solar energy in most LDCs is
    a. oil.
    b. hydropower.
    * c. fuelwood.
    d. coal.
    e. wind.

29. The most important supplement to solar energy in the United States is
    * a. oil.
    b. hydropower.
    c. fuelwood.
    d. coal.
    e. wind.

30. With 4.7% of the world's population, the U.S. uses about ___% of the world's commercial energy.
    a. 4.7
    b. 10
    c. 20
    * d. 25
    e. 35

31. Which of the following is an example of low-quality energy?
    a. electricity
    * b. heat in the ocean
    c. nuclei of uranium-235
    d. coal
    e. oil

32. High-quality energy is needed to do all of the following *except*
    a. run electric lights.
    b. run electric motors.
    c. run electric appliances.
    d. run a car.
    * e. heat the White House.

3-4 PHYSICAL AND CHEMICAL CHANGES AND THE LAW OF CONSERVATION OF MATTER

33. Which of the following statements is *not* an example of a physical change?
    a. Confetti is cut from pieces of paper.
    b. Water evaporates from a lake.
    c. Ice cubes are formed in the freezer.
    d. Gold earrings are melted down and recast into new works of art.
    * e. A plant converts carbon dioxide into carbohydrate.

34. All of the following statements can be concluded from the law of conservation of matter *except*
    a. We can't throw anything away because there is no away.
    * b. We'll eventually run out of matter if we keep consuming it at current rates.
    c. There will always be pollution of some sort.
    d. Everything must go somewhere.
    e. Things we throw away just change form.

35. To become a more sustainable society,
    a. "wastes" need to be viewed as "resources."
    b. resources need to be conserved.
    c. pollution needs to be prevented.
    d. the flow of matter needs to be reduced.
    * e. all of the above.

3-5 NUCLEAR CHANGES

36. Which of the following involves changes of mass into energy?
    a. chemical changes
    b. energy changes
    c. physical changes
    * d. nuclear changes
    e. biological changes

37. Nuclear fission is best described as a
    a. chemical change.
    b. physical change.
    c. nuclear change with spontaneous release of fast-moving particles and/or high-energy radiation from unstable isotopes.
    * d. nuclear change with release of energy from splitting of isotopes with large mass numbers.
    e. nuclear change with release of energy from combining two nuclei of isotopes of light elements.

38. Nuclear fusion is best described as a
   a. chemical change.
   b. physical change.
   c. nuclear change with spontaneous release of fast-moving particles and/or high-energy radiation from unstable isotopes.
   d. nuclear change with release of energy from splitting of isotopes with large mass numbers.
 * e. nuclear change with release of energy from combining two nuclei of isotopes of light elements.

39. Natural radioactive decay is best described as a
   a. chemical change.
   b. physical change.
 * c. nuclear change with spontaneous release of fast-moving particles and/or high-energy radiation from unstable isotopes.
   d. nuclear change with release of energy from splitting of isotopes with large mass numbers.
   e. nuclear change with release of energy from combining two nuclei of isotopes of light elements.

40. All of the following are given off by natural radioactivity *except*
   a. alpha particles.
 * b. delta rays.
   c. gamma rays.
   d. beta particles.
   e. b and c.

41. The two most common types of ionizing particles emitted by radioactive isotopes are
   a. gamma and alpha particles.
   b. gamma and beta particles.
 * c. alpha and beta particles.
   d. electrons and protons.
   e. protons and neutrons.

42. Isotopes of the same element differ from each other in their number of
   a. electrons.
   b. alpha particles.
 * c. neutrons.
   d. ions.
   e. protons.

43. Which of the following statements is *true?*
   a. Exposure of a substance to alpha, beta, or gamma radiation makes it radioactive.
   b. All isotopes are radioactive.
 * c. Radioactive isotopes give off radiation at a fixed rate.
   d. Only naturally occurring substances are radioactive.
   e. Radioactive substances originate from volcanoes.

44. Which of the following types of radiation is *most* penetrating?
   a. ultraviolet
   b. alpha
   c. beta
 * d. gamma
   e. X-ray

45. To decay to what is considered to be a safe level, a sample of radioisotope should be stored in a safe enclosure for approximately _____ half-lives.
   a. 2
   b. 5
 * c. 10
   d. 20
   e. 50

46. During a nuclear-fission reaction, each fission releases two or three
   a. protons and energy.
   b. electrons and energy.
 * c. neutrons and energy.
   d. protons, requiring the input of energy.
   e. electrons, requiring the input of energy.
   f. neutrons, requiring the input of energy.

47. In conventional nuclear-fission reactors, the fuel is
   a. natural gas.
 * b. uranium-235.
   c. alpha particles.
   d. beta particles.
   e. gamma rays.

48. The explosion of an atomic bomb is best described as
   a. uncontrolled radioactive decay.
 * b. uncontrolled nuclear fission.
   c. uncontrolled nuclear fusion.
   d. controlled nuclear fission.
   e. controlled nuclear fusion.
   f. controlled radioactive decay.

49. The generation of energy in a nuclear electric powerplant comes from
   a. uncontrolled radioactive decay.
   b. uncontrolled nuclear fission.
   c. uncontrolled nuclear fusion.
 * d. controlled nuclear fission.
   e. controlled nuclear fusion.
   f. controlled radioactive decay.

50. Which of the following would be the third step in the production of electricity from a nuclear powerplant?
   a. spinning turbines
   b. heat
   c. fission
 * d. high-pressure steam
   e. electricity

51. Nuclear fusion requires temperatures of at least _____ degrees centigrade.
   a. 100
   b. 100 thousand
 * c. 100 million
   d. 100 billion
   e. 100 trillion

52. The source of energy in the sun and stars is
    a. chemical change.
    b. physical change.
    c. nuclear change with spontaneous release of fast-moving particles and/or high-energy radiation from unstable isotopes.
    d. nuclear change with release of energy from splitting of isotopes with large mass numbers.
  * e. nuclear change with release of energy from combining two nuclei of isotopes of light elements.

53. Thermonuclear weapons get their energy from
    a. uncontrolled radioactive decay.
    b. uncontrolled nuclear fission.
  * c. uncontrolled nuclear fusion.
    d. controlled nuclear fission.
    e. controlled nuclear fusion.
    f. controlled radioactive decay.

54. Which of the following statements about nuclear fusion is *false?*
    a. During nuclear fusion, two nuclei of isotopes of light elements are forced together at high temperatures until they fuse and release energy.
    b. Fusion is the source of energy in the sun.
    c. High-temperature fusion is much harder to initiate but releases more energy per unit of fuel than fission.
    d. Hydrogen bombs use nuclear fusion.
  * e. Controlled nuclear fusion reactors are being tested in California and will probably be ready to come on-line by the year 2000.

55. Which of the following is a product of a fusion reaction?
    a. alpha particles
    b. deuterium
    c. tritium
  * d. helium
    e. heavy water

3-6   THE FIRST AND SECOND LAWS OF ENERGY

56. Which of the following statements is *false?*
    a. Energy can be converted from one form to another.
  * b. Energy and matter can generally be converted into each other.
    c. Energy input always equals energy output.
    d. The laws of thermodynamics can be applied to living systems.
    e. We can't get something for nothing in terms of energy quantity.

57. The first law of energy tells us that
    a. doing work always creates heat.
    b. altering matter is the best source of energy.
    c. energy cannot be recycled.
  * d. it takes energy to get energy.
    e. entropy tends to increase.

58. Which of the following statements does *not* apply to the second law of energy?
    a. Energy conversion results in lower-quality energy.
  * b. Energy can be neither created nor destroyed.
    c. Energy conversion results in more-dispersed energy.
    d. Heat is usually given off from energy conversions.
    e. We cannot recycle high-quality energy to do useful work.

59. Energy input is
    a. usually greater than energy output.
    b. always greater than energy output.
  * c. always equal to energy output.
    d. usually less than energy output.
    e. always less than energy output.

60. The energy lost by a system is
    a. usually found.
    b. equal to the energy the system creates.
  * c. converted to lower-quality energy.
    d. returned to the system eventually.
    e. converted into an equal amount of matter.

61. Which of the following energy sources has the lowest quality?
    a. high-velocity water flow
    b. fuelwood
    c. food
  * d. dispersed geothermal energy
    e. Saudi Arabian oil deposits

62. In an energy transformation, some of the energy usually ends up as
  * a. heat energy that flows into the environment.
    b. mechanical energy that performs useful work.
    c. chemical energy that performs useful work.
    d. electrical energy that performs useful work.
    e. potential energy that flows into the environment.

63. The matter and energy laws tell us that we can recycle
    a. both matter and energy.
    b. neither matter nor energy.
  * c. matter but not energy.
    d. energy but not matter.
    e. none of the above.

64. In any heat-to-work conversion, the quality of the energy available after the work is performed will always be _____ the initial energy quality.
    a. equal to
    b. higher than
    c. equal to or higher than
  * d. lower than
    e. equal to or lower than

65. Earth's supply of concentrated usable energy is being steadily
  * a. depleted.
    b. replenished.
    c. converted to more usable forms.
    d. converted to higher-quality forms.
    e. exported to space.

66. Which of the following statements is *not* an observation derived from applying the second law of thermodynamics to living systems?
    a. Life is a creation and maintenance of ordered structures.
    b. High-quality energy sources are required to maintain life.
    c. Living things give off heat.
  * d. Cooking foods turns them into high-quality energy sources.
    e. The more energy we use, the more disorder we create in the environment.

67. A throwaway society sustains economic growth by
    a. minimizing the rate of energy resource use.
    b. minimizing the rate of energy and matter resource use.
    c. maximizing the rate of energy resource use.
  * d. maximizing the rate of energy and matter resource use.
    e. maximizing the rate of matter resource use.

68. Which of the following statements is the most logical way to cope with the problem of limitations imposed by the three basic physical laws governing matter?
  * a. Use and waste less energy and matter.
    b. Shift to nonpolluting nuclear fusion power.
    c. Increase the output of low-quality heat.
    d. Increase the input of high-quality energy.
    e. Decrease the use of low-quality matter.

69. Which of the following statements about a matter-recycling society is *false?*
    a. The goal of a matter-recycling society is to allow economic growth to continue without depleting matter resources and without producing excessive pollution and environmental degradation.
    b. One limit of a matter-recycling society is dependence on high-quality energy to recycle materials.
    c. A matter-recycling society is limited by the environment's capacity to absorb and disperse waste heat and to dilute and degrade waste matter.
  * d. A matter-recycling society becomes independent of high-quality matter because materials can continue to be recycled indefinitely.
    e. The main purpose of a matter-recycling society is to give us more time to shift to a sustainable-Earth society.

70. Sustainable-Earth societies would do all of the following *except*
    a. use energy more efficiently.
    b. shift to perpetual and renewable energy sources.
    c. recycle and reuse most matter that is now discarded.
  * d. create goods with a short life cycle to increase recycling.
    e. slow human population growth and reduce poverty.

CHAPTER 4
ECOSYSTEMS AND HOW THEY WORK

1.   Wind carries
    a.   nutrients.
    b.   soot.
    c.   DDT and PCBs.
    d.   particles from volcanic eruptions.
  * e.   all of the above.

4-1   LIFE AND EARTH'S LIFE-SUPPORT SYSTEMS

2.   The basic unit of life is the
    a.   nucleotide.
    b.   mitochondrion.
  * c.   cell.
    d.   tissue.
    e.   organ.
    f.   system.

3.   All of the following are characteristic of life forms *except*
  * a.   highly diffuse internal structure and organization.
    b.   the ability to capture and transform matter and energy from the environment.
    c.   the ability to reproduce.
    d.   the ability to adapt to external change by mutations.
    e.   the ability to maintain favorable internal conditions in spite of external changes, if not
        overwhelmed.

4.   Biodiversity emerges from
    a.   mutations.
    b.   natural selection.
    c.   extinction.
    d.   evolution.
  * e.   all of the above.

5.   The thin, gaseous layer of air around the planet is called the
  * a.   atmosphere.
    b.   lithosphere.
    c.   stratosphere.
    d.   hydrosphere.
    e.   troposphere.

6.   All physical forms of water (solid, liquid, and gas) make up the
    a.   atmosphere.
    b.   lithosphere.
    c.   biosphere.
  * d.   hydrosphere.
    e.   troposphere.

7.   Fossil fuels and minerals are found in the
    a.   atmosphere.
  * b.   lithosphere.
    c.   biosphere.
    d.   hydrosphere.
    e.   troposphere.

8. Submarines explore the
    a. atmosphere.
    b. lithosphere.
    c. biosphere.
*   d. hydrosphere.
    e. troposphere.

9. Geologists find rock and soil samples in the
    a. atmosphere.
*   b. lithosphere.
    c. biosphere.
    d. hydrosphere.
    e. troposphere.

10. Children fly kites in the
    a. stratosphere.
    b. lithosphere.
    c. biosphere.
    d. hydrosphere.
*   e. troposphere.

11. Ecosphere is the same as
    a. atmosphere.
    b. lithosphere.
*   c. biosphere.
    d. hydrosphere.
    e. troposphere.

12. Life on Earth depends on interaction of gravity and
    a. one-way flow of energy.
    b. cycling of energy.
    c. one-way flow of matter.
    d. cycling of matter.
*   e. a and d.
    f. b and c.

13. Energy
    a. recycles through the ecosystem.
*   b. flows in only one direction.
    c. is used over and over again.
    d. tends to be concentrated by living organisms.
    e. originates in the center of the earth.

14. The sun is composed primarily of
*   a. hydrogen.
    b. helium.
    c. heavy metals.
    d. ions.
    e. neutrons.

15. Which of the following statements is *false?*
    a. About one-third of the solar energy hitting the earth is immediately reflected back
        to space.
    b. A spectrum of electromagnetic radiation emanates from the sun.
*   c. About one-third of the solar energy hitting the earth warms the land and lower
        atmosphere, runs cycles of matter, and generates winds.
    d. Less than 1% of sunlight is captured via photosynthesis.
    e. Nuclear fusion is the source of energy radiating from the sun.

16. When incoming solar radiation is converted to heat, it may be trapped in the atmosphere by all of the following *except*
    a. water vapor.
    b. carbon dioxide.
    c. methane.
  * d. nitrogen gas.
    e. ozone.

17. Which of the following statements is *true*?
    a. The earth's elements generally do not occur in a form useful to living organisms.
    b. The elements and compounds of important nutrients are continually cycled in complex paths.
    c. Nutrients are made available to living organisms by geological, biological, and chemical processes.
    d. Air, water, and land are reservoirs for biologically important compounds.
  * e. All of the above.

18. All of the following are elements involved in major biogeochemical cycles *except*
    a. nitrogen.
  * b. calcium.
    c. sulfur.
    d. oxygen.
    e. phosphorous.

4-2   ECOSYSTEM COMPONENTS

19. Ecology is the study of
    a. how atoms make up the environment.
    b. how humans affect the environment.
  * c. how organisms interact with each other and their nonliving environment.
    d. how energy runs the environment.
    e. how societies pass laws to protect the environment.

20. The goal of ecology is to
    a. eliminate pollution.
    b. eliminate environmental degradation.
    c. trace flow of energy through the environment.
  * d. learn about connections in nature.
    e. identify all the organisms in the world.

21. Which of the following includes all of the others?
    a. species
    b. population
    c. community
    d. organism
  * e. biome

22. A group of individuals of the same species occupying a given area at the same time is called a
    a. species.
  * b. population.
    c. community.
    d. genus.
    e. subspecies.

23. A community of living organisms interacting with one another and the physical and chemical factors of their nonliving environment is called
    a. a species.
*   b. an ecosystem.
    c. a population.
    d. a lithosphere.
    e. a biosphere.

24. The primary factor determining the types and abundance of life in a particular land area is
*   a. climate.
    b. longitude.
    c. weather.
    d. soil.
    e. oxygen.

25. Which term includes the others?
    a. species diversity
    b. genetic diversity
*   c. biological diversity
    d. ecological diversity

26. Simple cells without a distinct nucleus are called
*   a. prokaryotic.
    b. eukaryotic.
    c. akaryotic.
    d. bacterial.
    e. protistan.

27. To which of the following kingdoms do diatoms and amoebas belong?
    a. bacteria
*   b. protists
    c. plants
    d. animals
    e. fungi

28. To which of the following kingdoms do mushrooms and molds belong?
    a. bacteria
    b. protists
    c. plants
    d. animals
*   e. fungi

29. Which of the following are examples of deciduous plants?
*   a. maples and oaks
    b. algae and seaweed
    c. bacteria
    d. pines and cedars
    e. mosses and ferns

30. Which of the following are examples of evergreen plants?
    a. maples and oaks
    b. algae and seaweed
    c. bacteria
*   d. pines and cedars
    e. mosses and ferns

31. Succulent plants are most likely to be found in
    a. aquatic habitats.
    b. cold ecosystems.
    c. high altitudes.
    * d. deserts.
    e. tropical rain forests.

32. The service *least* likely to be performed by the insect family is
    a. plant reproduction.
    b. plant pollination.
    c. turning the soil.
    * d. chemosynthesis.
    e. decomposing dead tissues.

33. Which adjective *least* applies to insects?
    a. adaptable
    b. abundant
    * c. unnecessary
    d. diverse
    e. invincible

34. All of the following are abiotic factors *except*
    a. light.
    b. temperature.
    c. pH.
    d. size of soil particles.
    * e. bacteria.

35. The most inclusive components of the biotic portion of an ecosystem are
    * a. producers, consumers, and decomposers.
    b. primary and secondary consumers.
    c. herbivores, carnivores, and omnivores.
    d. all nonliving chemicals or matter.
    e. detritivores.

36. Autotrophs
    a. carry on chemosynthesis.
    b. are known as producers.
    c. carry on photosynthesis.
    d. can live without heterotrophs.
    * e. all of the above.

37. Photosynthesis
    a. converts glucose into energy and water.
    b. requires the combustion of carbon.
    c. produces carbon dioxide and oxygen gas.
    * d. yields glucose and oxygen gas as products.
    e. requires carbon dioxide and nitrogen gas.

38. The conversion of solar energy into chemical energy occurs in
    * a. photosynthesis.
    b. food chains.
    c. chemosynthesis.
    d. heterotrophic organisms.
    e. decomposers.

39. Chemosynthesis could utilize
    a. sunlight.
    b. carbon dioxide.
  * c. hydrogen sulfide.
    d. ammonia.
    e. water.

40. Organisms that feed on plants are called
    a. detritus feeders.
    b. omnivores.
    c. carnivores.
  * d. herbivores.
    e. decomposers.

41. Organisms that feed on both plants and animals are called
    a. detritus feeders.
  * b. omnivores.
    c. carnivores.
    d. herbivores.
    e. decomposers.

42. All of the following are consumers *except*
    a. herbivores.
    b. carnivores.
    c. omnivores.
  * d. producers.
    e. decomposers.

43. The organisms that are classified as primary consumers are the
    a. detritivores.
    b. omnivores.
    c. carnivores.
  * d. herbivores.
    e. producers.

44. Aerobic respiration requires
    a. glucose and carbon dioxide.
  * b. glucose and oxygen.
    c. oxygen and water.
    d. carbon dioxide and water.
    e. carbon dioxide and energy.

45. An ecosystem can survive without
    a. producers.
  * b. consumers.
    c. decomposers.
    d. b and c.
    e. all of the above.

46. Which of the following statements is *true*?
    a. Energy cycles and nutrients cycle.
    b. Energy cycles and nutrients flow.
  * c. Energy flows and nutrients cycle.
    d. Energy flows and nutrients flow.
    e. None of the above.

47. The range of tolerance is controlled by
    a. the species involved.
    b. the individual's reproductive state.
    c. the stage in the life cycle.
    d. b and c.
  * e. all of the above.

48. Which of the following statements is *false?*
    a. The existence, abundance, and distribution of a species in an ecosystem are determined by whether the levels of one or more physical or chemical factors fall within the range tolerated by a species.
    b. Organisms can adapt to slowly changing new conditions by acclimation.
    c. Too much or too little of any abiotic factor can limit or prevent growth of a population of a species in an ecosystem even if all other factors are at or near the optimum range of tolerance.
  * d. There is no such thing as too much fertilizer.
    e. Temperature, water, light, and soil nutrients can be limiting factors in terrestrial ecosystems.

49. The factor *least* likely to limit growth of a population in a land ecosystem is
  * a. salinity.
    b. temperature.
    c. water.
    d. soil nutrients.
    e. light.

50. In aquatic ecosystems, the factor *least* likely to limit growth of a population is
    a. salinity.
    b. temperature.
    c. sunlight.
    d. dissolved oxygen.
    e. nutrient availability.
  * f. precipitation.

4-3   CONNECTIONS: ENERGY FLOW IN ECOSYSTEMS

51. Complex feeding patterns for consumers in an ecosystem are called
  * a. food webs.
    b. food chains.
    c. trophic levels.
    d. pyramids of energy.
    e. biomass.

52. Most of the energy input in a food chain is
    a. in the form of heat.
    b. converted to biomass.
    c. recycled as it reaches the chain's end.
  * d. degraded to low-quality heat.
    e. used to recycle detritus.

53. The shorter the food chain the
  * a. smaller the loss of usable energy.
    b. fewer the number of organisms supported.
    c. lower the net primary productivity.
    d. smaller the gross primary productivity.
    e. greater the heat loss.

54. Which of the following statements is *false?*
    a. There is no waste in functioning natural ecosystems.
    b. Food webs reflect the complexity of real ecosystems better than food chains.
    c. Producers belong to the first trophic level.
    d. Primary consumers are on a lower trophic level than secondary consumers.
  * e. Decomposers are on the highest trophic level.

55. Which of the following would be considered a tertiary consumer?
    a. phytoplankton
    b. zooplankton
    c. fish
    d. jellyfish
  * e. osprey

56. The amount of energy transferred from an organism on one trophic level to the next trophic level is about _____%.
    a. 1
  * b. 10
    c. 25
    d. 33
    e. 50

57. In ecological pyramids, which group is most likely to form the base?
    a. tertiary consumers
    b. secondary consumers
    c. primary consumers
    d. omnivores
  * e. producers

58. Which term includes the others?
    a. biomass pyramids
    b. energy pyramids
    c. numbers pyramids
  * d. ecological pyramids
    e. none of the above

59. Net primary productivity
    a. is the rate at which producers manufacture chemical energy through photosynthesis.
    b. is the rate at which producers use chemical energy through respiration.
    c. is the rate of photosynthesis plus the rate of respiration.
    d. can be thought of as the basic food source for decomposers in an ecosystem.
  * e. is usually reported as the energy output of a specified area of producers over a given time.

60. Which of the following ecosystems has the lowest level of kilocalories per square meter per year?
    a. savanna
    b. tropical rain forest
    c. agricultural land
    d. lakes and streams
  * e. open ocean

61. Which of the following ecosystems has the highest net primary productivity?
    a. agricultural land
    b. open ocean
    c. temperate forest
  * d. swamps and marshes
    e. lakes and streams

62. Which of the following statements is *false?*
    a. Biomass is the organic matter synthesized by producers.
    b. Energy pyramids show why a larger human population can be supported if people eat grains rather than animals.
  * c. Plants from swamps and marshes are a good alternative food for the growing human population.
    d. Clearing tropical forests rapidly depletes soil nutrients.
    e. Longer food chains have a greater cumulative loss of usable high-quality energy.

4-4   CONNECTIONS: MATTER CYCLING IN ECOSYSTEMS

63. The key component of nature's thermostat is
    a. oxygen.
  * b. carbon dioxide.
    c. glucose.
    d. methane.
    e. hydrogen.

64. All of the following increase the amount of carbon dioxide in the atmosphere *except*
    a. respiration.
  * b. photosynthesis.
    c. combustion.
    d. decomposition.
    e. volcanic eruptions.

65. Transfer of carbon among organisms depends primarily on
    a. fuel combustion and decomposition.
  * b. photosynthesis and cellular respiration.
    c. soil bacteria and precipitation.
    d. volcanic activity and organic decay.
    e. condensation and circulation.

66. Of the following carbon-based compounds, which would have the slowest turnover rate?
  * a. calcium carbonate shells
    b. wood in a tree
    c. protein in a cow
    d. DNA in a bacterium
    e. carbon dioxide in the atmosphere

67. The two ways in which humans have most interfered with the carbon cycle are
    a. removal of forests and aerobic respiration.
    b. aerobic respiration and burning fossil fuels.
    c. respiration and photosynthesis.
  * d. burning fossil fuels and removal of forests.
    e. combustion and causing volcanic eruptions.

68. The most common gas in the atmosphere is
  * a. nitrogen.
    b. carbon dioxide.
    c. oxygen.
    d. hydrogen.
    e. water vapor.

69. Nitrogen is a major component of all of the following *except*
    a. proteins.
    b. nucleic acids.
    c. ammonia.
    * d. groundwater.
    e. nitrates.

70. Nodules containing nitrogen-fixing bacteria would be expected to occur on the roots of
    a. pine.
    b. roses.
    * c. legumes.
    d. grasses.
    e. mosses.

71. Nitrogen fixation is accomplished by
    a. legumes.
    * b. cyanobacteria.
    c. algae.
    d. protozoa.
    e. worms in the ocean trenches.

72. The form of nitrogen most usable to plants is
    a. ammonia.
    b. nitrogen gas.
    c. proteins.
    * d. nitrates.
    e. nucleic acids.

73. Which of the following statements about human alteration of the nitrogen cycle is false?
    a. Large quantities of nitric oxide are released into the atmosphere when fuel is burned.
    b. Nitric oxide can be converted in the atmosphere to nitric acid, which contributes to acid deposition.
    c. Nitrous oxide, released by bacterial action on commercial inorganic fertilizers, is a heat-trapping gas.
    d. Nitrate and ammonium ions are depleted from the soil by harvesting nitrogen-rich crops.
    * e. Eating protein puts "dead ends" in the nitrogen cycle.

74. All of the following human behaviors substantially affect the nitrogen cycle in aquatic systems *except*
    a. use of nitrogen-based detergents.
    b. use of nitrogen fertilizers.
    c. runoff from feedlots.
    d. addition of sewage to aquatic systems.
    * e. runoff from salt-treated icy highways.

75. Which one of the following is *not* one of the common phosphorous reservoirs in the ecosystem?
    a. water
    b. organisms
    * c. atmosphere
    d. rocks
    e. soil

76. To which of the following cycles is guano an important component?
    * a. phosphorous
    b. carbon
    c. hydrologic
    d. sulfur
    e. nitrogen

77. All of the following are sources of phosphorous *except*
    a. inorganic fertilizer.
    b. runoff of animal wastes from feedlots.
    c. detergents.
    * d. acid rain.
    e. ocean sediments.

78. The major plant nutrient most likely to be a limiting factor is
    * a. phosphorous.
    b. calcium.
    c. nitrogen.
    d. potassium.
    e. magnesium.

79. The primary location of sulfur is the
    a. troposphere.
    b. stratosphere.
    c. biosphere.
    d. hydrosphere.
    * e. lithosphere.

80. In the atmosphere, sulfur dioxide is a major contributor to
    a. global warming.
    * b. acid rain.
    c. ozone depletion.
    d. accidental death of wildlife.
    e. algae blooms.

81. About _____ of all sulfur reaching the atmosphere comes from human activities.
    a. 1/2
    * b. 1/3
    c. 1/4
    d. 1/5
    e. 1/6

82. All of the following human activities intervene in the sulfur cycle *except*
    a. smelting sulfur compounds of metallic minerals.
    b. burning sulfur-containing oil.
    c. burning sulfur-containing coal.
    * d. burning sulfur-containing wood.
    e. refining petroleum.

83. The hydrologic cycle refers to the movement of
    a. hydrogen.
    b. oxygen.
    * c. water.
    d. hydrocarbons.
    e. rain.

84. The hydrologic cycle is driven primarily by
  *   a.   solar energy and gravity.
      b.   solar energy and the moon.
      c.   solar energy and mechanical energy.
      d.   mechanical and chemical energy.
      e.   chemical energy and gravity.

85. Humans strongly affect the hydrologic cycle through all of the following *except*
      a.   water withdrawal in heavily populated areas.
      b.   clearing vegetation for agriculture.
  *   c.   boiling water.
      d.   paving roads and parking lots.
      e.   putting up buildings.

4-5 ROLES AND INTERACTIONS OF SPECIES IN ECOSYSTEMS

86. The *least* likely contributor to the decline in numbers of amphibians is
      a.   loss of habitat or fragmentation.
      b.   prolonged drought.
      c.   overhunting.
  *   d.   increase in predator populations.
      e.   pollution.

87. Characteristics which make amphibians particularly sensitive to pollution include
      a.   living part of their life cycle in water and part on land.
      b.   soft, permeable skin.
      c.   insect diet.
      d.   sensitivity to ultraviolet radiation.
  *   e.   all of the above.

88. Which of the following would serve as an indicator species with reference to DDT?
      a.   alligator
  *   b.   bald eagle
      c.   squid
      d.   earthworm
      e.   giraffe

89. Which of the following is a keystone species?
  *   a.   alligator
      b.   robin
      c.   orchid
      d.   opossum
      e.   crow

90. Which of the following statements about alligators is *false*?
      a.   Alligators have no natural predators other than humans.
      b.   Alligators dig deep depressions, which collect fresh water during dry spells.
      c.   The comeback of the alligator is an important success story for wildlife conservation.
  *   d.   Alligators are antagonistic to game fish such as bream and bass.
      e.   Large alligator nesting mounds also serve as nest sites for birds such as herons and egrets.

91. An organism's niche is analogous to its
      a.   address.
  *   b.   occupation.
      c.   food source.
      d.   trash dump.
      e.   power plant.

92. An organism is classified as a specialist or generalist based on
      a. its range of tolerance.
      b. its habitat.
      c. its major source of food.
      d. its limiting factors.
  *   e. all of the above.

93. Specialist species
      a. are very adaptable.
      b. tolerate a wide range of environments.
  *   c. are more likely to become extinct.
      d. eat a wide variety of food.
      e. have broad niches.

94. Which of the following is a specialist?
      a. cockroach
      b. fly
      c. human
  *   d. giant panda
      e. white-tailed deer

95. Which of the following statements is *false?*
      a. Symbiotic relationships describe two or more organisms living together to the benefit of
          one or both.
      b. The fundamental niche of a species is the full range of physical, chemical, and biological
          factors it could use if there were no competition.
      c. The competitive exclusion principle states that no two species with the same
          fundamental niche can indefinitely occupy the same habitat.
  *   d. Interspecific competition is competition between two members of the same species.
      e. Resource partitioning limits competition by two species using the same scarce resource
          at different times, in different ways, or in different places.

96. When two species occupy a similar fundamental niche, the competitive exclusion principle
      can be insured through
      a. exploitation competition.
      b. interference competition.
      c. resource partitioning.
  *   d. all of the above.
      e. none of the above.

97. Which of the following predators avoid competition by being active at different times?
      a. lions and tigers
      b. hummingbirds and bees
  *   c. hawks and owls
      d. zebras and antelopes
      e. coyotes and wolves

98. Which of the following broadly defined categories includes the others?
      a. carnivore—prey
      b. parasite—host
      c. herbivore—plant
  *   d. predator—prey
      e. none of the above

99. A shark is *least* likely to be killed
    * a.   by a predator.
      b.   for sport.
      c.   out of fear.
      d.   for food.
      e.   for economic gain.

100. Which of the following statements about sharks is *false*?
      a.   Sharks have an acute sense of smell.
    * b.   Sharks hunt humans.
      c.   Sharks sense weak electrical impulses.
      d.   Sharks have an acute sense of hearing.
      e.   Sharks help control the populations of other predators.

101. Sharks help humans by
      a.   providing people with food.
    * b.   distinguishing humans from sea lions, their typical prey.
      c.   providing chemicals from their cartilage used to fight cancer.
      d.   their blood, used in AIDS research.
      e.   providing chemicals from their cartilage used in production of artificial skin for burn victims.

102. The obvious relationship demonstrated by a food chain is
      a.   competition.
    * b.   predation.
      c.   parasitism.
      d.   mutualism.
      e.   commensalism.

103. The relationship between clownfish and sea anemones is
      a.   competition.
      b.   predation.
      c.   parasitism.
      d.   mutualism.
    * e.   commensalism.

104. The relationship between oxpeckers and a rhino is
      a.   competition.
      b.   predation.
      c.   parasitism.
    * d.   mutualism.
      e.   commensalism.

105. The relationship between mistletoe and pecan trees is
      a.   competition.
      b.   predation.
    * c.   parasitism.
      d.   mutualism.
      e.   commensalism.

106 The relationship in which one organism benefits and the other is neither benefitted nor harmed is called
      a.   competition.
      b.   predation.
      c.   parasitism.
      d.   mutualism.
    * e.   commensalism.

107. The relationship in which both organisms benefit is called
    a. competition.
    b. predation.
    c. parasitism.
  * d. mutualism.
    e. commensalism.

108. Forms of nondestructive behavior between organisms include all of the following *except*
    a. sharing resources by hunting at different times.
    b. sharing resources by looking for food in different places.
  * c. parasitism.
    d. mutualism.
    e. commensalism.

CHAPTER 5
ECOSYSTEMS: WHAT ARE THE MAJOR TYPES AND WHAT CAN HAPPEN TO THEM?

1.   Jim Callender, a California rancher, used all of the following methods to restore a
     marsh *except*
     a.   replanting tules and bulrushes.
     b.   hollowing out low areas.
*   c.   planting slow-maturing hardwoods.
     d.   building up islands.
     e.   reintroducing plants needed by birds.

5-1   BIOMES: LIFE ON LAND

2.   Climate is the general pattern of weather over a _____-year period.
     a.   10
     b.   20
*   c.   30
     d.   50
     e.   100

3.   The two most important factors determining the climate of an area are
     a.   temperature and ocean currents.
     b.   precipitation and light.
*   c.   temperature and precipitation.
     d.   light and temperature.
     e.   precipitation and ocean currents.

4.   Large ecological regions with characteristic types of natural vegetation are called
     a.   ecosystems.
     b.   communities.
     c.   populations.
*   d.   biomes.
     e.   ecospheres.

5.   The distribution of the desert, grassland, and forest biomes is determined principally by
     a.   temperature.
*   b.   precipitation.
     c.   latitude.
     d.   sunlight.
     e.   none of the above.

6.   The limiting factor that controls the vegetative character of a biome is
     a.   light.
*   b.   precipitation.
     c.   nutrients.
     d.   animal species.
     e.   soil type.

7.   The biome most likely to be found on the top of a very high tropical mountain is the
     a.   desert.
*   b.   tundra.
     c.   grassland.
     d.   temperate deciduous forest.
     e.   taiga.

8. You are going on a scientific expedition from the equator to the North Pole. As you leave the boreal forest behind, you anticipate next exploring
    a. gases captured in the ice.
    b. the fall leaves of New England.
    c. patterns of cone design in coniferous trees.
    d. stratified layers of the tropical forest.
    *   e. the role of lichens and mosses in boggy ecosystems.

9. Which of the following is *not* characteristic of desert plants?
    a. widespread, shallow root systems
    b. deep root systems
    *   c. large leaves that droop in the bright sunlight
    d. succulent leaves or stems
    e. wax-coated leaves

10. Which of the following would be the least common adaptation of animals to living in a desert?
    a. live underground during the heat of the day
    b. have thick outer coverings to minimize water loss
    *   c. drink and store large amounts of water
    d. become dormant during periods of extreme heat or drought
    e. be active at night

11. Which of the following is *not* characteristic of the kangaroo rat?
    a. excretes dry feces
    b. excretes almost solid urine
    *   c. is an imported pest from Australia
    d. is a nocturnal animal that stays in its burrow throughout the day
    e. gets the water it needs from its own metabolism and the food it eats

12. The fragility of the desert ecosystem is indicated by
    a. the slow growth rate of plants.
    b. low species diversity.
    c. shortages of water.
    d. long regeneration time from vegetation destruction.
    *   e. all of the above.

13. Off-road vehicles are a major threat to
    a. grassland.
    *   b. desert.
    c. tundra.
    d. taiga.
    e. tropical rain forest.

14. Through your binoculars you observe migrating wildebeests and antelopes. You are most likely in a
    a. desert.
    b. tropical grassland during wet season.
    *   c. tropical grassland during dry season.
    d. temperate grassland during wet season.
    e. temperate grassland during dry season.

15. Which of the following is *not* a temperate grassland?
    a. steppes
    b. veld
    *   c. taiga
    d. pampas
    e. prairie

16. The _____ is located in Europe.
    a. veld
  * b. steppe
    c. savanna
    d. pampas
    e. prairie

17. _____ can convert grasslands into desert.
    a. Overgrazing
    b. Mismanagement
    c. Wind erosion
    d. Occasional prolonged drought
  * e. All of the above

18. Recurring fires in summer and fall are most typical of
    a. desert.
    b. topical grassland.
  * c. temperate grassland.
    d. polar grassland.
    e. coniferous forest.

19. Permafrost and lichens are characteristic of the _____ biome.
    a. tropical savanna
  * b. arctic tundra
    c. taiga
    d. thorn woodland
    e. desert

20. Through your binoculars you observe a pack of wolves stalking a caribou separated from its herd. Geese take to the air, departing the boggy scene. You are most likely in
    a. desert.
    b. topical grassland.
  * c. temperate grassland.
    d. polar grassland.
    e. coniferous forest.

21. Arctic tundra is perhaps Earth's most fragile biome because of
    a. low rate of decomposition.
    b. shallow soil.
    c. slow growth rate of plants.
  * d. all of the above.

22. The primary limiting factor of the rain forest is
    a. water.
  * b. nutrients.
    c. temperature.
    d. light.
    e. wind.

23. A mature ____ has the greatest species diversity of all terrestrial biomes.
    a. tundra
  * b. tropical rain forest
    c. taiga
    d. temperate deciduous forest
    e. prairie

24. Which of the following is *not* appropriate to use in describing a tropical rain forest?
   a. humid
 * b. rich soil
   c. stratified
   d. diversity
   e. consistent temperatures

25. Plants of the tropical forest are *least* likely to be pollinated by
   a. beetles.
   b. butterflies.
 * c. wind.
   d. bats.
   e. birds.

26. Which of the following biomes would *not* be considered particularly fragile?
 * a. temperate deciduous forest
   b. desert
   c. tropical rain forest
   d. tundra
   e. none of the above; all are fragile

27. Which of the following is *false?* Deciduous forests
   a. change significantly during four distinct seasons.
   b. are dominated by a few species of broadleaf trees.
   c. have trees that survive winter by dropping their leaves.
 * d. have nutrient-poor soil.
   e. that originally grew in the United States have mostly been cleared.

28. Which of the following does *not* belong with the others?
   a. taiga
 * b. steppes
   c. boreal forest
   d. northern coniferous forest
   e. none of the above; they all belong together

29. Cone-bearing trees are characteristic of the
 * a. taiga.
   b. tropical rain forest.
   c. temperate deciduous forest.
   d. savanna.
   e. prairies.

30. Trees with needlelike leaves that are kept year round are especially abundant in which biome?
   a. tundra
   b. tropical rain forest
 * c. coniferous forest
   d. temperate deciduous forest
   e. none of the above

31. Which of the following is *least* descriptive of coniferous forest?
   a. carpet of needles on forest floor
   b. long, cold, dry winter
   c. short summer
 * d. high species diversity
   e. nutrient-poor soil

32.   The ocean zone that covers the continental shelf is the
       a.   estuary.
   *   b.   coastal zone.
       c.   littoral zone.
       d.   benthic zone.
       e.   none of the above.

33.   The open sea has ____% of the ocean's surface area and ____% of its plant and animal life.
       a.   95 . . . 5
   *   b.   90 . . . 10
       c.   90 . . . 25
       d.   80 . . . 25
       e.   75 . . . 25

34.   Oceans
       a.   play a major role in controlling climate by distributing solar heat.
       b.   function to dilute and disperse human wastes.
       c.   participate in biogeochemical cycles.
       d.   provide habitats for organisms.
   *   e.   do all of the above.

35.   Which of the following choices is *false*? Oceans are important because they
       a.   regulate climates.
       b.   provide a source of many natural resources, such as minerals and fossil fuels.
   *   c.   are one of the most highly productive ecosystems in the world on a unit area basis.
       d.   participate in the biogeochemical cycles.
       e.   play a major role in the hydrologic cycle.

36.   The oceans
       a.   contain 50% of this planet's water.
       b.   cover about 90% of the planet's surface.
   *   c.   contain vast amounts of dissolved CO_2.
       d.   provide niches for six million species.
       e.   have unchanging currents that distribute heat.

37.   The deepest part of the ocean is the
   *   a.   abyssal zone.
       b.   euphotic zone.
       c.   estuary zone.
       d.   bathyal zone.
       e.   estuarine zone.

38.   Most photosynthesis in the open sea occurs in the
   *   a.   euphotic zone.
       b.   abyssal zone.
       c.   bathyal zone.
       d.   coastal zone.
       e.   estuarine zone.

39.   The twilight zone of the sea is the
       a.   euphotic zone.
       b.   abyssal zone.
   *   c.   bathyal zone.
       d.   coastal zone.
       e.   estuarine zone.

40. The ecosystems with the world's highest net primary productivities per unit area are found in the
    a. euphotic zone.
    b. abyssal zone.
    c. bathyal zone.
    * d. coastal zone.
    e. benthic zone.

41. Which of the following trees is characteristic of tropical coastal wetlands?
    a. cypress
    b. coconut
    * c. mangrove
    d. palm
    e. baobab

42. The relationship demonstrated by coral polyps and algae is
    a. predation.
    b. commensalism.
    c. parasitism.
    * d. mutualism.
    e. interspecific competition.

43. The *least* appropriate use of coastal wetlands is for
    a. spawning and nursery grounds.
    * b. condominiums and disposal of landfill waste.
    c. food production.
    d. recreational diving.
    e. filtering pollutants and increasing water quality.

44. The *most* threatened ecosystems in the coastal zone are
    a. trenches.
    b. estuaries.
    * c. coral reefs.
    d. coastal wetlands.
    e. inland wetlands.

45. Thriving coral reefs require
    a. cloudy water.
    b. cool water.
    c. fairly deep water.
    d. salinity that fluctuates with the tides.
    * e. dissolved oxygen and nutrients.

46. All of the following threaten the survival of coral reefs *except*
    a. hurricanes and El Niño–Southern Oscillations.
    b. eroded soil from deforestation and poor land management.
    c. global warming and chemical pollution.
    * d. predation by sharks.
    e. oil spills and limestone mining.

47. Estuaries and coastal wetlands are important for all of the following reasons *except*
    a. spawning and nursery grounds for marine fish and shellfish.
    b. filtering out waterborne pollutants from swimming and wildlife areas.
    c. breeding grounds for waterfowl.
    * d. providing coral for limestone production and the tourist trade.
    e. providing food to people in coastal LDCs.

48. Which of the following ecosystems is least likely to be found in a temperate coastal wetland?
    a. bay
    b. salt flat
    c. mud flat
    * d. mangrove swamp
    e. salt marsh

49. During the past 200 years, about _____ of the area of estuaries and coastal wetlands in the United States has been destroyed or damaged.
    a. one-quarter
    b. one-third
    * c. one-half
    d. two-thirds
    e. three-quarters

50. Lakes that have few minerals and low productivity are referred to as
    a. autotrophic.
    b. eutrophic.
    * c. oligotrophic.
    d. mesotrophic.
    e. heterotrophic.

51. Lakes with large nutrient supplies are called
    a. autotrophic.
    * b. eutrophic.
    c. oligotrophic.
    d. mesotrophic.
    e. heterotrophic.

52. In lakes, large numbers of decomposers are found in the
    a. limnetic zone.
    * b. benthic zone.
    c. littoral zone.
    d. profundal zone.
    e. estuarine zone.

53. In lakes, the nutrient-rich water near the shore is part of the
    a. limnetic zone.
    b. benthic zone.
    * c. littoral zone.
    d. profundal zone.
    e. estuarine zone.

54. In lakes, the open-water surface layer is called the
    * a. limnetic zone.
    b. benthic zone.
    c. littoral zone.
    d. profundal zone.
    e. estuarine zone.

55. Fish adapted to cool, dark water are found in the zone of lakes called the
    a. limnetic zone.
    b. benthic zone.
    c. littoral zone.
    * d. profundal zone.
    e. estuarine zone.

56. In a stratified lake, a high concentration of dissolved oxygen is most likely found in the
    a. hypolimnion.
    b. benthic zone.
    c. thermocline.
    d. abyssal zone.
  * e. epilimnion.

57. During spring and fall turnovers,
    a. nutrients sink down from the top.
    b. nutrients are brought up from the bottom to the top.
    c. dissolved oxygen goes from top to bottom.
    d. dissolved oxygen goes from bottom to top.
    e. a and d.
  * f. b and c.

58. The highest level of dissolved oxygen is most likely to be found in the _____ phase of a river system.
  * a. first
    b. second
    c. third
    d. pro-
    e. meta-

59. A river is most likely to be wide and deep at its _____ phase.
    a. first
    b. second
  * c. third
    d. pro-
    e. meta-

60. A mix of warm-water and cold-water fish are most likely to be found in the _____ phase of a river.
    a. first
  * b. second
    c. third
    d. pro-
    e. meta-

61. Waterfalls are most likely to be found in the _____ phase of a river.
  * a. first
    b. second
    c. third
    d. pro-
    e. meta-

62. Inland wetlands include all of the following *except*
    a. bogs.
    b. wet arctic tundra.
    c. marshes.
    d. swamps.
  * e. estuaries.

63. Inland wetlands are valuable for providing
    a. wildlife habitat.
    b. improved water quality.
    c. regulated stream flow.
    d. recharged groundwater supplies.
  * e. all of the above.

64. Inland wetlands are often lost to
    a. croplands.
    b. mining.
    c. urban development.
    d. oil and gas exploration.
  * e. all of the above.

5-3   RESPONSES OF LIVING SYSTEMS TO ENVIRONMENTAL STRESS

65. The maintenance of favorable internal conditions despite fluctuations in external conditions
    is called
    a. evolution.
    b. coevolution.
    c. natural selection.
  * d. homeostasis.
    e. mutation.

66. The ability of a population to maintain a certain size is known as
    a. stability.
    b. inertia.
  * c. constancy.
    d. resilience.
    e. none of the above.

67. Which of the following characteristics of an organism includes the other three?
  * a. stability
    b. inertia
    c. constancy
    d. resilience
    e. none of the above

68. Which of the following characteristics refers to the ability of an organism to return to its
    former condition after a period of stress?
    a. stability
    b. inertia
    c. constancy
  * d. resilience
    e. none of the above

69. Which of the following refers to the ability of a living system to resist being disturbed or
    altered?
    a. stability
  * b. inertia
    c. constancy
    d. resilience
    e. none of the above

70. Tropical rain forests and California redwood forests are characterized by all of the
    following *except*
  * a. high resilience.
    b. high diversity.
    c. high stability.
    d. high inertia.
    e. none of the above; all are characteristics of the forests.

71. Grasslands can be destroyed by
    a. drought.
    b. fire.
    c. mowing.
  * d. plowing.
    e. all of the above.

72. Changes such as drought and earthquake are considered to be
  * a. natural and catastrophic.
    b. natural and gradual.
    c. human-caused and catastrophic.
    d. human-caused and gradual.
    e. human-caused and moderate.

73. Changes such as depletion of underground aquifers and waterlogging of irrigated soils are considered to be
    a. natural and catastrophic.
    b. natural and gradual.
    c. human-caused and catastrophic.
  * d. human-caused and gradual.
    e. human-caused and moderate.

74. Signs of ill health in stressed ecosystems include all of the following *except*
    a. increase in numbers of insect pests.
    b. drop in primary productivity.
    c. drop in species diversity.
  * d. increase in number of indicator species.
    e. presence of contaminants.

5-4   POPULATION RESPONSES TO STRESS: POPULATION DYNAMICS

75. In response to environmental conditions, populations undergo changes in
    a. age distribution.
    b. size.
    c. density.
    d. dispersion.
  * e. all of the above.

76. The curve that depicts the growth of a population that is limited by a definite carrying capacity is shaped like the letter ____.
    a. J
    b. L
    c. M
  * d. S
    e. U

77. The changes in population size, density, dispersion, and age structure are known as
    a. succession.
    b. demography.
  * c. population dynamics.
    d. biotic potential.
    e. evolution.

78. The biotic potential of a population
    * a. is the maximum reproductive rate of a population.
    b. is the current rate of growth of a population.
    c. is an expression of how many offspring survive to reproduce.
    d. can be determined only by studying an age structure diagram.
    e. determines the fitness of a population.

79. Population size is governed by
    a. births.
    b. deaths.
    c. immigration.
    d. emigration.
    * e. all of the above.

80. Biotic potential is determined by
    a. reproductive age span.
    b. litter size.
    c. how many offspring survive to reproductive age.
    d. how often reproduction occurs.
    * e. all of the above.

81. Which of the following factors leads to an increase in environmental resistance?
    a. ability to compete for resources
    b. ability to resist disease and parasites
    * c. a specialized niche
    d. high reproductive rate
    e. ability to migrate

82. Which of the following factors leads to an increase in biotic potential?
    a. too much or too little light
    b. low reproductive rate
    c. too many competitors
    * d. optimal level of critical nutrients
    e. insufficient ability to hide from predators

83. A population will increase if
    a. natality decreases.
    b. mortality increases.
    * c. the biotic potential increases.
    d. the environmental resistance increases.
    e. emigration increases.

84. Carrying capacity refers to
    a. reproductive rate.
    b. interaction of natality and mortality.
    * c. the maximum size of population the environment will support.
    d. the proportion of males to females.
    e. litter size.

85. Carrying capacity is determined by
    a. climate.
    b. migration.
    c. predation.
    d. interspecific competition.
    * e. all of the above.

86. The curve depicting the growth of a population that is limited by a definite carrying capacity is shaped like the letter ____.
    a. J
    b. L
    c. M
  * d. S
    e. W

87. A population crash occurs when
    a. a population approaches its carrying capacity.
    b. environmental resistance comes into play gradually.
    c. resources are essentially unlimited.
  * d. a population overshoots the carrying capacity and the environmental pressures and shortfalls begin to exert their effects.
    e. resistance to disease develops.

88. The human population of Ireland experienced a major crash as a result of
    a. the plague.
  * b. the potato fungus.
    c. an influenza epidemic.
    d. a prolonged drought.
    e. the Crusades.

89. Humans have extended Earth's carrying capacity for the human species by
    a. controlling many diseases.
    b. using energy resources at a rapid rate.
    c. using material resources at a rapid rate.
    d. increasing food production.
  * e. all of the above.

90. Density-dependent population controls include all of the following *except*
    a. disease.
  * b. human destruction of habitat.
    c. parasitism.
    d. competition for resources.
    e. predation.

91. A population of mice undergoing crowding is most likely to show
    a. increased sexual activity.
    b. increased milk production in nursing females.
    c. increased number of offspring per litter.
  * d. increased cannibalism and killing of the young.
    e. decreased spontaneous abortions.

92. Density-independent population controls include all of the following *except*
    a. drought.
    b. fire.
  * c. resource competition.
    d. unfavorable chemical changes in the environment.
    e. unseasonal temperature changes.

93. All of the following are general types of population change curves *except*
  * a. explosive.
    b. stable.
    c. cyclic.
    d. irruptive.
    e. none of the above.

94. Which of the following terms *best* describes the type of population change you would expect to find for a monkey in an undisturbed section of the Brazilian rain forest?
    a. explosive
  * b. stable
    c. cyclic
    d. irruptive
    e. none of the above

95. Which of the following terms *best* describes the type of population change you would expect to find for rabbits and coyotes in an undisturbed habitat?
    a. explosive
    b. stable
  * c. cyclic
    d. irruptive
    e. none of the above

96. Which of the following terms *best* describes the type of population change you would expect to find for a muskrat population in a state that has just outlawed trapping?
    a. explosive
    b. stable
    c. cyclic
  * d. irruptive
    e. none of the above

97. An r-strategist generally
    a. has a low biotic potential.
  * b. is small and short-lived.
    c. gives much parental care to its offspring.
    d. survives to reproduce.
    e. lives in a stable environment.

98. A K-strategist generally
    a. is opportunistic.
    b. exhibits "boom-and-bust" cycles.
    c. has populations that rise quickly, then crash.
    d. generally lives in a rapidly changing environment.
  * e. has populations that follow an S-shaped growth curve.

99. K-strategists
    a. have high genetic diversity.
    b. are more responsive to environmental changes than r-strategists.
    c. exhibit fast rates of evolution.
  * d. are generally less adaptable to change than r-strategists.
    e. have large gene pools.

100. Which of the following is an r-strategist?
    a. human
  * b. insect
    c. rhinoceros
    d. saguaro cactus
    e. eagle

101. Which of the following is a K-strategist?
    a. algae
    b. bacteria
  * c. geese
    d. rodents
    e. annual plants

102. Which of the following *best* describes the survivorship curve you would expect to find for a mountain gorilla?
    * a. late loss (type I)
      b. constant loss (type II)
      c. early loss (type III)
      d. no loss (type IV)
      e. cyclical loss (type V)

103. Which of the following *best* describes the survivorship curve you would expect to find for a fish?
      a. late loss (type I)
      b. constant loss (type II)
    * c. early loss (type III)
      d. no loss (type IV)
      e. cyclical loss (type V)

5-5   RESPONSES TO CHANGING CONDITIONS: THE RISE OF LIFE ON EARTH

104. Evidence for the evolution of life comes from
      a. chemical experiments.
      b. fossils.
      c. chemical analysis of ancient rocks and core samples.
      d. analysis of genetic makeup of fossil specimens.
    * e. all of the above.

105. The fossil record is uneven because
      a. some lifeforms left no fossils.
      b. some fossils have decomposed.
      c. some fossils have yet to be found.
    * d. all of the above.
      e. none of the above.

106. The distance of the Earth from the sun
      a. ensures that our climate will be too hot for evolution to continue.
      b. ensures that our climate will be too cold for evolution to continue.
      c. ensures that the Earth will overheat of its own accord long before global warming from human causes could happen.
    * d. has created a temperature just right for the evolution of life that is dependent upon water.
      e. is getting larger and threatens more ice ages to come.

107. The size of the Earth
      a. keeps the iron-nickel core molten.
      b. allows transfer of geothermal energy to the surface of the planet.
      c. gives it enough mass to gravitationally hold the atmosphere.
    * d. all of the above.
      e. none of the above.

108. The four elements that form the major components of living organisms include all of the following *except*
    * a. calcium.
      b. carbon.
      c. hydrogen.
      d. oxygen.
      e. nitrogen.

109. The gas that is *least* likely to have formed Earth's primitive atmosphere is
    a. methane.
    b. ammonia.
  * c. oxygen.
    d. water vapor.
    e. carbon dioxide.

110. The source of energy that probably contributed *least* to the synthesis of biological chemicals on primitive Earth is
    a. ultraviolet light.
  * b. hydropower.
    c. radioactivity.
    d. cosmic rays.
    e. lightning.

111. The *most* likely sequence for the evolution of life is
    a. organic chemicals—inorganic chemicals—proto-cells—prokaryotic cells—eukaryotic cells—photosynthetic cells.
    b. inorganic chemicals—organic chemicals—prokaryotic cells—proto-cells—eukaryotic cells—photosynthetic cells.
  * c. inorganic chemicals—organic chemicals—proto-cells—prokaryotic cells—eukaryotic cells—photosynthetic cells.
    d. inorganic chemicals—organic chemicals—proto-cells—eukaryotic cells—prokaryotic cells—photosynthetic cells.
    e. inorganic chemicals—organic chemicals—proto-cells—prokaryotic cells—photosynthetic cells—eukaryotic cells.

112. The oxygen revolution was *most* likely brought about by
    a. lightning creating ozone.
    b. ultraviolet light breaking down ozone.
    c. chemosynthetic bacteria.
  * d. photosynthesizing bacteria.
    e. all of the above acting synergistically.

113. As the oxygen revolution transpired, harmful ultraviolet radiation was filtered out by
  * a. ozone.
    b. carbon dioxide.
    c. oxygen.
    d. methane.
    e. nitrogen.

5-6 POPULATION RESPONSES TO STRESS: EVOLUTION, ADAPTATION, AND NATURAL SELECTION

114. A change in the genetic composition of a population is known as
    a. emigration.
    b. mutation.
    c. natural selection.
  * d. evolution.
    e. immigration.

115. A gene pool is
    a. the collection of genes being used in the human genome project.
    b. the genetic composition of an organism.
  * c. the genetic composition of a population.
    d. the genetic composition of a community.
    e. the habitat where evolving organisms live.

116. Darwin proposed that the mechanism for most evolutionary change is
    a. inheritance of acquired characteristics.
    b. synthetic selection.
  * c. natural selection.
    d. supernatural selection.
    e. none of the above.

117. Adaptations may be
    a. behavioral.
    b. structural.
    c. physiological.
  * d. all of the above.

118. An example of a structural adaptation is
    a. resource partitioning.
  * b. protective cover.
    c. ability to hibernate.
    d. ability to produce predator poisons.
    e. mutualism.

119. An example of a physiological adaptation is
    a. resource partitioning.
    b. protective cover.
  * c. ability to hibernate.
    d. gripping mechanisms.
    e. mutualism.

120. Which of the following is *false?* Coevolution
    a. occurs when interacting species exert selective pressures on each other.
    b. occurs between plants and the herbivores who eat them.
    c. occurs between predator and prey.
    d. may play a role in the evolution of camouflage.
  * e. leads to competitive relationships.

121. Coevolution involves the interaction of
    a. plants and herbivores.
    b. pollinators and flowers.
    c. parasites and hosts.
    d. predator and prey.
  * e. all of the above.

122. Approximately ____% of the species that have ever lived are now extinct.
    a. 10–20
    b. 30–50
    c. 60–80
    d. 85–90
  * e. 98–99

5-7 COMMUNITY-ECOSYSTEM RESPONSES TO STRESS

123. The World Health Organization in Borneo (Brunei) initiated spraying to
  * a. reduce the incidence of malaria.
    b. control sylvatic plague.
    c. eliminate small lizards.
    d. kill the caterpillars found in the thatched roofs.
    e. control the cat population.

124. The World Health Organization controlled sylvatic plague in Borneo (Brunei) by
    a. spraying with pesticide.
  * b. parachuting healthy cats into the area.
    c. introducing lizards.
    d. burning all the huts that housed infected people.
    e. all of the above.

125. Which of the following statements about ecological succession is *false?*
    a. It is initiated by pioneer species.
    b. It often involves organisms changing the environment so that they can no longer thrive and must be replaced by others.
    c. It generally results in more complex, more self-sustaining stages than the preceding ones.
  * d. It is an orderly progression from pioneer species to climax species indicative of the region being studied.
    e. None of the above.

126. Which of the following would exhibit primary succession?
  * a. rock exposed by a retreating glacier
    b. an abandoned farm
    c. a forest that had been clear-cut
    d. a newly created lake
    e. all of the above

127. Which of the following would undergo secondary succession?
    a. cooled volcanic lava
    b. a new sandbar exposed by the ocean
  * c. a heavily polluted stream that has been cleaned up
    d. a bare rock outcrop
    e. all of the above

128. In immature ecosystems
    a. the species diversity is high.
    b. the decomposers are numerous.
    c. there are many specialized niches.
  * d. the food webs are simple.
    e. nutrients are efficiently recycled.

129. In mature ecosystems
    a. most plants are annuals.
    b. species diversity is low.
  * c. the efficiency of energy use is high.
    d. the efficiency of nutrient recycling is low.
    e. there are mostly generalized ecological niches.

5-8   HUMAN IMPACTS ON ECOSYSTEMS: WORKING WITH NATURE

130. Which of the following is *not* true?
    a. Humans simplify ecosystems to use them or their products.
    b. Unwanted pioneer plants are called weeds.
  * c. Once an effective insecticide is found, it can be used indefinitely.
    d. The first law of ecology is that we can never do just one thing.
    e. None of the above.

131. Which of the following is *false?* Compared to a natural ecosystem, a simplified human system
    a.   consumes energy from fossil fuels rather than the sun.
    b.   depletes rather than creates fertile soil.
    c.   uses and contaminates rather than cycles and purifies water.
    d.   produces pollutants that must be cleaned up at considerable expense rather than filtering and detoxifying for free.
  * e.   is usually capable of self-renewal instead of requiring maintenance.

132. Which of the following statements is *true?*
  * a.   A pond will eventually undergo succession to become a terrestrial ecosystem.
    b.   Humans usually make ecosystems more complex.
    c.   Ecological succession is a natural process upon which human beings have no influence.
    d.   Pioneer species will not invade a human-managed ecosystem.
    e.   None of the above.

133. Garrett Hardin's first law of ecology is
    a.   ecological succession cannot be prevented.
    b.   natural ecosystems cannot be controlled by artificial means.
  * c.   we can never do just one thing.
    d.   there is no such place as "away."
    e.   entropy wins.

134. Nature's secrets for sustainable living include all of the following *except*
    a.   rely on solar energy.
    b.   rely on renewable resources.
    c.   maintain biodiversity.
    d.   keep population size appropriate for environmental conditions.
  * e.   eat, drink, and make merry, for tomorrow we die!

CHAPTER 6
THE HUMAN POPULATION: GROWTH, URBANIZATION, AND REGULATION

1. Since 1960, Thailand's
   a. population growth rate fell from 3.2% to 1.4%.
   b. population growth rate fell from 5.2% to 1.2%.
   c. per capita income doubled.
   d. per capita income was cut in half.
   * e. a and c.
   f. b and d.
   g. a and d.
   h. b and c.

2. Thailand's population growth rate has changed because of
   a. a government-supported family-planning program.
   b. religious community support.
   c. openness of the Thai people to new ideas.
   d. the nonprofit Population and Community Development Association.
   * e. synergistic interaction of all of the above.

3. The family planning campaign in Thailand included
   a. free vasectomies on the king's birthday.
   b. condom distribution at public events.
   c. birth control carts at bus stations.
   d. insurance support for taxi drivers who distributed birth control pills.
   * e. all of the above.

6-1 FACTORS AFFECTING HUMAN POPULATION SIZE

4. Population fluctuations depend upon
   a. births.
   b. deaths.
   c. migrations.
   * d. all of the above.
   e. none of the above.

5. Between 1965 and 1993,
   a. the population base and growth rate increased.
   * b. the population base increased while the growth rate declined.
   c. the population base decreased while the growth rate increased.
   d. the population base and growth rate decreased.
   e. the population base and growth rate stayed constant.

6. For every death, there are (is) about
   a. 1 birth.
   b. 2 births.
   * c. 3 births.
   d. 4 births.
   e. 5 births.

7. Currently, the annual growth rate of the world's population is about
   a. 1.0%.
   b. 1.3%.
   * c. 1.6%.
   d. 1.9%.
   e. 2.1%.

8. The maximum recorded annual rate of human population growth was
   a. 1.4%.
   b. 1.6%.
   c. 1.8%.
 * d. 2.0%.
   e. 2.2%.

9. The greatest human population growth rate occurred in the middle
   a. 1940s.
   b. 1950s.
 * c. 1960s.
   d. 1970s.
   e. 1980s.

10. Which of the following statements about the world's population is *false?*
    a. Sixty percent of the world's population is Asian.
    b. The most rapid population growth is taking place in the LDCs.
    c. We will add 1 billion more people in the next decade.
  * d. When the total fertility rate drops below the replacement level fertility, the population
       is balanced and stops growing.
    e. The impact of exponential population growth on population size is much greater in
       countries with a large existing population base.

11. Which of the following countries would produce the greatest rise in population size from
    experiencing a growth rate of 1.2%?
    a. country A, with a population of 100,000
    b. country B, with a population of 1 million
    c. country C, with a population of 10 million
    d. country D, with a population of 100 million
  * e. country E, with a population of 1 billion

12. Which of the following contributes the most to total population size in one year?
    a. a country of 1 million people with a growth rate of 4%
    b. a country of 1.5 million people with a growth rate of 3%
    c. a country of 5 million people with a growth rate of 2.5%
    d. a country of 100 million people with a growth rate of 2%
  * e. a country of 500 million people with a growth rate of 1.5%

13. The crude birth rate is calculated by dividing the annual number of births by
    a. 50.
    b. 100.
    c. 500.
  * d. 1,000.
    e. 10,000.

14. The crude birth rate for the world is
    a. 21.
  * b. 27.
    c. 35.
    d. 43.
    e. 50.

15. The crude death rate for the world is
    a. 5.
  * b. 9.
    c. 15.
    d. 20.
    e. 25.

16. China and India have a combined population that accounts for ____% of the total population of the world.
    a. 20
    b. 26
    c. 32
    * d. 38
    e. 45

17. The crude death rate is highest in
    * a. Africa.
    b. Latin America.
    c. Asia.
    d. Europe.
    e. North America.

18. The crude birth rate is highest in
    * a. Africa.
    b. Latin America.
    c. Asia.
    d. Europe.
    e. North America.

19. The actual average replacement-level fertility for the world as a whole is slightly higher than
    a. 1.5 children per couple.
    * b. 2.0 children per couple.
    c. 2.5 children per couple.
    d. 3.0 children per couple.
    e. 3.5 children per couple.

20. The most useful measure of fertility for projecting future population change is the
    a. replacement-level fertility.
    b. one-year future fertility level.
    * c. total fertility rate.
    d. birth rate.
    e. ZPG rate.

21. The total fertility rate
    a. is the most useful measure of fertility for projecting future population change.
    b. is an estimate of the average number of children a woman will have during childbearing years.
    c. is 3.3 children per woman for the world as a whole.
    d. is expected to remain about 1.8 in MDCs and drop to 2.3 in LDCs by 2025.
    * e. is all of the above.

22. The birth rate in the United States has been at or below replacement level for
    a. 5 years.
    b. 10 years.
    c. 15 years.
    * d. 20 years.
    e. 25 years.

23. The total fertility rate in the United States reached a peak
    a. in the late 1700s when the United States was dominated by an agricultural economy.
    b. after the Civil War, in the 1870s.
    c. after World War I, in the mid-1920s.
    * d. after World War II, in the mid-1950s.
    e. during the golden 1980s.

24. All of the following factors contributed to a declining total fertility rate in the United States *except*
    a. widespread use of effective birth control methods.
    * b. illegality of abortion.
    c. rising costs of raising a family.
    d. increases in the average age of marriage.
    e. an increasing number of women in the work force.

25. The United States has not reached zero population growth because of
    a. the large number of baby-boomer women still in their childbearing years.
    b. increase in unmarried women having children.
    c. high levels of immigration.
    * d. all of the above.
    e. none of the above.

26. Of the following forms of birth control, the *second* most effective is
    a. total abstinence.
    b. condom (good brand).
    * c. IUD plus spermicide.
    d. diaphragm plus spermicide.
    e. rhythm method.

27. Of the following forms of birth control, the *least* effective is
    a. IUD plus spermicide.
    b. oral contraceptive.
    c. condom (good brand).
    d. diaphragm plus spermicide.
    * e. rhythm method.

28. An unreliable form of birth control is
    a. cervical cap.
    * b. douche.
    c. condom.
    d. spermicide.
    e. vaginal sponge impregnated with spermicide.

29. Social factors affecting birth and fertility rates include
    a. change in attitudes toward large families.
    b. average levels of education and affluence.
    c. urbanization.
    d. average marriage age.
    * e. all of the above.

30. Economic factors affecting birth and fertility rates include
    a. rising costs of having children.
    b. importance of children as part of the family labor force.
    c. employment opportunities for women.
    d. availability of pension plans.
    * e. all of the above.

31. Which of the following would decrease the likelihood of a couple having a child?
    a. The child is part of the family labor pool.
    b. Contraceptives are not available.
    c. They have no public or private pension.
    d. There is a high infant mortality rate.
    * e. Women have many opportunities to participate in the work force.

32. The drop in the human death rate is due to
    a. increased food productivity and better distribution.
    b. improved personal hygiene and sanitation.
    c. improved water supplies.
    d. antibiotics and immunization.
  * e. all of the above.

33. Two useful indicators of overall health in a country or region are
    a. birth rate and death rate.
    b. replacement-level fertility rate and total fertility rate.
  * c. life expectancy and infant mortality rate.
    d. life expectancy and death rate.
    e. birth rate and infant mortality rate.

34. Infant mortality rate refers to the number of children per 1,000 that die
    a. before birth.
    b. in their first month.
    c. in the first half year of life.
  * d. by their first birthday.
    e. by the time they are age 5.

35. In which of the following countries do citizens have the greatest life expectancy?
  * a. Japan
    b. Sweden
    c. the United States
    d. Portugal
    e. Germany

36. Between 1900 and 1993, the average life expectancy in the United States has increased
    a. 21 years.
    b. 27 years.
  * c. 38 years.
    d. 40 years.
    e. 43 years.

37. Between 1965 and 1993, the world's infant mortality rate dropped by about
    a. 10-15%.
    b. 20-25%.
  * c. 30-35%.
    d. 40-45%.
    e. 50-55%.

38. All of the following reasons help explain why the United States has one of the highest
    infant mortality rates of developed countries *except*
    a. lack of health care for children of the poor after birth.
  * b. the older age of pregnant women as a result of many women delaying having children.
    c. lack of adequate prenatal care.
    d. high birth rate for teenage women.
    e. drug addiction among pregnant women.

39. Which of the following statements about teenage sex and sex education in the United States is *false*?
    a. Polls show that 85% of Americans favor sex education in schools.
  * b. About 10,000 U.S. high schools provide free contraceptives at high school health clinics with parental consent.
    c. A small but highly vocal group of people oppose sex education because they believe it will increase sexual activity.
    d. Many analysts urge that U.S. children be made aware of the values of abstinence and types of contraceptives by age 12.
    e. Teenagers in the United States aren't more sexually active than those in other MDCs, but they are less likely to take precautions to prevent pregnancy.

40. Which of the following statements about U.S. teenage pregnancy is *false*?
    a. About 1 million U.S. teenagers become pregnant each year.
    b. About 83% of teenage pregnancies are unplanned.
    c. Of unplanned teenage pregnancies, about 40% end in abortion.
    d. Teenage pregnancies cost state and federal governments about $21 billion per year.
  * e. Babies born to teenagers are more likely to be high-weight babies than those born to other women.

41. Projections of total U.S. population by the year 2100, assuming different average total fertility rates, life expectancies, and net legal immigration rates, range from
    a. about 210 million to 310 million.
    b. about 200 million to 250 million.
    c. about 250 million to 350 million.
  * d. about 275 million to 300 million.
    e. about 190 million to 270 million.

42. Over the last 200 years, the rate of natural increase in MDCs
    a. oscillated wildly.
    b. increased greatly.
    c. remained fairly stable.
    d. decreased.
  * e. peaked during the Industrial Revolution and then decreased.

43. Over the last 200 years, the rate of natural increase in LDCs
    a. oscillated wildly.
  * b. increased greatly.
    c. remained fairly stable.
    d. decreased.
    e. peaked during the Industrial Revolution and then decreased.

6-2 POPULATION AGE STRUCTURE

44. The age structure of a population is the number or percentage of
    a. females age 14 years or less.
    b. females age 15 to 44.
    c. males age 15 to 44.
  * d. persons of each sex at each age level.
    e. persons of each sex aged 15 to 64.

45. Population age structure diagrams can be divided into all of the following categories *except*
  * a. infant.
    b. preproductive.
    c. reproductive.
    d. postproductive.
    e. none of the above; there are four categories.

46. Age structure diagrams
    a. show only two age groups: reproductive and not reproductive.
    b. show the number of males and females in the infant category.
    c. are strictly for present use and do not provide insight into future trends.
  * d. are useful for comparing one population with another.
    e. should not be considered a valid way of explaining the concept of momentum.

47. Countries that have achieved ZPG have an age structure that
    a. forms an inverted pyramid.
    b. has a broad-based pyramid.
  * c. shows little variation in population by age.
    d. has a large preproductive population.
    e. has a large infant population.

48. Rapidly growing countries have an age structure that
    a. forms an inverted pyramid.
  * b. has a broad-based pyramid.
    c. shows little variation in population by age.
    d. has a large preproductive population.
    e. has a large infant population.

49. Which of the following implies the greatest built-in momentum to maintain population growth?
    a. a large population size
    b. a large number of people age 29 to 44
    c. a large number of people under age 34
  * d. a large number of people under age 15
    e. a large number of people age 15 to 25

50. The most important *generalization* that can be drawn from the author's description of baby boomers is
    a. baby boomers make up nearly half of all adult Americans.
    b. baby boomers will play an increasingly important role in deciding who gets elected and what laws are made between now and 2030.
    c. Many baby boomers may delay retirement because of improved health and the need to build up adequate retirement funds.
    d. The economic burden of helping support retired baby boomers will fall on the baby-bust generation.
  * e. Indentations and expansions in the age structure create a number of social and economic changes that ripple through a society for decades.

51. The present age structure of the U.S. population implies that in the 1990s there will be
    a. fewer young adults.
    b. many more young adults.
  * c. many more middle-aged people.
    d. fewer elderly people.
    e. fewer young adults and elderly people.

52. Which of the following statements about Japan's population is *false?*
    a. Between 1949 and 1956, Japan's population growth rate was cut in half.
    b. Japan has liberal abortion laws and offers access to family planning.
    c. Japan has one of the world's lowest infant death rates.
    d. Japan's universal health insurance and pension systems used about 41% of Japan's national income in 1993.
  * e. The aging of Japan's population has encouraged investment in automation and encouraged women to stay at home.
    f. Japan is becoming increasingly dependent on illegal immigrants to keep its economy going.

53. Economic, environmental, and social conditions are affected by
    a. population density.
    b. population growth rate.
    c. age structure.
    d. population distribution in rural and urban areas.
    * e. all of the above.

6-3   POPULATION DISTRIBUTION: URBANIZATION AND URBAN PROBLEMS

54. In the text, an urban area is any town, village, or city with a population of at least
    a. 1,000.
    b. 1,500.
    * c. 2,500.
    d. 5,000.
    e. 10,000.

55. At present, _____% of the people on Earth live in urban environments.
    a. 22
    b. 32
    * c. 42
    d. 52
    e. 62

56. A megacity contains _____ people.
    a. 1 million
    b. 3 million
    c. 5 million
    * d. 10 million
    e. 50 million

57. Urban growth in LDCs is the result of
    a. rapid population growth.
    b. rural poverty.
    c. increased mechanization of agriculture.
    d. better social services for urban dwellers.
    * e. all of the above.

58. Which of the following statements is *false?*
    * a. The shift in poverty is moving rapidly from the city to the country.
    b. Urbanization is occurring at a faster rate in LDCs than in MDCs.
    c. Urbanization varies throughout the world but is increasing everywhere.
    d. The general population growth contributes to urban growth.
    e. The United Nations estimates that 18% of the world's population lives in crowded, often wretched, inhumane conditions in squatter settlements and slums.

59. It is projected that by 2025, about _____ of the world's people will live in urban areas.
    a. 40%
    b. 50%
    * c. 60%
    d. 70%
    e. 80%

60. By the year 2000, it is expected that there will be ____ megacities.
    a. 12
    * b. 26
    c. 43
    d. 55
    e. 100

61. The most populous city ever to exist is
    a. Calcutta.
    b. Tokyo.
 * c. Mexico City.
    d. Singapore.
    e. New York City.

62. Which of the following statements about Mexico City is *true?*
    a. There is work for all the residents.
    b. It is safe to walk the city streets.
 * c. Air pollution is severe.
    d. All barrios have running water and electricity.
    e. Human waste is recycled as fertilizer in the surrounding countryside.

63. Mexico City's air quality is due to
    a. topography.
    b. large population.
    c. industrialization and large numbers of motor vehicles.
    d. lack of pollution prevention measures.
 * e. all of the above.

64. Currently about _____ of Americans live in metropolitan areas.
    a. one-fourth
    b. one-third
    c. one-half
    d. two-thirds
 * e. three-fourths

65. Problems faced by numerous cities in industrialized countries include all of the following *except*
    a. deteriorating services.
    b. decaying infrastructure.
 * c. massive starvation.
    d. environmental degradation.
    e. neighborhood collapse.

66. The major population shift in the United States from 1800 to the mid-1900s was
    a. from West to East.
    b. from central cities to suburbs and smaller cities.
    c. from North and East to South and West.
    d. from rural to suburbs.
 * e. from rural to mostly large central cities.

67. Because of the large creation of new jobs, the 1970s saw a U.S. population shift
    a. from West to East.
 * b. from central cities to suburbs and smaller cities.
    c. from North and East to South and West.
    d. from rural to suburbs.
    e. from rural to mostly large central cities.

68. The major trend of migration starting in 1980 in the United States is
    a. from West to East.
    b. from central cities to suburbs and smaller cities.
 * c. from North and East to South and West.
    d. from rural to suburbs.
    e. from rural to mostly large central cities.

69. All of the following modes of transportation make sense in compact cities *except*
    a. buses.
    b. commuter trains.
 * c. cars.
    d. walking.
    e. bicycles.

70. Where land has been plentiful and inexpensive as cities grew, the most-used form of transportation is
    a. buses.
    b. bicycles.
    c. commuter trains.
    d. walking.
 * e. cars.

71. In the United States, ____% of all urban transportation is by car.
    a. 50
    b. 80
    c. 86
    d. 92
 * e. 98

72. With 4.7% of the people in the world, the United States has ___% of the cars.
    a. 15
    b. 20
    c. 25
 * d. 35
    e. 40

73. In the United States, one of every _____ dollars spent is connected to the automobile.
    a. 4
 * b. 6
    c. 8
    d. 10
    e. 20

74. Advantages of automobiles include all of the following *except*
    a. freedom to go where you want when you want.
    b. adventure.
 * c. energy efficiency.
    d. provision of jobs.
    e. none of the above; all are advantages.

75. Which of the following statements is *false*?
    a. About 18 million people have been killed by motor vehicles.
    b. Each year, cars and trucks kill as many people worldwide as were killed during the bombing of Hiroshima and Nagasaki.
 * c. Constructing more roads reduces automobile travel and congestion.
    d. More Americans have been killed by automobiles than in all U.S. wars.
    e. One-half of all the people in the world will be involved in an automobile accident before they die.

76. The first automobile was built by
    a. Henry Ford.
 * b. Karl Benz.
    c. Studebaker.
    d. General Motors.
    e. none of the above.

77. Which of the following statements about disadvantages of cars is *false*?
    a. Driving cars is one of the most dangerous things we do in our daily lives.
    b. Cars are the largest source of air pollution.
    c. Cars have reduced use of mass transit.
  * d. Cars are a major cause of consolidated cities.
    e. Cars are a major cause of oil addiction in MDCs.

78. Which of the following has a faster average speed?
  * a. horse-drawn carriages in Manhattan in 1907
    b. cars in the Manhattan streets
    c. traffic in London
    d. cars in Tokyo
    e. Paris traffic

79. Cities depend upon areas outside their boundaries for all of the following *except*
    a. food and water.
  * b. governance and banking institutions.
    c. energy.
    d. minerals.
    e. waste disposal.

80. Which of the following cities does the textbook say is a sustainable city?
  * a. Davis, California
    b. Pittsburgh, Pennsylvania
    c. Des Moines, Iowa
    d. Jackson, Mississippi
    e. Seattle, Washington

81. Davis, California has instituted all of the following policies *except*
    a. building codes that encourage use of solar energy for heating space and water.
    b. planting of shade trees.
  * c. development of surrounding farmland.
    d. many bicycle paths.
    e. mix of homes for low, medium, and high incomes.

82. All of the following contribute to air quality problems in cities *except*
    a. lack of trees and other natural vegetation.
    b. urban heat islands.
    c. dust domes.
  * d. fewer clouds.
    e. releases from cars and industries.

83. Urban resource problems include
    a. lack of water and rapid water runoff.
    b. water pollution.
    c. solid waste disposal.
    d. excessive noise.
  * e. all of the above.

84. All of the following noises can reach the human pain threshold *except*
    a. a jet takeoff.
  * b. a cotton spinning machine.
    c. live rock music.
    d. sirens at close range.
    e. a subway.

85.   Individuals interested in creating sustainable urban areas might consider all of the following land use policies *except*
    a.   relying on comprehensive, regional ecological land-use planning and control to regulate speed and nature of economic development.
    b.   repairing and revitalizing existing cities.
*   c.   establishing formal public gardens with exotic plants to demonstrate biodiversity.
    d.   establishing greenbelts.
    e.   planting large numbers of trees.

86.   Ecologically sound development can be discouraged by
    a.   requiring environmental impact analysis for private and public projects.
    b.   conservation easements.
    c.   land trusts.
*   d.   taxing land on the basis of the economically highest potential use.
    e.   all of the above.

87.   Land-use planning is *not* widely used because
    a.   there is economic and political pressure for short-term economic gain.
    b.   political officials focus on short-term goals.
    c.   it is difficult to get different jurisdictions to agree.
    d.   ecological land-use planning and implementation may be costly in the short run.
*   e.   all of the above.

88.   Which of the following is *not* an appropriate method of land-use control?
    a.   conservation easements
    b.   land trusts
*   c.   subsidy of farmers who use marginal land that is highly susceptible to erosion
    d.   taxation of land on the basis of its current use
    e.   purchase of land development rights

89.   Local governments can control land use by
    a.   requiring building permits.
    b.   zoning.
    c.   limiting services.
    d.   discouraging ecologically unsound development.
*   e.   all of the above.

90.   Conservation easements
    a.   are political payoffs.
*   b.   are tax breaks to individuals who use land only for specific purposes.
    c.   purchase land for preservation.
    d.   give subsidies to specific groups.
    e.   require environmental impact statements.

91.   Of the following land-use and maintenance strategies to improve the quality of urban life, the *least* desirable would be
    a.   well-planned, compact growth.
    b.   regional ecological land-use planning.
    c.   giving squatters legal title to land they have lived on.
*   d.   planting lawns with tropical forest plants to encourage biodiversity.
    e.   building wildlife corridors.

92. The quality of urban life could be improved by all of the following transportation strategies *except*
    a. charging single-occupant vehicles higher fees for tolls and parking.
    b. raising gasoline taxes and car registration fees.
    c. establishing a high-speed rail trust fund.
    d. using gas taxes for mass transit.
  * e. charging bikers for the construction of bike paths.

93. The quality of urban life could be improved by all of the following energy efficiency strategies *except*
    a. retrofitting public buildings to use renewable energy resources.
    b. writing building codes that require energy efficiency.
    c. imposing taxes on citizens who buy gas guzzling vehicles.
    d. passing laws to improve car gas mileage standards.
  * e. passing ordinances that ban solar collectors because they decrease property values.

94. The quality of urban life could be improved by all of the following renewable resource strategies *except*
    a. establishing neighborhood water recycling plants.
    b. enacting building codes that require water conservation.
    c. installing water meters.
  * d. growing food in large, maximum-efficiency farms.
    e. emphasizing sustainable organic agricultural methods.

95. The quality of urban life would be improved *least* by
    a. imposing pollution taxes.
    b. giving economic incentives to businesses that emphasize pollution prevention.
    c. establishing composting centers.
  * d. recycling and reusing 5% of urban solid wastes.
    e. enacting strict noise control laws.

96. According to David Orr, good ecological design includes all of the following qualities *except*
    a. right scale.
  * b. complexity.
    c. efficient use of resources.
    d. a close fit between means and ends.
    e. durability.

6-5   SOLUTIONS: INFLUENCING POPULATION SIZE

97. All of the following countries accept large numbers of immigrants *except*
    a. United States.
  * b. Mexico.
    c. Canada.
    d. Australia.

98. Population size can be influenced by
    a. controlling migration.
    b. economic development.
    c. family planning.
    d. economic rewards and penalties.
    e. empowering women.
  * f. all of the above.

99. Which factor accounts for the greatest increase to the population of the United States?
    a. illegal immigrants
    b. legal immigrants
    c. illegitimate births
  * d. legal births
    e. multiple births

100. Between 1960 and 1993, immigration into the United States
    a. rapidly decreased after steadily increasing.
    b. was about 1 million legal immigrants per year.
    c. was responsible for 10% of population growth.
  * d. was approximately twice that of all other nations combined.
    e. was not significant in the U.S. population growth rate.

101. Which of the following statements about U.S. immigration is *false?*
    a. Since 1820, the United States admitted almost twice as many immigrants and refugees as all other countries combined.
    b. Between 1960 and 1993, the United States let in about twice as many immigrants as all other countries combined.
    c. Between 200,000 to 500,000 people enter the United States illegally each year.
    d. Soon immigration is expected to be the primary factor increasing the population of the United States.
  * e. Immigration increases U.S. population size considerably less than teenage pregnancy.

102. Today, about _____ of the world's population live in countries with fertility reduction programs.
    a. 25%
    b. 50%
    c. 73%
  * d. 83%
    e. 93%

103. Proponents of population regulation say that
  * a. giving preference to skilled immigrants could be a brain drain on LDCs.
    b. people are the world's most valuable resource.
    c. if a person produces more than he consumes he is an asset.
    d. human ingenuity permits continued improvement in humanity's lot.
    e. all of the above.

104. The people who oppose population regulation are *least* likely to say that
    a. lack of a free and productive economic system in LDCs is the primary cause of poverty and despair.
    b. people are the world's most valuable resource for finding solutions to our problems.
    c. aging societies will be less innovative and dynamic.
    d. population regulation is a violation of religious beliefs and an intrusion into personal privacy and freedom.
  * e. people overpopulation in LDCs and consumer overpopulation in MDCs are threats to earth's life-support systems.

105. The people who support population regulation say that
    a. billions more people on the earth will intensify many environmental and social problems.
    b. it is unethical for us not to control births, because the alternative is a sharp rise in death rates and future human misery.
    c. technological innovation, not simply population size, is the key to military and economic power.
    d. we only have the freedom to produce as many children as we want as long as this does not reduce the quality of other people's lives, now and in the future, by impairing the ability of the Earth to sustain life.
  * e. all of the above are true.

106. All of the following ways of controlling human population size have large constituencies *except*
    a. increase economic development.
    b. increase family planning.
    c. bring about socioeconomic change.
  * d. increase the death rate.
    e. none of the above; all have large constituencies.

107. The term *demographic transition* refers to
    a. a requirement for a population to reach a specific size before it becomes stable.
    b. the slowing down in the growth of a population as it approaches the carrying capacity.
  * c. the decline in death rates followed by decline in birth rates when a country becomes industrialized.
    d. the decline in death rates followed by a decline in birth rates that occurred when the germ theory of disease was discovered.
    e. migration from overpopulated countries to new colonies.

108. The change that takes place in a demographic transition occurs when
    a. one-third of the population is below 15 years of age.
    b. the birth rate drops below the death rate.
  * c. the economic development of a country changes the population growth pattern.
    d. either immigration or emigration changes the population growth pattern.
    e. family planning is acceptable to a country's culture.

109. The demographic transition model explains why
    a. death rates rise in industrializing nations.
    b. industrialization leads to population growth.
    c. development requires large populations.
  * d. industrializing LDCs have high birth rates.
    e. population growth in MDCs goes down and then up.

110. During demographic transitions, birth rates of a population are high during the
    a. preindustrial and industrial stages.
    b. postindustrial and transitional stages.
    c. industrial and postindustrial stages.
  * d. preindustrial and transitional stages.
    e. preindustrial and postindustrial stages.

111. In the demographic transition model, ZPG in a country is likely to occur during
    a. the industrial stage.
  * b. the postindustrial stage.
    c. the transitional stage.
    d. the preindustrial stage.
    e. none of the above.

112. In the demographic transition model, death rates fall while birth rates remain high during
    a. the preindustrial stage.
    b. the industrial stage.
    c. the postindustrial stage.
*   d. the transitional stage.
    e. none of the above.

113. In the demographic transition model, birth and death rates are high during
*   a. the preindustrial stage.
    b. the industrial stage.
    c. the postindustrial stage.
    d. the transitional stage.
    e. none of the above.

114. Some experts fear that the LDCs lack sufficient ___ to allow the demographic transition to occur.
    a. people
*   b. capital
    c. cooperation
    d. commitment
    e. all of the above

115. The LDCs may not be able to complete the transition to the industrial stage because of
    a. the competition of the MDCs and other LDCs.
    b. a lack of capital.
    c. lack of education and technical skill to produce high-technology products.
    d. a lack of natural resources.
*   e. all of the above.

116. Family planning programs have been successful in reducing population growth in
    a. Brazil.
    b. Bangladesh.
    c. Indonesia.
    d. India.
*   e. a and c.

117 Basic family planning policy in most countries includes all of the following *except*
*   a. limiting families to two children each.
    b. employing contraceptive methods.
    c. helping parents space births as desired.
    d. helping parents regulate family size.
    e. providing information about prenatal care.

118. Benefits of family planning include
    a. saving government money by reducing the need for social services.
    b. preventing pregnancy-related deaths.
    c. reducing total fertility rate.
    d. increased use of contraceptives among the world's married women.
*   e. all of the above.

119. Governments try to reduce population growth with all of the following *except*
    a. paying couples who agree to use contraceptives.
    b. paying couples who agree to be sterilized.
    c. penalizing couples who have more than a set number of children (usually one or two).
*   d. providing needed health care and food allotments to those who have more than a certain number of children.
    e. reducing job opportunities for those who have more than a certain number of children.

120. Which of the following statements about population regulation are true?
    a. Programs that give economic rewards can be psychologically coercive for the poor.
    b. Programs that employ economic penalties can be psychologically coercive for the poor.
    c. Programs that withhold food or increase the cost of having children penalize the children.
    d. a and c.
  * e. All of the above.

121. Economic rewards and penalties in population control strategies work best if they
    a. nudge rather than push people to have fewer children.
    b. are not retroactive.
    c. reinforce existing customs and trends.
    d. increase a poor family's income or land.
  * e. include all of the above.

122. Women tend to have fewer children when
    a. they live in democratic societies.
    b. their rights are not suppressed.
    c. they have access to education.
    d. they have access to jobs outside the home.
  * e. all of the above.

123. Which of the following statements about women's employment/economic status is *false?*
    a. Women do more than half of the work gathering fuelwood.
    b. Women do more than half of the work involved in producing food.
  * c. Women have more than half of the world's assets.
    d. Women provide more health care than all of the world's organized health services.
    e. Women do most of the world's domestic work.

124. Women possess _____% of the world's land.
  * a. 1
    b. 10
    c. 25
    d. 50
    e. 75

125. Women receive _____% of the world's income.
    a. 1
  * b. 10
    c. 25
    d. 50
    e. 75

6-6   CASE STUDIES: POPULATION CONTROL IN INDIA AND CHINA

126. India
  * a. had the world's first national family planning program.
    b. has the world's most successful family planning program.
    c. has the world's only national family planning program.
    d. has the world's largest family planning program.
    e. has no family planning program.

127. Which of the following statements about India's population is *false?*
  * a. India's age structure is moving from a pyramid to a more vertical distribution.
    b. India's people are among the poorest in the world.
    c. Almost one-third of the present population goes hungry.
    d. Life expectancy is 59 years.
    e. Infant mortality is 91 deaths per 1,000 live births.

128. India's family planning has yielded disappointing results for all of the following reasons *except*
    a. poor planning and bureaucratic inefficiency.
  * b. government information about advantages of small families.
    c. extreme poverty.
    d. low status of women.
    e. too little financial support.

129. ____ of all Indian children die before their fifth birthday.
  * a. One-third
    b. One-fourth
    c. One-fifth
    d. One-seventh
    e. One-tenth

130. China's population policy has included all of the following *except*
    a. encouraging later marriages.
    b. expanding educational opportunities.
    c. urging couples to have no more than one child.
    d. training local people to carry on the family planning program.
  * e. encouraging contraceptive use but banning abortion.

131. China's population control program does all of the following *except*
    a. employ freely available contraceptives.
    b. employ compulsory measures.
  * c. emphasize huge family planning centers.
    d. offer economic incentives.
    e. encourage couples to postpone marriage.

132. Perhaps the most important feature of China's population control program that could be transferred to other countries is
    a. focusing control efforts on males.
  * b. localizing the program rather than forcing people to travel to distant centers.
    c. requiring one of the parents to be sterilized when a couple has two children.
    d. encouraging couples to postpone marriage.
    e. giving couples economic incentives to have fewer children.

133. Which of the following statements is *false?*
    a. The carrying capacity refers to the maximum population a given environment can support.
    b. The cultural capacity is always greater than simple carrying capacity.
    c. Robert Malthus focused on human carrying capacity 200 years ago.
  * d. It is possible to accurately determine the carrying capacity of the Earth.
    e. There is considerable negative reaction to proposals that make international philanthropy conditional upon stopping population growth by poor, overpopulated nations.

CHAPTER 7
ENVIRONMENTAL ECONOMICS AND POLITICS

1.   Which of the following questions would the author of your text consider *least* important?
     a.   How can we grow without plundering the planet?
   * b.   To grow or not to grow?
     c.   How can we build sustainable societies?
     d.   How can we grow as if Earth matters?
     e.   How can we grow and still leave a flourishing Earth to future generations?

2.   All of the following are primarily economic decisions *except*
     a.   deciding what goods and services to produce.
     b.   deciding how to produce goods and services.
     c.   deciding how many goods and services to produce.
   * d.   deciding to write letters to congresspersons about which goods you think should receive a Green Seal.
     e.   deciding how to distribute goods and services.

7-1   ECONOMIC SYSTEMS AND ENVIRONMENTAL PROBLEMS

3.   Which of the following categories includes the others?
     a.   human capital
     b.   natural resources
   * c.   economic resources
     d.   manufactured capital
     e.   none of the above

4.   Which of the following is *not* classified as an economic resource?
   * a.   time
     b.   labor
     c.   capital or intermediate goods
     d.   natural resources
     e.   none of the above; all are economic resources

5.   In a pure command (centralized planning) economic system, decisions are made by
     a.   the markets.
   * b.   the government.
     c.   past customs and experience.
     d.   none of the above.
     e.   all of the above.

6.   Economic decisions in a pure capitalistic economy are made by
   * a.   the markets.
     b.   the government.
     c.   past customs and experience.
     d.   none of the above.
     e.   all of the above.

7.   *Capitalistic system* is another name for
     a.   the traditional system.
     b.   the pure command system.
   * c.   the pure market system.
     d.   the mixed economic system.
     e.   none of the above.

8. The pure market system is based on all of the following *except*
   a. the flow of economic goods and money between households and businesses.
 * b. obligation to provide a safe workplace.
   c. freedom of choice.
   d. pure competition.
   e. balance of supply and demand.

9. In a pure capitalistic system, a company's only obligation is to
   a. provide taxes to the government.
   b. supply needed goods and services.
   c. provide jobs and safe workplaces.
 * d. produce the highest short-term profits for owners and stockholders.
   e. protect the environment.

10. If price, supply, and demand are the only factors involved, the demand and supply curves
    a. run parallel to each other.
    b. are reciprocals of each other.
 * c. intersect at the market equilibrium.
    d. are straight lines that run in opposite directions.
    e. are inverted S-shaped curves.

11. The *most* common economic system is
    a. the pure market economic system.
    b. the pure command economic system.
    c. socialism.
 * d. the mixed economic system.
    e. the traditional economic system.

12. Which of the following statements is *false?*
    a. Economists believe that unlimited economic growth is possible and desirable.
    b. Economists believe that resources are infinite in supply or that substitutes can be found.
 * c. Economists believe in S-shaped rather than J-shaped growth curves.
    d. Economists believe that the world has an essentially infinite capacity to absorb, dilute, or degrade wastes.
    e. Economists often represent pure capitalism as a flow of economic goods between households and businesses operating essentially independently of the ecosphere.

13. All of the following would be counted as part of the gross national product *except*
    a. a new car purchase.
    b. government purchases.
 * c. home health care.
    d. dental services.
    e. a cable installation.

14. Which of the following statements about economic growth is *false?*
    a. Economic growth increases the flow of matter and energy resources through society.
    b. Economic growth is generally seen as good, limitless, and necessary to maximize wealth and power.
    c. Economic growth is generally measured by gross national product (GNP).
 * d. Real GNP is more accurate than GNP because it takes the discount rate into account.
    e. The average person's slice of the economic pie is generally reported as real GNP per capita.

15. GNP is generally accepted as a measure of
    a. the quality of life.
    * b. the status of economic progress.
    c. social welfare.
    d. the quality of the environment.
    e. none of the above.

16. GNP does not measure the well-being of a society because it does not include
    a. differences between beneficial and harmful services.
    b. how the resources are divided.
    c. some positive effects on society.
    d. subtraction of natural resource depletion and degradation.
    * e. all of the above.

17. The use of gross national product as an indicator of economic well-being
    a. encourages a throwaway economy.
    b. creates a false sense of resources.
    c. is an inaccurate measure of quality of life.
    d. cannot predict ecological bankruptcy.
    * e. all of the above.

18. NEW (net economic welfare) measures
    * a. the quality of life.
    b. the GNP corrected for inflation.
    c. only the negative features of economic growth.
    d. the growth of the private sector only.
    e. GNP per capita.

19. The real net economic welfare calculated for the United States since 1940 has risen at _____ the rate of the GNP.
    a. one-fourth
    b. one-third
    * c. one-half
    d. twice
    e. the same rate as

20. The reason that net economic welfare is not widely used as an economic indicator is
    a. it is not widely recognized.
    b. it is not corrected for inflation.
    * c. it is hard to determine the cost of the "bad" features of economic growth.
    d. it tends to make people appear richer than they are.
    e. it hasn't been published in an academic journal.

21. Social and environmental indicators
    * a. may consider the value of goods and services based on natural resource depletion.
    b. are frequently used by politicians.
    c. are noncontroversial.
    d. are not as accurate as GNP in assessing the quality of the environment.
    e. cannot be calculated.

22. All of the following are social and/or environmental indicators of quality of life *except*
    a. net economic welfare (NEW).
    * b. real gross national product (real GNP).
    c. index of sustainable economic welfare (ISEW).
    d. net national product (NNP).
    e. none of the above; all are indicators.

23. Which of the following indicators is incorrectly matched with the factor(s) it takes into account?
  * a. real GNP per person—inflation, population growth, resource depletion
    b. net economic welfare (NEW)—pollution
    c. net national product (NNP)—resource degradation and depletion
    d. index of sustainable economic welfare (ISEW)—income distribution, depletion of nonrenewables, pollution

24. Which of the following is *not* an external cost of driving a domestic car?
    a. pollution
  * b. cost of manufacture
    c. highway accidents
    d. increased price of lumber, paper, and textbooks
    e. noise

7-2 SOLUTIONS: USING ECONOMICS TO IMPROVE ENVIRONMENTAL QUALITY

25. Full-cost pricing
  * a. requires government action.
    b. reduces final cost of goods.
    c. would result in redirecting economic growth and would drop the NEW.
    d. would make the market price higher than the true cost of a product or service.
    e. can be done voluntarily.

26. Full-cost pricing
  * a. involves making the market price approach the true cost of an economic good.
    b. fails to include the cost of pollution.
    c. omits the costs of taxes.
    d. increases the chance for environmental degradation.
    e. is a simple change to make.

27. Which of the following statements is *false*?
    a. Full-cost pricing would make pollution prevention more profitable.
    b. Full-cost pricing makes consumers more aware of the true cost of goods and services.
    c. Real market prices would not always be cheaper if external costs were internalized.
    d. Full-cost pricing is in opposition to the throwaway society.
  * e. Fuel-inefficient cars would still have competitive prices if their external costs were internalized.

28. Full-cost pricing is difficult to implement for all of the following reasons *except*
    a. producers of harmful and wasteful products might go bankrupt.
    b. producers of harmful and wasteful products might lose subsidies.
  * c. it is difficult to externalize internal costs.
    d. it is difficult to place a price tag on external costs.
    e. none of the above.

29. Which of the following statements is *false*?
    a. Cost-benefit analyses have a built-in bias against the future.
    b. Conservationists generally place a higher value on the future value of resources than economists do.
  * c. In cost-benefit analysis it is easy to determine who pays the costs and who gets the benefit.
    d. Many costs and benefits cannot be labeled with a price tag.
    e. In cost-benefit analyses, there is often disagreement about the level of the discount rate.

30. Which statement about cost-benefit analysis is *true*?
   * a. In cost-benefit analyses, the future value of a resource is usually discounted.
   b. Cost-benefit analyses clearly assess all the winners and losers in any situation.
   c. Cost-benefit analyses are based on well-established values.
   d. Cost-benefit analyses employ only objective criteria that can be easily assessed.
   e. Cost-benefit analysis of a particular manufacturing process would be the same throughout the world.

31. Cost-benefit estimates
   * a. analyze data to determine the most economically efficient course of action.
   b. provide bottom-line numbers.
   c. are not useful and should not be made.
   d. are objective assessments of the real costs of doing business.
   e. always have the same discount rate.

32. Which of the following statements is the most important thing to keep in mind about economics and the environment?
   a. Cost-benefit analyses will answer our questions about options to choose.
   b. Businesses will voluntarily produce less pollution.
   * c. We need to achieve a balance between the costs of cleaning up pollution and the costs of a polluted environment.
   d. It's important not to hold up new chemical products by requiring them to be tested for environmental safeness before release into the marketplace.
   e. The only way to get a cleaner environment is to reward those companies who use pollution control devices.

33. For most pollutants, economists say our goal should *not* be zero pollution because
   a. it is too costly.
   b. everything we do produces some pollution.
   c. nature can handle some of the pollution.
   d. some businesses would go bankrupt.
   * e. all of the above.

34. The graph of the cost of removing pollutants is
   a. an ascending straight line.
   b. a descending straight line.
   c. a normal distribution (first going up, then coming down).
   d. a horizontal line.
   * e. a J-shaped curve showing exponential growth.

35. In 1992, 13 nations of _____ made a commitment to achieve zero discharge of persistent toxic substances.
   * a. Europe
   b. Asia
   c. Africa
   d. South America
   e. Central America

36. The pollution control approach that does the least to internalize external costs is
   a. tradable rights.
   b. withdrawing harmful subsidies.
   * c. subsidizing beneficial actions.
   d. green taxes and user fees.
   e. pollution prevention bonds.

37. Using tax dollars to subsidize businesses that install pollution control equipment is an example of
   a. marketing pollution rights.
   b. marketing resource use rights.
   * c. rewarding beneficial actions.
   d. making harmful actions illegal.
   e. taxing harmful actions.

38. Selling the right to harvest timber up to a sustainable yield level would be an example of
   a. tradable rights.
   b. withdrawing harmful subsidies.
   c. subsidizing beneficial actions.
   * d. green taxes and user fees.
   e. pollution-prevention bonds.

39. The pollution control strategy *least* likely to confer a disadvantage in the international marketplace is
   a. tradable rights.
   b. green taxes.
   * c. subsidizing beneficial actions.
   d. user fees.
   e. pollution-prevention bonds.

40. Administrative costs are highest for
   a. tradable rights.
   b. withdrawing harmful subsidies.
   c. subsidizing beneficial actions.
   d. green taxes and user fees.
   * e. regulation.

41. All of the following economic approaches contribute to government revenues *except*
   a. tradable rights.
   b. withdrawing harmful subsidies.
   c. user fees.
   d. green taxes and user fees.
   * e. pollution-prevention bonds.

42. The discount rate is
   a. a social indicator.
   b. an environmental indicator.
   c. an economic indicator.
   * d. an ethical decision about responsibility to future generations.
   e. a cost-benefit decision about responsibility to the GNP.

7-3 SOLUTIONS: REDUCING POVERTY

43. Poverty is usually defined as
   a. not being able to buy shoes.
   * b. not being able to meet one's basic economic needs.
   c. earning less than $100 per year.
   d. eating an unbalanced diet.
   e. none of the above.

Environmental Economics and Politics   183

44. Currently, one in every _____ persons on the planet lives in poverty in an LDC.
    a. nine
    b. seven
  * c. five
    d. four
    e. three

45. Poverty causes
    a. preventable health problems.
    b. unsustainable use of potentially renewable resources.
    c. premature deaths.
    d. increasing birth rates.
  * e. all of the above.

46. The trickle-down theory describes how
    a. waterfalls flow during the fall season.
    b. toxic wastes filter down in landfills.
    c. water percolates through loam soils.
  * d. economic growth is supposed to increase the number of jobs and help the poor help
       themselves.
    e. the World Bank makes investments.

47. Since 1950, the gap between the rich and the poor has
    a. disappeared.
  * b. increased.
    c. stayed the same.
    d. decreased.
    e. increased briefly, then leveled off.

48. Mahatma Ghandi's concept of *antyodaya* was
    a. nonviolent protest.
  * b. putting the poor and their environment first, not last.
    c. improvement of economic conditions through sit-down strikes.
    d. full employment.
    e. none of the above.

49. Critics of past development efforts say that improvements can be made by all of the
    following methods *except*
    a. protecting those who are living sustainable lifestyles.
    b. fostering individual and community self-reliance.
    c. involving local residents.
  * d. bringing in expert opinion from MDCs.
    e. learning from cultures with sustainable lifestyles and disseminating what works.

50. Governments of LDCs can help eliminate poverty by all of the following methods *except*
    a. shifting more of the national budget to aid the poor.
    b. giving villages and urban poor title to common lands and to crops and trees they plant
       on common lands.
  * c. redistributing some of the land to the large landowners who have shown they use it
       most efficiently.
    d. extending full human rights to women.
    e. increasing spending on health, education, family planning, and sanitation.

51. MDCs can help address poverty, thereby investing in global environmental and economic security by all of the following methods *except*
    a. forgiving of the present debt owed by LDCs to MDCs in exchange for agreement by the governments of LDCs to increase expenditure for rural development, family planning, health care, education, land distribution, and protection of wilderness areas.
    * b. insisting that LDCs stop ruining their natural resource base and that they pay off their debt to MDCs to establish their reliability and encourage more MDC investment in their economies.
    c. increasing nonmilitary aid given by MDCs to LDCs.
    d. lifting trade barriers that hinder the export of commodities from LDCs to MDCs.
    e. tracking the flight and concealment of capital from LDCs to MDCs.
    f. developing sustainable-Earth economies in both MDCs and LDCs.

52. Which of the following is *not* an instance that illustrates the effect GATT can have on a country's environmental policies?
    a. Canada had to relax its pesticide regulations to bring them in line with the United States.
    b. British Columbia ended a tree-planting program because the United States claimed it was an unfair subsidy to Canada's timber industry.
    c. The Canadian timber industry is urging its government to challenge a U.S. law requiring the use of recycled fiber in newsprint.
    * d. Japan had to lower its automobile mpg standards to match the United States.
    e. A GATT panel of judges ruled that the U.S. requirement that tuna imported to the United States had to be caught in dolphin-sparing ways was an unfair and illegal trade barrier.

53. Environmentalists generally believe all of the following *except*
    a. trade restrictions that put LDCs at a disadvantage should be lifted.
    b. countries should not be restricted in setting higher environmental standards by other countries.
    c. countries should not be restricted in setting higher consumer and worker safety standards by other countries.
    d. countries should not be restricted in setting higher resource depletion standards by other countries.
    * e. a trade agreement should be judged on how it benefits environment, workers, and the richest 40% of humanity.

7-4   SOLUTIONS: MAKING THE TRANSITION TO A SUSTAINABLE-EARTH ECONOMY

54. A sustainable-Earth economy might do all of the following *except*
    a. discourage nondegradable products by requiring a "cradle to grave" environmental audit for all economic goods.
    b. discourage use of fossil fuels by taxing gasoline to internalize the external costs.
    c. encourage pollution prevention by removing government subsidies of pollution-producing economic enterprises.
    d. discourage resource waste by discouraging policies that deplete Earth's natural resource capital.
    * e. encourage use of carbon-dioxide-free nuclear power by decreasing the amount of citizen participation in the licensing process.

55. A sustainable-Earth economy would *most* discourage use of
    a. solar energy.
    b. small-scale hydropower.
    c. wind power.
    d. geothermal energy.
    * e. fossil fuels.

56. A sustainable-Earth economy determines progress by all of the following factors *except*
    a. pollution prevention.
    b. the amount of recycling and reuse of nonrenewable materials.
  * c. the bottom line.
    d. the sustainable use of renewable resources.
    e. energy efficiency.

57. Sustainable-Earth economic actions that are in the hands of the consumer include buying
    a. only what is needed.
    b. products that are durable and reusable.
    c. products made from recycled materials.
    d. products with the least packaging.
    e. products with a Green Seal.
  * f. all of the above.

58. A shift to a sustainable-Earth economy could be stimulated by
    a. removing subsidies to Earth-degrading activities.
    b. taxing Earth-degrading activities.
    c. subsidizing Earth-sustaining activities.
    d. improving environmental management in business.
  * e. all of the above.

59. Which of the following is *not* among the major causes of job loss in the United States?
    a. unsustainable use of potentially sustainable resources
    b. automation
    c. rapid depletion of nonrenewable resources
  * d. environmental protection
    e. more efficient and innovative competitors

60. Which of the following is *not* among the major causes of job loss in the United States?
    a. cheaper labor in other countries
  * b. worker safety regulations
    c. decline of unsustainable "sunset" industries
    d. decreased research and development by government and business.
    e. failure to modernize or to invest in "sunrise" industries

61. A Bureau of Labor study found that _____ of job loss was linked to environmental causes.
    a. 0.01%
  * b. 0.1%
    c. 1%
    d. 10%
    e. 15%

62. Which of the following is *least* likely to be a growth business of the future?
    a. environmental monitoring
    b. energy-efficient products
  * c. coal production
    d. solar-cell technology
    e. bike-path construction

63. Giving tax breaks and low-interest loans for improving energy efficiency for homes would
    a. reduce energy waste.
    b. reduce air pollution.
    c. create jobs.
    d. help the lower and middle classes.
  * e. all of the above.

64. Ways to ease the transition from the current economy to a sustainable-Earth economy include all of the following *except*
    a. providing tax breaks to make it more profitable for companies to keep or hire more workers instead of automating.
    b. using incentives to encourage location of sunrise industries in hard-hit communities and helping such areas diversify their economic base.
    c. providing income and retraining assistance for workers displaced from environmentally destructive businesses.
  * d. changing as quickly as possible so that change is not dragged out.
    e. none of the above.

65. Good environmental management includes all of the following *except*
    a. making a commitment to quality.
  * b. using built-in obsolescence as a guiding principle.
    c. recycling and reusing resources at sustainable rates.
    d. protecting biodiversity.
    e. preventing pollution and waste.

66. All of the following describe Germany's efforts *except*
    a. German companies sell the world's cleanest and most efficient gas turbines.
    b. Germany reduced energy waste and air pollution by requiring large and medium industries and utilities to use cogeneration.
    c. Germany developed bar-coded disassembly of used cars.
    d. Germany is the world leader in solar-cell technology and hydrogen fuel.
  * e. Germany is the world leader in percent of GNP invested in future growth.

7-5 POLITICS AND ENVIRONMENTAL POLICY: PROBLEMS AND SOLUTIONS

67. The process by which individuals and groups try to influence the decisions and policies of governments is
    a. economics.
    b. resource management.
  * c. politics.
    d. ethics.
    e. visioning.

68. The most powerful thing that decision makers do is
    a. compromise.
  * b. develop and adopt a budget.
    c. enact laws.
    d. establish policies.
    e. talk to the media.

69. Jim Hightower would do all of the following *except*
    a. go to the people.
    b. introduce a new ethic of environmental responsibility.
    c. green the economy.
    d. plan for global ecological security.
  * e. set a personal example of acting globally while thinking locally.
    f. none of the above.

70. Which is *not* one of the three branches of the U.S. government?
    a. judicial
    b. executive
    c. legislative
  * d. long-range planning board
    e. none of the above; there are four branches of the U.S. government

71.	In the U.S. political system, legislative decision makers are theoretically responsive to
  a.	the executive and judicial branches.
  b.	labor and industry.
  c.	decision implementers.
  d.	the quality of life.
* e.	all of the above.
  f.	none of the above; everyone in the political system operates in a compartmentalized way.

72.	Individuals and organized groups influence and change government policies in constitutional democracies by all of the following methods *except*
  a.	voting and carrying out grass-roots activities.
  b.	contributing time and money to candidates running for office.
  c.	lobbying and writing elected representatives to pass certain laws.
  d.	using the formal education system and the media to influence public opinion.
* e.	voting directly on legislation through electronic mail.

73.	Incremental rather than revolutionary change in constitutional democracies is encouraged by
  a.	conflicts among interest groups.
  b.	conflicting information from experts.
  c.	distribution of power among different branches of government.
  d.	distribution of power among federal, state, and local authorities.
* e.	all of the above.

74.	Environmental legislation in the United States
  a.	sets standards for pollution.
  b.	encourages resource conservation.
  c.	requires comprehensive evaluation of environmental impact before an activity is undertaken.
  d.	protects certain ecosystems or species.
  e.	screens new substances.
* f.	all of the above.

75.	One approach used by environmental legislation in the United States is
  a.	banning pollution.
  b.	establishing recycling standards and procedures for communities to follow.
  c.	establishing at least one "forever free" area for each type of ecosystem in the United States.
* d.	requiring comprehensive evaluation of environmental impact before an activity is undertaken.
  e.	putting the responsibility for proof of environmental safety for new chemicals in the hands of the chemical producers.

76.	Environmental regulations may be set using the general principle of
  a.	no-risk.
  b.	no unreasonable risk.
  c.	risk-benefit balancing.
  d.	standards based on best-available technology.
  e.	cost-benefit balancing.
* f.	all of the above.

77.	The Delaney clause is an example of
* a.	no-risk.
  b.	no unreasonable risk.
  c.	risk-benefit balancing.
  d.	standards based on best-available technology.
  e.	cost-benefit balancing.

78. The National Environmental Policy Act requires
    a. screening of new chemical substances to determine if they are a hazard.
    b. protection of certain areas, resources, or species.
  * c. comprehensive evaluation of the environmental impact statement of an activity before it is undertaken.
    d. encouragement of resource conservation.
    e. taxes to be levied on polluters.
    f. all of the above.

79. According to your text, history shows that significant change
    a. is cyclical.
    b. is exponentially decreasing.
  * c. is from the bottom up.
    d. is from the top down.
    e. occurs at a constant rate.

80. Effective environmental leadership
    a. leads by working within the system.
    b. leads by example.
    c. leads by challenging the system.
    d. involves each individual using his or her preferred style.
  * e. all of the above.

7-6 ENVIRONMENTAL AND ANTI-ENVIRONMENTAL GROUPS

81. The motto of the environmental movement is
    a. Go Green!
  * b. Think Globally, Act Locally!
    c. Think Locally, Act Globally!
    d. Think and Act!
    e. You are What You Eat, Breathe, and Drink!

82. Compared to national- and state-level environmental organizations, grass-roots groups
    a. are less willing to compromise.
    b. are fighting immediate threats to their families' lives.
    c. want pollution stopped and prevented.
    d. "draw the line across the ground of their homes and their beings."
  * e. all of the above.

83. The world's largest environmental organization is
  * a. Greenpeace.
    b. the Sierra Club.
    c. the Nature Conservancy.
    d. the Audubon Society.
    e. the National Wildlife Federation.

84. Mainstream environmental groups
    a. are active primarily at the national level.
    b. often form coalitions to work together on issues.
    c. have been a major force in passing environmental laws.
    d. can end up spending too much time in fundraising.
  * e. all of the above.

85. Grass-roots environmental groups
    a. have been a major force in passing environmental laws.
    b. often form coalitions to work together on issues.
*   c. are organized to protect themselves from environmental damage at the local level.
    d. are active primarily at the national level.
    e. can lose touch with their goals.

86. A pro-environmental position is held by
    a. the National Wetlands Council.
    b. the U.S. Council for Energy Awareness.
    c. America the Beautiful.
    d. the Information Council on the Environment.
*   e. none of the above.
    f. all of the above.

87. A pro-environmental position is held by
    a. the National Wetlands Council.
    b. the U.S. Council for Energy Awareness.
    c. the Information Council on the Environment.
*   d. Earth First!
    e. all of the above.

88. Corporations with poor environmental management skills might
    a. adopt the latest environmental slogans while carrying on business as usual.
    b. fund and develop environmental curricula that promote industry positions.
    d. brand environmentalists as radical and anti-American.
    e. circumvent local, state, and national environmental and resource use laws and regulations by having laws and regulations established by international commissions dominated by multinational companies and elites, under the guise of promoting free trade.
*   f. all of the above.
    g. none of the above; these strategies would be unethical.

89. In the U.S. political system, legislative decision makers are theoretically responsive to
    a. the executive and judicial branches.
    b. labor and industry.
    c. decision implementers.
    d. the quality of life.
*   e. all of the above.
    f. none of the above; everyone in the political system operates in a compartmentalized way.

90. Environmental careers
    a. include solid and hazardous waste management.
    b. include sustainable forestry.
    c. include ecology and population dynamics.
    d. include environmental economics.
*   e. include all of the above.
    f. do not exist.

7-7   CONNECTIONS AND SOLUTIONS: GLOBAL ENVIRONMENTAL POLICY

91. Today, approximately _____ countries have environmental protection agencies.
    a. 15
    b. 25
    c. 75
*   d. 115
    e. 150

92. International treaties cover all of the following subjects *except*
    a. endangered species protection.
    * b. deep sea mining.
    c. acid deposition and ozone depletion.
    d. ocean pollution.
    e. reducing biodiversity loss.

93. Successes of the 1992 Rio Earth Summit included all of the following *except*
    a. a convention on climate change.
    b. an action plan for developing the planet sustainably during the 21st century.
    c. a broad statement of principles on forest protection.
    * d. an agreement to slow human population growth.
    e. a convention on protecting biodiversity.

94. Failures of the 1992 Rio Earth Summit included all of the following *except*
    a. lack of targets and timetables for stabilizing carbon dioxide emissions.
    b. failure of the United States to sign the biodiversity treaty.
    c. general failure to address population issues.
    * d. failure to establish a U.S. Commission on Sustainable Development to implement and oversee the agreements.
    e. lack of a treaty on reversing desertification.

95. Non-Governmental Organizations (NGO) at the 1992 Rio Earth Summit formed new global networks linked by
    a. television.
    b. fax machines.
    c. computer networks.
    d. electronic conferences.
    * e. all of the above.

96. Global environmental protection could be improved by all of the following *except*
    * a. streamlining the role and budget of the United Nations Environmental Programme.
    b. having heads of all countries meet together in an annual Earth Summit.
    c. requiring data collection and regular reporting and publication of compliance progress by countries signing environmental treaties.
    d. creating an Environmental Security Council as part of the United Nations.
    e. facilitating transfer of environmental and resource-efficient technologies to LDCs.

CHAPTER 8
RISK, TOXICOLOGY, AND HUMAN HEALTH

1. Which of the following statements is *false*?
   a. Smoking tobacco causes more death and suffering among adults than does any other environmental factor.
   b. Over 1,000 Americans die each day because of disease caused by smoking.
   c. Daily, the death toll from smoking is equivalent to three jumbo jet crashes.
   d. Annually, smoking kills more than illegal drugs, alcohol, automobile accidents, suicide, and homicide combined.
   * e. None of the above; all statements are true.

2. Which of the following statements is *false*?
   a. If an adolescent smokes more than one cigarette, he or she has an 85% chance of becoming a smoker.
   b. Inhaling passive smoke increases the chance of respiratory infections.
   c. Cigarette smoking is the single most preventable major cause of death and suffering among U.S. adults.
   * d. America's economic benefits to tobacco farmers far exceed tobacco's harmful costs to the American public.
   e. Smoking increases the chance of heart disease, lung cancer, emphysema, and stroke.

3. Of the following, the leading cause of death in the United States is
   * a. tobacco use.
   b. alcohol use.
   c. suicide.
   d. accidents.
   e. hard drug use.

8-1   TYPES OF HAZARDS

4. A hazard is a source of risk and refers to any substance or action that can cause
   a. injury.
   b. disease.
   c. economic loss.
   d. environmental damage.
   * e. all of the above.

5. Four types of hazards discussed in Chapter 8 include all of the following *except*
   a. biological hazards.
   b. physical hazards.
   * c. global hazards.
   d. chemical hazards.
   e. cultural hazards.

6. All of the following are considered to be cultural hazards *except*
   a. smoking.
   * b. drought.
   c. drugs.
   d. diet.
   e. working and living conditions.

7. All of the following are considered to be biological hazards *except*
   a. pollen.
   b. parasites.
   * c. diet.
   d. bacteria.
   e. viruses.

8. All of the following are considered to be physical hazards *except*
    * a. driving.
    b. hurricanes.
    c. landslides.
    d. ionizing radiation.
    e. drought.

8-2 CHEMICAL HAZARDS AND TOXICOLOGY

9. The harmfulness of a chemical is determined by
    a. the body's detoxification system.
    b. the size of the dose.
    c. how often exposure occurs.
    d. the vulnerability of the people exposed.
    * e. all of the above.

10. An acute effect from chemical exposure
    a. might be kidney or liver damage.
    b. is a long-lasting effect.
    c. can result from a single large dose or many doses over a long period of time.
    * d. is an immediate reaction.
    e. is none of the above.

11. The principal types of chemical hazards include all of the following *except*
    a. toxic and hazardous chemicals.
    b. mutagens.
    c. teratogens.
    * d. zymogens.
    e. carcinogens.

12. Carcinogens cause
    a. genetic defects.
    b. birth defects.
    * c. cancer.
    d. chronic health effects.
    e. acute health effects.

13. Cancer can be caused by
    a. genetic factors.
    b. occupational and environmental pollutants.
    c. cigarette smoke.
    d. diet.
    * e. all of the above.

14. Of the following environmental and lifestyle factors, the one that contributes most significantly to cancer is
    a. genetic factors.
    b. occupational pollutants.
    * c. cigarette smoke.
    d. diet.
    e. all of the above.

15. A recommended diet for Americans is *least* likely to limit
    a. protein.
    b. fat and cholesterol.
    * c. whole grains.
    d. alcohol.
    e. sodium.

16. Toxic substances
    * a. are fatal to humans in low doses.
      b. cause birth defects.
      c. are harmful because they are flammable, explosive, irritating to skin or lungs, or cause allergic reactions.
      d. cause mutations.
      e. cause or promote the growth of cancer.

17. Mutagens
      a. are fatal to humans in low doses.
      b. cause birth defects.
      c. are harmful because they are flammable, explosive, irritating to skin or lungs, or cause allergic reactions.
    * d. cause mutations.
      e. cause or promote the growth of cancer.

18. Teratogens
      a. are fatal to humans in low doses.
    * b. cause birth defects.
      c. are harmful because they are flammable, explosive, irritating to skin or lungs, or cause allergic reactions.
      d. cause mutations.
      e. cause or promote the growth of cancer.

19. All of the following are known teratogens *except*
      a. thalidomide.
    * b. carbon monoxide.
      c. lead.
      d. PCBs.
      e. arsenic.

20. Hazardous substances
      a. are fatal to humans in low doses.
      b. cause birth defects.
    * c. are harmful because they are flammable, explosive, irritating to skin or lungs, or cause allergic reactions.
      d. cause mutations.
      e. cause or promote the growth of cancer.

21. Which of the following statements *best* reflects the nature of natural and synthetic chemicals?
      a. Natural chemicals are safe and synthetic chemicals are harmful.
      b. Synthetic chemicals are safe and natural chemicals are harmful.
      c. Natural and synthetic chemicals are safe.
      d. Natural and synthetic chemicals are harmful.
    * e. Chemicals, regardless of their origin, may be safe or harmful.

22. Animal testing
      a. extrapolates from low-dose levels to high-dose levels.
    * b. extrapolates from test animals to humans.
      c. is noncontroversial.
      d. usually involves low dose levels.
      e. is relatively inexpensive.

23. The problem with using animal testing to determine dose levels for humans is
    a. uncertainty about the accuracy of a threshold dose-response model.
    b. uncertainty about the accuracy of a linear dose-response model.
    c. human metabolic processes differ from those of test animals.
    d. extrapolation from high-dose to low-dose levels is uncertain and controversial.
  * e. all of the above.

24. Determining the level at which a substance poses a health threat is done by
    a. animal tests.
    b. epidemiological studies.
    c. bacterial tests.
    d. tissue cultures.
  * e. all of the above.

25. The Ames test is used to demonstrate that a particular
    a. organism causes a disease.
  * b. chemical causes a mutation.
    c. chemical causes a birth defect.
    d. chemical causes cancer.
    e. chemical causes allergic reactions.

26. The study of the pattern of a disease in a population is called
    a. pathology.
    b. oncology.
  * c. epidemiology.
    d. ecology.
    e. bacteriology.

27. Epidemiological studies are limited by
    a. difficulty in selecting a sample because people have been exposed to high levels of many hazardous substances.
  * b. difficulty in establishing cause and effect because people have been exposed to many different toxic agents.
    c. prohibitive costs.
    d. taking at least a decade to complete.
    e. all of the above.

8-3   PHYSICAL HAZARDS: EARTHQUAKES AND VOLCANIC ERUPTIONS

28. Which of the following statements is *true?*
    a. Focus and epicenter are synonymous.
    b. The epicenter must be located on the fault line.
    c. Fault lines are always vertical.
  * d. The focus is the point of initial movement of an earthquake.
    e. The epicenter is the point in the outer core directly below the focus.

29. The strength of an earthquake is measured on the ____ scale.
  * a. Richter
    b. Miller
    c. Zambini
    d. Geiger
    e. pH

30. An earthquake reported as magnitude 9 would be considered
    a. insignificant.
    b. minor.
    c. damaging.
    d. destructive.
    * e. great.

31. Secondary effects from earthquakes include all of the following *except*
    a. urban fires.
    * b. shaking and permanent vertical displacement of the ground.
    c. damaged coastal areas.
    d. mass wasting.
    e. flooding.

32. Signals that may indicate the onset of an earthquake (precursor phenomena) include
    a. changes in electrical and magnetic properties of rock.
    b. slight tilting of rock.
    c. concentration of radon in groundwater.
    d. unusual animal behavior.
    * e. all of the above.

33. Tsunamis are
    a. underground caverns.
    * b. large, seismic ocean waves.
    c. atolls formed by volcanoes found in the South Pacific.
    d. sites of meteor impacts.
    e. hurricanes.

34. Ejecta
    * a. is debris released from a volcano.
    b. are substances injected into faults to relieve pressure.
    c. is material released from rifts on the floor of the ocean.
    d. is the depressed region inside the cone of an inactive volcano.
    e. is the material found in geological intrusions.

35. Zones around the Mt. St. Helens eruption include all of the following *except* the
    a. tree-removal zone.
    * b. radioactive zone.
    c. seared zone.
    d. tree-down zone.

36. Prediction of volcanic activity is enhanced by precursor phenomena such as
    a. tilting of the cone.
    b. increased seismic activity.
    c. changes in gas composition.
    d. changes in magnetic properties.
    * e. all of the above.

37. A positive result of volcanic activity is the
    a. creation of majestic mountains and lakes.
    b. production of geysers and hot springs.
    c. generation of fertile soil.
    d. geothermal energy.
    * e. all of the above.

38. Vectors are
    a. dose levels.
    b. agents of infection.
  * c. agents of disease transmission.
    d. parasitic organisms.
    e. antibiotics.

39. Vectors of disease include
    a. blood.
    b. air and water.
    c. food.
    d. insects.
  * e. all of the above.

40. All of the following are transmissible diseases *except*
  * a. diabetes.
    b. schistosomiasis.
    c. measles.
    d. elephantiasis.
    e. sexually transmitted diseases.

41. All of the following are transmissible diseases *except*
    a. measles.
    b. sleeping sickness.
  * c. lung cancer.
    d. malaria.
    e. sexually transmitted diseases.

42. All of the following are nontransmissible diseases *except*
    a. heart disease.
    b. cancer.
  * c. sexually transmitted diseases.
    d. malnutrition.
    e. chronic respiratory diseases.

43. Primary preventive health care that could make significant improvements in LDCs at low cost include all of the following *except*
    a. contraceptives.
    b. postnatal care.
    c. antibiotics for infections.
  * d. cataract surgery.
    e. clean drinking water.

44. An epidemiologic transition
    a. is the same as the demographic transition.
    b. happens before the demographic transition.
  * c. happens when mortality is caused more from chronic diseases than infectious diseases of childhood.
    d. happens when mortality is caused more from infectious diseases of childhood than chronic diseases of adulthood.
    e. none of the above.

45. The leading causes of death in the MDCs are
   a. bacteria.
   b. viruses.
   c. vectors.
 * d. environmental and lifestyle factors.
   e. wars.

46. In the United States, ____% of health care dollars is spent on preventive medicine.
 * a. 5
   b. 15
   c. 25
   d. 35
   e. 50

47. About ____% of the world's population live in malaria-prone regions.
   a. 30
 * b. 40
   c. 50
   d. 60
   e. 70

48. Malaria is caused by
   a. viruses.
   b. bacteria.
 * c. *Plasmodium* parasites.
   d. parasitic worms.
   e. algae.

49. Malaria is spread by
 * a. *Anopheles* mosquitos.
   b. flies.
   c. worms.
   d. snails.
   e. a and c.

50. During the mid-1900s, the spread of malaria decreased sharply from
   a. draining swamplands.
   b. draining marshes.
   c. spraying breeding areas with DDT and other pesticides.
   d. using drugs to kill the parasites in the bloodstream.
 * e. all of the above.

51. The reason for the increased incidence of malaria since 1970 has been
   a. organisms causing malaria developed resistance to the drugs.
   b. resistance of the vectors to insecticides.
   c. reduced budgets for malaria control because of the belief that the disease was under control.
   d. increase in number of irrigation ditches where vectors breed.
 * e. all of the above.

52. The American Medical Association has called for
    a. a total ban on cigarette advertising in the United States and a tax on cigarettes at about $2.20 a pack.
    b. prohibition of the sale of cigarettes to anyone under 21 and a ban on all cigarette vending machines.
    c. classification of nicotine as a drug and elimination of all federal subsidies to U.S. tobacco farmers.
    d. prohibition of government officials pressuring other countries to accept U.S. tobacco products.
  * e. all of the above.

53. The American Medical Association has called for
    a. pressuring other countries to accept U.S. tobacco products so U.S. citizens don't use them.
    b. eliminating the tax on cigarettes.
  * c. classification of nicotine as a drug.
    d. instating new subsidies to tobacco farmers.
    e. prohibition of sale of cigarettes to anyone under 18.

8-5   RISK ANALYSIS

54. Formal risk assessment
    a. evaluates the nature and severity of risks.
    b. is significant in risk management and risk communication.
    c. is difficult and imprecise.
    d. is controversial.
  * e. all of the above.

55. Calculating the risk of a new technological system
    a. requires assessment of technological reliability.
    b. requires careful economic analysis.
    c. requires assessment of human reliability in operating the system.
    d. is impossible.
  * e. a and c.

56. In analyzing the risks of new, complex technological systems, it is important to remember
    a. technological reliability is almost always much lower than human reliability.
  * b. poor management, poor training, and poor supervision can increase the chances of human error.
    c. reliable computer programs can replace humans in the safe and proper operation of complex systems.
    d. a system is the sum of its parts.
    e. all of the above.

57. A desirability quotient
    a. is the ratio of economic needs to economic wants.
    b. is a measure of the strength of advertising.
  * c. is the ratio of societal benefits to societal risks of a new technology.
    d. is the sum of all the benefits of a new technology.
    e. is the ratio of negatives to positives.

58. Which of the following has the highest desirability quotient?
  * a. diagnostic X rays
    b. nuclear power plants
    c. automobiles
    d. coal-burning power plants
    e. genetic engineering

59. Which of the following has the least certain desirability quotient?
    a. diagnostic X rays
    b. nuclear power plants
    c. computer technology
    d. coal-burning power plants
  * e. genetic engineering

60. Genetic engineering is an activity with a(n) ____ desirability quotient.
    a. very large
    b. large
    c. very small
    d. small
  * e. uncertain

61. Which of the following statements is *true?*
    a. There are no known risks involved in genetic engineering.
    b. The only specific way that human beings will benefit from genetic engineering is the development of pharmaceuticals.
  * c. Genetically engineered products can be patented and produce enormous profits.
    d. Only bacteria can be used in biotechnology.
    e. All of the potential gains from genetic engineering are still in the planning stages.

62. Genetic engineering might be used in the future to develop plants with
    a. higher yields.
    b. greater resistance to disease.
    c. greater resistance to drought.
    d. better balance of nutrients.
  * e. all of the above.

63. Genetic engineering has already been used to fight
    a. diabetes.
    b. hemophilia.
    c. some forms of cancer.
    d. heart attacks.
  * e. all of the above.

64. Critics of genetic engineering are concerned about
    a. the production of a "superorganism" that could cause unpredictable consequences.
    b. mutations that change the form and behavior of the genetically engineered product.
    c. the escape and rapid spread of genetically engineered organisms.
    d. greed rather than ecological wisdom operating unless there are strict controls.
  * e. all of the above.

65. Problems with risk assessment include
    a. some technologies benefit one group and harm another group.
    b. people making risk assessments vary in their emphasis on long-term vs. short-term risks and benefits.
    c. there may be conflict of interest in those carrying out the risk assessment and review of the results.
    d. consideration of cumulative impacts of various risks rather than consideration of each impact separately.
  * e. all of the above.

66. Risk management involves trying to answer all of the following questions *except:*
    a. Which of the risks have top priority?
    b. How reliable is the risk-benefit analysis or risk assessment?
    c. How much risk is acceptable?
  * d. Is it morally responsible to develop this risk?
    e. How much money will it take to reduce each risk to an acceptable level?
    f. How will the risk management plan be communicated to the public, monitored, and enforced?

67. The general public perceives a technology or product as a greater risk than the experts do when it
    a. is relatively familiar and simple.
  * b. is involuntarily thrust upon the public rather than an individual choice.
    c. is viewed as beneficial and necessary rather than unnecessary.
    d. involves a large number of deaths spread out over a long period of time rather than from a single catastrophic accident.
    e. involves equitable distribution of risks.

PART III. RESOURCES: AIR, WATER, SOIL, MINERALS, AND WASTES

CHAPTER 9
AIR

1. In Chapter 9, the use of lichens indicates that
    a. nature wins over the technological fix.
    b. a technological fix can solve an environmental problem.
    * c. nature and technology together can be used to solve an environmental problem.
    d. neither nature nor technology contribute to environmental problem-solving.
    e. global warming is in process.

9-1   THE ATMOSPHERE

2. The innermost layer of the atmosphere is known as the
    a. ionosphere.
    * b. troposphere.
    c. stratosphere.
    d. stratopause.
    e. lithosphere.

3. The troposphere, which contains 75% of the mass of Earth's air, extends ___ miles above sea level.
    a. 5
    * b. 11
    c. 19
    d. 25
    e. 34

4. If Earth were an apple, the troposphere would be the thickness of
    a. the seed.
    b. the core.
    c. the fleshy part of the apple that you eat.
    * d. the skin.
    e. the stem.

5. The most common gas in the atmosphere is
    a. argon.
    b. oxygen.
    * c. nitrogen.
    d. water vapor.
    e. carbon dioxide.

6. The concentration of carbon dioxide in the atmosphere is ____%.
    a. 36
    b. 3.6
    c. 0.36
    * d. 0.036
    e. 0.0036

7. The atmospheric gas that shows considerable variation (about 5%) between the tropics and the poles is
    a. nitrogen.
    b. argon.
    c. carbon dioxide.
    * d. water vapor.
    e. oxygen.

8.	The atmosphere's second layer above Earth is called the
	a.	ionosphere.
	b.	troposphere.
*	c.	stratosphere.
	d.	stratopause.
	e.	lithosphere.

9.	Harmful ultraviolet radiation can cause
	a.	skin cancer.
	b.	eye cataracts.
	c.	immune system damage.
	d.	sunburn.
*	e.	all of the above.

10.	The atmospheric gas that filters out harmful ultraviolet radiation is
	a.	oxygen.
*	b.	ozone.
	c.	nitrogen.
	d.	carbon dioxide.
	e.	CFCs.

11.	Human activities produce _____ as much carbon dioxide as produced by nature.
	a.	one-third
*	b.	one-fourth
	c.	one-half
	d.	twice
	e.	three times

12.	Humans alter the carbon cycle *least* by
	a.	burning fossil fuels.
	b.	clearing forests.
	c.	exponential population growth.
	d.	unsustainable economic growth.
*	e.	exhaling.

13.	Humans alter natural energy flows by
	a.	creating heat islands.
	b.	burning fossil fuels.
	c.	creating dust domes.
	d.	clearing forests.
*	e.	all of the above.

14.	Human activities produce ____ as much nitrogen oxide and gaseous ammonia as produced by nature.
	a.	one-third
	b.	one-fourth
	c.	one-half
	d.	twice
*	e.	three times

15.	Humans alter the nitrogen cycle by
	a.	burning fossil fuels.
	b.	exhaling.
	c.	methane release from solid waste dumps.
	d.	using nitrogen fertilizers.
*	e.	a and d.
	f.	all of the above.

16. Human activities release _____ as much sulfur (primarily sulfur dioxide) into the atmosphere as natural processes.
     a. one-third
     b. one-fourth
     c. one-half
  * d. twice
     e. three times

17. Humans alter the sulfur cycle by
     a. refining petroleum.
     b. causing volcanic eruptions.
     c. burning coal and oil.
     d. producing toxic wastes.
  * e. a and c.
     f. all of the above.

9-2 OUTDOOR AIR POLLUTION: POLLUTANTS, SMOG, AND ACID DEPOSITION

18. Each of the following is one of the five primary outdoor pollutants *except*
     a. carbon monoxide.
  * b. photochemical oxidants.
     c. nitrogen oxides.
     d. sulfur oxides.
     e. suspended particulate matter.

19. All of the following are volatile organic compounds (VOCs) *except*
     a. methane.
     b. chlorofluorocarbon.
  * c. carbon monoxide.
     d. benzene.
     e. bromine-containing halons.

20. All of the following are photochemical oxidants *except*
  * a. dioxin.
     b. hydroxyl radicals.
     c. peroxyacyl nitrates (PANs).
     d. aldehydes, such as formaldehyde.
     e. ozone.

21. Suspended particulate matter includes
     a. asbestos.
     b. droplets of liquids.
     c. nitrate and sulfate salts.
     d. dust, soot, and pollen.
     e. lead, arsenic, and cadmium.
  * f. all of the above.

22. An example of a primary pollutant is
     a. PANs (peroxyacyl nitrates).
  * b. sulfur dioxide.
     c. nitric acid.
     d. hydrogen peroxide.
     e. aldehydes.

23. An example of a secondary air pollutant is
    a. volatile organic compounds (VOCs).
    b. suspended particulate matter.
 * c. nitric acid.
    d. carbon monoxide.
    e. nitrogen oxides.

24. Which of the following statements is *false?*
    a. Stationary sources of air pollution include power and industrial plants.
    b. Mobile sources of air pollution are motor vehicles.
 * c. Human contributions to air pollution are smaller than those from volcanoes.
    d. Burning fossil fuels is the primary contributor to air pollution in MDCs.
    e. In congested Mexico City, the major cause of air pollution is motor vehicles.

25. Photochemical smog is formed when primary pollutants interact with
 * a. sunlight.
    b. water vapor.
    c. sulfur dioxide.
    d. oxygen.
    e. heat.

26. Photochemical smog is found in urban areas with many vehicles and a climate that is
    a. cool, wet, and cloudy.
    b. cool, dry, and sunny.
 * c. warm, dry, and sunny.
    d. warm, wet, and cloudy.
    e. warm, wet, and sunny.

27. Which of the following statements is *true?*
    a. Thermal inversion occurs when a layer of cold air prevents warm air from rising.
 * b. Thermal inversions enhance pollution problems.
    c. Thermal inversions last only a few minutes to a few hours.
    d. Normally, cool air near Earth's surface expands and rises, carrying pollutants higher into the troposphere.
    e. None of the above.

28. Which of the following statements is *false?*
    a. Industrial smog consists of a mixture of sulfur dioxide, suspended droplets of sulfuric acid, and a variety of suspended solid particles.
    b. Industrial smog is primarily a problem in the winter.
    c. In the United States, gray-air smog was a greater problem 30 years ago than it is now.
    d. Industrial smog originates with burning sulfur-containing coal and heavy oil.
 * e. All industrial countries now use modern technologies to prevent industrial smog.

29. A thermal inversion is the result of
    a. precipitation.
    b. cold air drainage.
 * c. a warm lid on top of stagnant air.
    d. a cold blanket of air that prevents warm air from rising.
    e. ocean currents.

30. The city in the United States that experienced one of the first thermal inversions, with high pollution that killed 20 people, was
    a. New York City, New York.
    b. Los Angeles, California.
 * c. Donora, Pennsylvania.
    d. Austin, Texas.
    e. Washington, D.C.

31. Which of the following areas would be least likely to have a thermal inversion?
    a. an area near the coast
    * b. an area in the central plains
    c. a valley surrounded by mountains
    d. the leeward side of a mountain range
    e. none of the above; all are vulnerable

32. The city in the United States distinguished by having the toughest pollution control program and the greatest air pollution problem is
    a. New York City, New York.
    b. Birmingham, Alabama.
    * c. Los Angeles, California.
    d. Boston, Massachusetts.
    e. Washington, D.C.

33. Acid deposition is properly defined as the ____ deposition of ____ pollutants onto Earth's surface.
    a. wet . . . primary
    b. wet . . . secondary
    c. dry . . . secondary
    d. wet and dry . . . primary
    * e. wet and dry . . . secondary

34. Tall chimneys
    a. are inexpensive ways to disperse pollution.
    b. carry the pollutants above any local inversion layer.
    c. are an output approach to pollution.
    d. increase pollution in downwind areas.
    * e. all of the above.

35. Acid deposition
    a. increases the mobility of aluminum ions.
    b. contaminates fish with methylmercury.
    c. leads to excessive nitrogen in the soil.
    d. damages statues, buildings, and car finishes.
    * e. all of the above.

36. In general, acid deposition has harmful effects when it falls below
    a. 12.5.
    b. 7.0.
    c. 6.7.
    * d. 5.1.
    e. 4.2.

37. Acid deposition can affect living organisms by
    a. killing fish, aquatic plants, and microorganisms in lakes and streams.
    b. damaging tree roots.
    c. making trees more susceptible to diseases.
    d. aggravating human respiratory diseases.
    * e. doing all of the above.

38. Acid deposition can effect nonliving components of the ecosystem by
    a. damaging statues, buildings, metals, and car finishes.
    b. leaching plant nutrients from the soil.
    c. releasing metal ions from the soil.
    * d. doing all of the above.

39. Acid deposition
    a. exhibits a threshold effect.
    b. can exceed the buffering of limestone and other alkaline minerals.
    c. may cause yearly damages between $6 billion and $10 billion.
    d. is often exported from one country to another.
  * e. all of the above.

40. Acid deposition is more of a problem in the _____ United States than in the other regions.
  * a. eastern
    b. western coastal
    c. Rocky Mountains of the
    d. Mississippi Valley of the
    e. Ohio Valley of the

9-3 INDOOR AIR POLLUTION

41. In 1990, out of 18 sources of cancer risk, the EPA rated indoor air pollution
  * a. first.
    b. second.
    c. third.
    d. fifth.
    e. last.

42. Individuals at greatest risk from indoor air pollution include
    a. nonsmokers.
    b. single women.
    c. white collar workers.
  * d. the young and the old.
    e. people with hereditary disorders.

43. "Sick building syndrome" is *least* likely to result in
    a. coughing and sneezing.
    b. chronic fatigue.
  * c. cancer.
    d. nausea.
    e. dizziness and headaches.

44. About _____% of all U.S. buildings are considered "sick" from indoor air pollutants.
    a. 5–10
    b. 10–20
  * c. 20–30
    d. 40–50
    e. 70–80

45. Radioactive _____ is a product of uranium decay and a serious indoor air pollutant.
  * a. radon
    b. radium
    c. plutonium
    d. lead
    e. arsenic

46. Which of the following is *not* a major indoor air pollutant?
    a. asbestos
    b. radon-222
  * c. sulfur dioxide
    d. cigarette smoke
    e. formaldehyde

47. In LDCs, the single largest cause of indoor air pollution is
   * a. tobacco smoke.
   b. sick building syndrome.
   c. formaldehyde from building materials.
   d. particulate matter from fires and stoves.
   e. radon gas from the ground.

48. Furniture stuffing, paneling, particle board, and foam insulation may be sources of
   a. chloroform.
   * b. formaldehyde.
   c. carbon monoxide.
   d. asbestos.
   e. nitrogen oxides.

49. Pipe insulation and vinyl ceiling and floor tiles may be sources of
   a. chloroform.
   b. formaldehyde.
   c. carbon monoxide.
   * d. asbestos.
   e. sulfur dioxide.

50. Faulty furnaces, unvented gas stoves and kerosene heaters, and wood stoves may be sources of
   a. chloroform.
   b. formaldehyde.
   * c. carbon monoxide.
   d. asbestos.
   e. nitrogen oxides.

51. All of the following commonly used household products can contribute to indoor air pollution *except*
   a. air fresheners.
   b. mothball crystals.
   * c. baking powder.
   d. paint strippers.
   e. aerosol sprays.

52. Radon-222 is
   a. a colorless, odorless, tasteless gas.
   b. a product of uranium decay.
   c. particularly concentrated in underground deposits of uranium, phosphate, granite, and shale.
   d. basically a problem in confined spaces, such as basements, and underground wells over radon-containing deposits.
   * e. all of the above.

53. The radiation produced by radon-222 and its damaging decay products is _____ radiation.
   * a. alpha
   b. beta
   c. gamma
   d. ultraviolet
   e. none of the above

54. In 1988, the EPA and the U.S. Surgeon General's Office recommended that everyone living in a detached house or the first three floors of an apartment building test for radon. By 1991, _____% of all households had been tested for radon.
     a. 5
   * b. 11
     c. 20
     d. 52
     e. 75

55. You have been looking for your first house for months. You find one in just the right neighborhood at just the right price for you. In the course of negotiations, you have a radon test done and find that the level is 15 picocuries. A reasonable course of action would be to
     a. get out of the housing market.
     b. back out of the deal quickly and look for another house.
   * c. make a purchase offer, but recognize you will need to make some changes over the course of a few years.
     d. make a purchase and move in happily ever after.
     e. none of the above.

56. You have been looking for your first house for months. You find one in just the right neighborhood at just the right price for you. In the course of negotiations, you have a radon test done and find that the level is 250 picocuries. A reasonable course of action would be to
     a. get out of the housing market.
   * b. back out of the deal quickly and look for another house.
     c. make a purchase offer, but recognize you will need to make some changes over the course of a few years.
     d. make a purchase and move in happily ever after.
     e. none of the above.

57. Which of the following statements is *false?*
     a. Houses with wells should have their well water tested for radon.
   * b. Gas-heated houses are more likely to have radon problems than those heated with electricity.
     c. Well-insulated airtight houses are more likely to have radon problems than those that leak heat.
     d. In Sweden, no house can be built until the lot has been tested for radon.
     e. Radon problems can be prevented in new housing by using solid concrete blocks or poured concrete walls and installing a heat-bonded nylon mat under the slab.

58. Prolonged exposure to asbestos is *least* likely to cause
     a. chronic breathing difficulties.
   * b. hardening of the arteries.
     c. lung cancer.
     d. cancer of the chest cavity lining.
     e. mesothelioma.

59. Out of the following groups, the group *least* likely to be exposed to asbestos fibers is
     a. auto mechanics.
   * b. battery manufacturers.
     c. insulators.
     d. shipyard employees.
     e. pipe fitters.

60. Asbestos has historically been used for
    a. fireproofing.
    b. insulation of pipes.
    c. decoration.
    d. soundproofing.
  * e. all of the above.

61. The least recommended method of asbestos treatment is
    a. sealing.
    b. containment.
    c. wrapping.
  * d. removal.

9-4 EFFECTS OF AIR POLLUTION ON LIVING ORGANISMS AND ON MATERIALS

62. Humans are protected from air pollution by
    a. sneezing and coughing.
    b. mucus capturing small particles.
    c. nasal hairs filtering out large particles.
    d. cilia transporting mucus to the mouth.
  * e. all of the above.

63. The people least vulnerable to air pollution are
    a. infants.
    b. elderly people.
  * c. adult males.
    d. people with heart and respiratory disease.
    e. pregnant women.

64. Pollutants that are especially hazardous because they can penetrate the lungs' natural defenses are
    a. large particles
  * b. small particles.
    c. ozone.
    d. acid rain.
    e. all of the above.

65. Years of smoking and exposure to air pollutants is *least* likely to cause
  * a. diabetes.
    b. asthma.
    c. bronchitis.
    d. emphysema.
    e. lung cancer.

66. The World Health Organization estimates that one in ____ people live in cities where outdoor air is unhealthy to breathe.
    a. two
  * b. four
    c. seven
    d. ten
    e. one hundred

67. Trees that are chronically exposed to gaseous pollutants (particularly ozone) are more likely to suffer from all of the following *except*
    a. excessive water loss.
    b. damage from disease.
    * c. reduced vitamin synthesis.
    d. reduced nutrient uptake.
    e. needle loss.

68. Waldsterben in Europe
    a. kills small rodents.
    b. increases plant resistance to drought and disease.
    * c. kills large forests.
    d. spreads bubonic plague.
    e. causes lakes to become sterile.

69. The greatest pollution damage to forests in Europe has occurred in
    a. Germany.
    b. Sweden.
    * c. Poland.
    d. Great Britain.
    e. France.

70. Air pollution in the United States has most seriously affected trees
    a. along the shores of lakes.
    b. lining major interstate highways.
    * c. on high-elevation slopes facing moving air masses.
    d. in the low-lying swamps in the southeast.
    e. forming windbreaks in the Central Plains states.

71. U.S. air pollution has most seriously affected trees in the
    * a. Appalachian Mountains.
    b. Rocky Mountains.
    c. Olympic Mountains.
    d. Sierra Nevada Mountains.
    e. Everglades.

72. Air pollution, mostly ozone, has reduced crop production by 5% to 10% especially in
    a. corn and wheat.
    b. tobacco and sorghum.
    c. soybeans and peanuts.
    d. alfalfa and cotton.
    * e. a and c.
    f. all of the above.

73. Acid shock that damages aquatic life in the Northern Hemisphere is the result of the sudden runoff of acid water with dissolved
    a. lead.
    b. chromium.
    c. fluorine.
    d. aluminum.
    * e. nitrates.

74. Acidified lakes can be neutralized by treating them with
    * a. large amounts of limestone.
    b. large amounts of vinegar.
    c. small amounts of baking soda.
    d. small amounts of drain cleaner.
    e. none of the above.

Air

75. Air pollutants are *least* likely to be responsible for
    a. increasing cleaning bills for clothing.
    b. ruining painted surfaces.
  * c. rusting metal surfaces.
    d. pitting marble statues.
    e. discoloring stained-glass windows.

9-5 SOLUTIONS: PREVENTING AND REDUCING AIR POLLUTION

76. National ambient air quality standards have been set for all of the following except
    a. sulfur oxides.
    b. nitrogen oxides.
  * c. carbon dioxide.
    d. ozone.
    e. hydrocarbons.

77. Of the following major outdoor air pollutants, the one which has *not* declined since 1970 is
    a. sulfur dioxide.
  * b. nitrogen oxides.
    c. suspended particles.
    d. lead.
    e. suspended particles.

78. Due to pollution prevention strategies, the outdoor air pollutant which has shown the steepest decline since 1970 is
    a. sulfur dioxide.
    b. nitrogen oxides.
    c. suspended particles.
  * d. lead.
    e. suspended particles.

79. National ambient air quality standards
    a. have been established for almost 100 air pollutants.
    b. must be met by 50 major U.S. metropolitan areas that are responsible for implementation plans.
    c. are established by Congress.
  * d. specify the maximum allowable level, averaged over a specific time period, for a certain outdoor air pollutant.
    e. include all of the above.

80. The Clean Air Act has been criticized for
    a. increasing fuel efficiency standards for cars and light trucks.
    b. classifying ash from municipal trash incinerators as hazardous waste.
  * c. giving municipal trash incinerators 30-year permits.
    d. setting standards for air pollution emissions from incinerators too high.
    e. setting recycling goals at an unattainable 50%.

81. In 1992, a car sold in California emitted _____ the pollution emitted by a new car in 1970.
    a. 1/2
    b. 1/4
  * c. 1/10th
    d. 1/25th
    e. 1/100th

82. Of the following motor vehicle fuels, the greatest polluter is
    * a. gasoline.
    b. hydrogen gas.
    c. alcohol.
    d. natural gas.
    e. none of the above; they all burn with no emissions.

83. All of the following are provisions of the California South Coast Air Quality Management District Council's proposals in 1989 to reduce ozone and smog in the Los Angeles area *except*
    * a. establish emission standards for trucks.
    b. ban drive-through facilities.
    c. increase parking fees and assess high fees for multivehicle families.
    d. require gas stations to use a hydrocarbon-vapor recovery system on gas pumps and sell alternative fuels.
    e. control or relocate petroleum refining, dry-cleaning, auto-painting, printing, and other highly polluting industries.

84. Indoor air pollution could be sharply reduced by
    a. modifying building codes to prevent radon infiltration.
    b. requiring exhaust hoods or vent pipes for stoves, refrigerators, or other appliances burning natural gas or other fossil fuels.
    c. setting emission standards for building materials.
    d. finding substitutes for potentially harmful chemicals.
    * e. all of the above.

85. One way to help protect the atmosphere would be to
    a. quickly burn all remaining fossil fuels to encourage faster change to alternative fuels.
    b. compartmentalize air pollution, water pollution, and energy policies so that each department has its own focus.
    c. emphasize local control and responsibility for air pollution.
    * d. control population growth.
    e. emphasize pollution control rather than pollution prevention.

CHAPTER 10
CLIMATE, GLOBAL WARMING, AND OZONE LOSS

1. Humans are now altering the chemical content of Earth's entire atmosphere _____ times faster than its natural rate of change over the past 100,000 years.
   - a. 2–10
   - * b. 10–100
   - c. 100–1000
   - d. 1000–10,000
   - e. 10,000–100,000

10-1   WEATHER AND CLIMATE: A BRIEF INTRODUCTION

2. Weather occurs
   - a. over the short term.
   - b. at a particular place.
   - c. at a particular time.
   - d. in the troposphere.
   - * e. all of the above.

3. Climate
   - a. describes long-term weather patterns.
   - b. describes seasonal variations.
   - c. describes weather extremes.
   - d. occurs over at least a 30-year period.
   - * e. includes all of the above.

4. Climate is the general pattern of weather over a period of at least
   - a. 10 years.
   - b. 20 years.
   - * c. 30 years.
   - d. 50 years.
   - e. 100 years.

5. The two most important factors in climate are
   - a. temperature and insulation.
   - b. precipitation and pressure.
   - c. humidity, clouds, and wind.
   - * d. temperature and precipitation.
   - e. wavelengths of light and atmospheric particulates.

6. Climate is influenced by
   - a. the amount of incoming solar radiation.
   - b. the rotation and revolution of Earth.
   - c. distribution of land and water and topography.
   - d. air and ocean currents.
   - * e. all of the above.

7. Which of the following statements is *false*?
   - a. The amount of solar energy reaching Earth's surface is dependent on latitude.
   - b. Hot air rises.
   - * c. Air at the lower latitudes tends to cool and fall or sink downward.
   - d. Cool air is denser than warm air.
   - e. Air is heated much more at the equator than at the poles.
   - f. None of the above; all are true.

8. Which of the following statements of cause and effect is *false?*
   a. The differential in solar energy striking the equator versus the poles sets up general global air circulation patterns.
 * b. Earth's rotating faster under air at the poles causes the prevailing winds.
   c. Earth's rotational tilt and revolution around the sun cause seasonal variations in temperature.
   d. Greenhouse gases let in the sun's ultraviolet radiation and trap infrared waves radiated from Earth.
   e. The warm Gulf Stream is largely responsible for the mild northwestern European climate.

9. There are _____ separate belts of moving air or prevailing winds.
   a. two
   b. four
 * c. six
   d. eight
   e. an indefinite number of

10. Which of the following wind patterns is found at the equator?
 * a. doldrums
   b. tradewinds
   c. easterlies
   d. southerlies
   e. westerlies

11. All of the following are greenhouse gases *except*
   a. carbon dioxide.
   b. water vapor.
   c. methane.
 * d. sulfur dioxide.
   e. nitrous oxide.

12. The term *greenhouse effect*
   a. describes occupational diseases of florists.
 * b. describes the trapping of heat energy in the troposphere by certain gaseous molecules.
   c. describes the trapping of heat energy in the stratosphere by nitrogen.
   d. describes efforts being made by the White House to support environmental legislation.
   e. makes Earth uninhabitable.

10-2 GLOBAL WARMING OR A LOT OF HOT AIR?

13. Earth's climate is the result of complex interactions between
   a. the sun and the atmosphere.
   b. the atmosphere and the oceans.
   c. the land and the biosphere.
   d. the sun and the oceans.
 * e. all of the above.

14. During Earth's past 800,000 years, glacial periods were about _____ interglacial periods.
   a. a hundred times longer than
 * b. ten times longer than
   c. the same length as
   d. ten times shorter than
   e. a hundred times shorter than

15. During the current interglacial period, mean surface temperatures have fluctuated _____ degrees centigrade.
    * a. 0.5–1.0
      b. 1.0–2.0
      c. 3.0–5.5
      d. 5.0–10.5
      e. 10.0–15.0

16. Estimated variations in Earth's mean surface temperature correlate closely with
      a. carbon dioxide and ozone.
      b. water vapor and nitrous oxide.
      c. CFCs and methane.
    * d. carbon dioxide and methane.
      e. nitrous oxide and sulfur dioxide.

17. All of the following are major greenhouse gases *except*
      a. chlorofluorocarbons (CFCs).
      b. carbon dioxide and water vapor.
    * c. sulfur dioxide.
      d. methane.
      e. nitrous oxide.

18. All of the following greenhouse gases have increased in recent decades *except*
      a. carbon dioxide.
      b. methane.
    * c. water vapor.
      d. nitrous oxide.
      e. chlorofluorocarbons (CFCs).

19. Increased greenhouse gases originate from
      a. burning fossil fuels.
      b. agriculture.
      c. use of CFCs.
      d. deforestation.
    * e. all of the above.

20. The fluctuations in mean global temperature measured over the last 100 years may be caused by
      a. enhanced greenhouse effect.
      b. volcanic eruptions.
      c. air pollution.
    * d. all of the above.
      e. none of the above.

21. Since 1860, mean global temperature has risen _____ degrees centigrade.
      a. 0.1–0.3
    * b. 0.3–0.6
      c. 0.6–1.0
      d. 1.0–1.5
      e. 1.5–2.0

22. Of the years from 1980 to 1992, _____ were among the hottest years in the 110-year history of land surface temperature measurements.
      a. two
      b. four
      c. six
    * d. eight
      e. 10

23. Warming or cooling by more than _____ degrees centigrade could be disastrous for Earth's ecosystems.
    a. 0.5
    b. 1.0
  * c. 2.0
    d. 3.0
    e. 4.0

24. Major climate models project all of the following *except*
    a. 1.5- to 5.5-degree centigrade rise in Earth's mean surface temperature.
  * b. an Earth warmer than at any time in the last 100,000 years.
    c. more warming of the air over land than over oceans.
    d. more warming in the Northern Hemisphere than in the Southern Hemisphere.
    e. a greater rise in middle and high latitudes than near the equator.

25. Major climate models project all of the following *except*
    a. drier soil in the middle latitudes.
  * b. particularly dry soil during the winter in the Southern Hemisphere.
    c. more extreme heat waves.
    d. more forest fires.
    e. an average sea level rise of 2 to 4 centimeters per decade.

26. We have the most certainty about
    a. variations in solar output.
    b. the role the oceans will play in global warming.
  * c. patterns of glacial and interglacial periods in Earth's history.
    d. the role of polar ice in global warming.
    e. the role of air pollution in global warming.

27. We are uncertain about
    a. how carbon dioxide will affect the rate of photosynthesis.
    b. how increased temperatures will affect insect populations.
    c. how increased food production will affect global warming.
    d. how gas trapped in the permafrost will affect global warming.
  * f. all of the above.

28. If the permafrost in the tundra melts, we could expect a major increase in the amount of _____ in the atmosphere.
  * a. methane
    b. ammonia
    c. chlorofluorocarbons
    d. carbon dioxide
    e. nitrous oxide

29. Evidence indicates that climate belts would shift toward the poles ____ miles for every 1 degree centigrade increase.
    a. 10–30
    b. 30–60
  * c. 60–90
    d. 90–120
    e. 120–150

30. Which of the following would *most* likely occur if global warming continues?
    a. Irrigated land would dry up.
    b. Countries near the poles would reap an agricultural bonanza.
    c. Wind patterns would probably remain the same.
  * d. Agricultural prices would rise.
    e. Tropical fruits will be grown in Canada.

31. Tree species typically move _____ mile(s) per decade.
    a. 1
  * b. 5
    c. 10
    d. 20
    e. 50

32. If climate belts move faster than trees migrate, there could be
    a. large forest diebacks.
    b. mass extinctions of species that couldn't migrate.
    c. massive fires.
    d. increased carbon dioxide from decomposition.
  * e. all of the above.

33. If global warming trends continue, sea levels might
    a. rise because of melting of polar sea ice.
    b. rise because of water expanding when heated.
    c. drop because of increased use of water for irrigation.
    d. drop because of increased evaporation.
  * e. a and b.
    f. c and d.

34. A modest rise in average sea level is *least* likely to cause
    a. flooding of low-lying cities.
    b. acceleration of coastal erosion.
  * c. exposure of coral reefs to more intense ultraviolet light.
    d. contamination of coastal aquifers with salt.
    e. inland migration of marshes.

35. If global warming trends continue, human health may be affected by
    a. flooding of sewage systems in coastal cities.
    b. spread of tropical diseases to temperate zones.
    c. disrupted food supplies.
    d. disrupted water supplies.
  * e. all of the above.

36. Which of the following statements about the potential effects of global warming on human health is *false*?
    a. Food and fresh water supplies are likely to be disrupted.
    b. People are likely to be displaced.
  * c. Insect-borne diseases are likely to decrease in today's temperate zones.
    d. Sanitation systems in coastal cities may be flooded.
    e. None of the above is false.

10-3 SOLUTIONS: DEALING WITH GLOBAL WARMING

37. Humans who wish to avoid the fate of the frog are *least* likely to propose
    a. improving energy efficiency.
    b. starting reforestation programs.
    c. replacing fossil fuels with renewable energy resources.
  * d. more research before taking action.
    e. halting deforestation.

38. The quickest way to reduce the buildup of carbon dioxide in the atmosphere is to
    a. switch from fossil fuels to nuclear fuels.
  * b. increase the efficiency of energy use.
    c. plant trees to trap more carbon dioxide.
    d. stop deforestation.
    e. do none of the above.

39. The fossil fuel that releases the *most* carbon dioxide per unit of energy is
    a. oil.
    b. natural gas.
  * c. coal.
    d. petroleum.
    e. none of the above; the law of conservation of energy says that each unit of fossil fuel releases the same amount of energy.

40. All of the following would be win/win pacts to reduce the threat of global warming *except*
    a. LDCs stop deforestation and MDCs shift to sustainable agriculture.
    b. LDCS protect biodiversity and MDCs transfer pollution prevention technologies.
    c. LDCS slow population growth and MDCs abandon use of ozone-depleting chemicals.
    d. LDCs enact fairer land distribution and MDCs forgive foreign debt.
  * e. LDCs phase out coal burning and MDCs deplete remaining fossil fuels as quickly as possible.

41. Prevention approaches to global warming include all of the following *except* to
  * a. increase beef production to strengthen public health.
    b. halt unsustainable deforestation.
    c. switch to sustainable agriculture.
    d. slow population growth.
    e. decrease poverty.

42. Of the following adjustments to global warming, the *least* effective would be to
    a. breed food plants that need less water and are more tolerant of salt.
    b. build dikes.
    c. move storage tanks of hazardous materials away from coastal areas.
    d. ban new construction on low-lying coastal areas.
  * e. establish large food supplies on barrier beaches.
    f. expand protected areas, connecting them by corridors.
    g. develop better water conservation strategies.

43. Things individuals can do to alleviate the potential threats of global warming include all of the following *except* to
    a. improve energy efficiency where they live.
  * b. use low-sulfur coal if they can't use perpetual or renewable energy sources.
    c. plant trees.
    d. reduce, reuse, and recycle.
    e. urge state and national legislators to sponsor bills aimed at greatly improving energy efficiency.

10-4 OZONE DEPLETION: SERIOUS THREAT OR HOAX?

44. Which of the following statements is *false?*
    a. You can thank photosynthetic bacteria for the ozone sunscreen.
    b. Chlorofluorocarbons are also known by their trademark: freons.
    c. CFCs are odorless and stable.
    d. CFCs are nonflammable, nontoxic, and noncorrosive.
    e. Under the influence of ultraviolet radiation, CFCs break down.
  * f. Fluorine atoms are most responsible for the breakdown of ozone to molecular oxygen.

45. Which of the following statements is *false?*
    a. Over 44 years passed from the first production of CFCs until the first awareness that they could cause environmental damage.
    b. CFCs are stable, odorless, nonflammable, nontoxic, and noncorrosive chemicals.
    c. CFCs are found in bubbles in styrofoam and insulation.
    * d. CFCs are important because they help screen out ultraviolet radiation from reaching Earth's surface.
    e. Bromine-containing compounds, called halons, are used in fire extinguishers.

46. Which of the following statements is *false?*
    a. CFCs are relatively unreactive compounds.
    * b. CFCs are heavy molecules that will sink in the atmosphere.
    c. Ultraviolet radiation will cause CFCs to break down and release chlorine.
    d. One chlorine molecule may convert 100,000 molecules of ozone to molecular oxygen.
    e. It is common for a CFC molecule to last 100 years in the atmosphere.

47. Which of the following statements is *false?*
    * a. The ozone hole is larger in the Northern Hemisphere than in the Southern Hemisphere.
    b. Up to 50% of the ozone over Antarctica is destroyed each year.
    c. The large annual decrease in ozone over the South Pole is caused by spinning vortices with clouds of ice crystals that have absorbed CFCs on their surfaces.
    d. The ozone hole over Antarctica appears to last longer each year.
    e. The annual polar vortex is weaker at the North Pole than at the South Pole.

48. Maduro and Schauerhammer claim that, of the following, chlorine is contributed to the atmosphere *least* by
    * a. CFCs.
    b. seawater.
    c. biomass burning.
    d. volcanoes.

49. Ozone-layer research scientists critique Maduro and Schauerhammer's claims with all of the following evidence *except*
    a. questioning extrapolations of the data.
    b. disputing that seawater chlorine reaches the stratosphere.
    c. claiming that only 20% of the chlorine from biomass burning reaches the stratosphere.
    d. measuring contributions of chlorine from eruption of a volcano in Mexico.
    * e. claiming that natural sources of chlorine do not affect stratospheric ozone.

50. CFCs are used for all of the following *except*
    a. coolants in refrigerators and air conditioners.
    b. propellants in aerosol spray cans.
    c. sterilants in hospitals.
    * d. fuels in camp stoves.
    e. fumigants for granaries.
    f. cleaners for electronic parts.
    g. creating bubbles in polystyrene foam.

51. CFCs are released into the atmosphere by all of the following *except*
    a. spray cans.
    b. discarded refrigerators.
    * c. burning of artificial logs in fireplaces.
    d. leaking of air conditioners.
    e. burning of plastic foam products.

52. Chemicals capable of destroying ozone include all of the following *except*
    a. chlorofluorocarbons.
  * b. formaldehyde used as a preservative.
    c. halons in fire extinguishers and crop fumigants.
    d. carbon tetrachloride used as a solvent.
    e. methyl chloroform used as a cleaning solvent for metals.

53. In the 1980s, researchers discovered a _____%-loss of ozone in the upper stratosphere over the Antarctic during the Antarctic springtime.
    a. 10
    b. 25
  * c. 50
    d. 75
    e. 90

54. Increases in ultraviolet radiation will cause an increase in all but which one of the following?
    a. skin cancers
  * b. yields of food crops
    c. eye cataracts
    d. suppression of the human immune system
    e. ozone at ground level

55. The world's highest rate of skin cancer is found in
    a. Sweden.
    b. the United States.
    c. the Commonwealth of Independent States.
    d. China.
  * e. Australia.

56. Human health problems closely associated with ozone depletion include all of the following *except*
    a. skin cancer.
    b. eye cataracts.
  * c. increased incidence of heart disease.
    d. suppression of the immune response.
    e. increased eye irritation.

10-5 SOLUTIONS: PROTECTING THE OZONE LAYER

57. All of the following are substitutes for CFCs *except*
    a. chemicals outside the fluorocarbon family that can be used as cleaning and blowing agents.
    b. hydrofluorocarbons.
  * c. halons.
    d. hydrochlorofluorocarbons.
    e. all of the above; there are no substitutes for CFCs.
    f. none of the above; they are all substitutes.

58. In 1987, 24 nations meeting in Montreal, Canada developed the Montreal Protocol to reduce production of
    a. carbon dioxide.
    b. nitrous oxide.
  * c. CFCs.
    d. toxic wastes.
    e. methane.

59. If all of the ozone-depleting substances were banned tomorrow, it would take ___ years for Earth to recover.
    a. 15
    b. 50
  * c. 100
    d. 300
    e. 500

60. To help protect the ozone layer, individuals should do all of the following *except*
    a. avoid purchasing products that contain CFCs.
  * b. buy halon fire extinguishers.
    c. pressure legislators to ban all uses of CFCs, halons, and methyl bromide by 1995.
    d. buy new refrigerators that use vacuum insulation and helium as a coolant.
    e. make sure their auto air conditioners are not leaking.

61. Ray Turner is known for his work in
    a. stopping use of CFCs in air conditioners.
    b. stopping use of CFCs in refrigerators.
  * c. substituting lemon juice for CFCs used to clean electronic circuit boards.
    d. stopping use of lasers.
    e. substituting harmless chemicals for CFCs used in fire extinguishers.

CHAPTER 11
WATER

1. Which of the following countries is mismatched with the headwaters of a river found in that country?
    a. Sudan–Nile River
    b. Ethiopia–Nile River
    * c. Jordan–Jordan River
    d. Turkey–Tigris River
    e. Turkey–Euphrates River

2. All of the following are downstream countries that could suffer from an upstream country withdrawing more water from a shared river *except*
    a. Egypt.
    b. Jordan.
    c. Israel.
    * d. Turkey.
    e. Syria.

3. There is potential for conflict over water resources among all of the following pairs of countries *except*
    a. Sudan and Egypt.
    b. Syria and Jordan.
    c. Syria and Israel.
    * d. Turkey and Egypt.
    e. Turkey and Iraq.

4. Resolution of water-supply conflicts in the Middle East will require
    a. reducing population growth.
    b. improving water efficiency.
    c. regional cooperation.
    d. none of the above.
    * e. all of the above.

5. Water is a vital resource for
    a. agriculture.
    b. manufacturing.
    c. recreation.
    d. transportation.
    * e. all of the above.

11-1 WATER'S IMPORTANCE AND UNIQUE PROPERTIES

6. Water covers about _____% of Earth's surface.
    a. 56
    b. 62
    * c. 71
    d. 79
    e. 83

7. Water
    a. helps to dilute pollutants.
    b. is essential to all life.
    c. maintains Earth's climate.
    d. makes up most of living organisms.
    * e. includes all of the above.

8. Which of the following statements is *false?*
   a. Water has a very high heat capacity and changes temperatures very slowly.
   b. Water helps distribute heat throughout the earth.
   * c. Water has a low heat of vaporization; that is, it evaporates very easily.
   d. Water functions well as a coolant.
   e. Water exists as a liquid over a wide temperature range.

9. Which of the following statements is *false?*
   a. Water is easily polluted because it is a good solvent.
   * b. Water has a low surface tension; it is not attracted and does not adhere to other molecules.
   c. Water exhibits capillarity, the ability to rise through small pores.
   d. Water expands when it freezes.
   e. None of the above is false.

10. Water
   a. contracts when it freezes and expands when it melts.
   b. reaches its maximum density when it freezes.
   * c. breaks rocks during weathering.
   d. is used as a lubricant in engines.
   e. includes all of the above.

11. Which of the following statements is *false?*
   * a. Water is one of the better-managed resources.
   b. We do not charge enough for water.
   c. We waste and pollute water.
   d. More than 50% of organisms' weight is water.
   e. None of the above is false.

11-2 SUPPLY, RENEWAL, AND USE OF WATER RESOURCES

12. Approximately ____% of Earth's water is fresh rather than salt water.
   * a. 3
   b. 10
   c. 12
   d. 21
   e. 24

13. Only about ____% of the world's total water supply exists as uncontaminated fresh water on or close to the surface and readily available for human use.
   a. 0.0003
   * b. 0.003
   c. 0.03
   d. 0.3
   e. 3.0

14. The hydrologic cycle will naturally purify and recycle fresh water as long as humans don't
   a. pollute the water faster than it is replenished.
   b. withdraw it from groundwater supplies faster than it is replenished.
   c. overload it with slowly degradable and nondegradable wastes.
   * d. do any of the above.
   e. worry about any of the above.

15. During which of the following does water move in a direction different from the others?
    a. percolation
*   b. transpiration
    c. infiltration
    d. precipitation
    e. surface runoff

16. There is ____ times as much groundwater as there is water in all the world's rivers and lakes.
    a. 12
*   b. 40
    c. 80
    d. 100
    e. 200

17. Porous water-saturated layers of underground rock are known as
*   a. aquifers.
    b. recharge areas.
    c. watersheds.
    d. runoff areas.
    e. cones of depression.

18. Which of the following statements is *false?*
    a. Recharging of water is a slow process.
    b. Fossil aquifers are nonrenewable resources on a human time scale.
    c. Aquifers could be called underground lakes.
*   d. Groundwater is stationary and does not move.
    e. The water table is the upper surface of the zone of saturation.

19. Which of the following leads to a cone of depression of the water table?
*   a. excessive withdrawal
    b. recharging
    c. percolation
    d. infiltration
    e. transpiration

20. In which of the following would large amounts of water be consumed?
    a. Water from a river cools a power plant and returns to the river.
    b. A teenager rinses a car with water and the water goes down the storm drain.
    c. When the washing machine wash cycle is over, the water goes down the drain.
*   d. Golf courses are watered in the middle of the day in Las Vegas, Nevada.
    e. All of the above.

21. Throughout the world, the most water is used for
*   a. irrigation.
    b. industrial processes.
    c. needs of animals and humans.
    d. transportation.
    e. power production.

22. Worldwide, energy production accounts for ____% of withdrawn water.
    a. 3
    b. 13
*   c. 23
    d. 33
    e. 43

23. There is _____ groundwater than (as) surface water.
    * a. 40 times more
    b. 10 times more
    c. the same amount of
    d. 10 times less
    e. 40 times less

24. The global increase in water withdrawal since 1950 is due largely to
    * a. increased population demands.
    b. climatic variation.
    c. demands of new species of crops.
    d. industrialization in LDCs.
    e. cooling towers for powerplants.

25. The world's largest water user is
    a. India.
    b. China.
    * c. the United States.
    d. the Commonwealth of Independent States.
    e. Canada.

26. Which of the following uses tends to consume the smallest amount of water?
    a. irrigation
    * b. rural domestic use
    c. industry
    d. power plant cooling

27. Averaged globally, about two-thirds of the water withdrawn each year is used for
    a. industrial processes.
    b. cooling towers of power plants.
    * c. irrigation of croplands.
    d. domestic use.
    e. sewage treatment.

11-3 WATER RESOURCE PROBLEMS

28. The natural hazard that does the most economic damage and harm to people worldwide is
    * a. drought.
    b. foods.
    c. volcanic eruptions.
    d. hurricanes.
    e. tornadoes.

29. Drought is intensified by all of the following *except*
    a. rapid population growth.
    b. overgrazing.
    c. desertification.
    * d. replacement of monocultures with natural grasslands.
    e. deforestation.

30. About _____ of the world's 214 major river systems are shared by two or more countries.
    a. 50
    b. 100
    c. 150
    d. 175
    * e. 200

31. The Mekong River is shared by all of the following countries *except*
    a. Thailand.
    * b. Pakistan.
    c. Cambodia.
    d. Laos.
    e. Vietnam.

32. Monsoon seasons can
    a. cause floods.
    b. wash away topsoil and crops.
    c. waterlog soils.
    d. leach soil nutrients.
    * e. do all of the above.

33. Benefits of floods include all of the following except
    a. provision of productive farmland downstream.
    b. refilling of wetlands.
    c. provision of important breeding and feeding grounds for wildlife.
    * d. filling up of soil air spaces to prevent oxidation of nutrients.
    e. recharging of groundwater supplies.

34. People like to settle on floodplains for all of the following reasons *except*
    a. good transportation.
    b. flat sites for buildings.
    * c. security.
    d. fertile soil.
    e. ample water for irrigation.

35. Of the following natural hazards, the one that causes the most deaths is
    * a. floods.
    b. volcanic eruptions.
    c. typhoons and hurricanes.
    d. earthquakes.
    e. gales and thunderstorms.

36. In India, 90% of the annual precipitation occurs from
    a. December to March.
    b. March to May.
    * c. June to September.
    d. September to December.
    e. none of the above.

37. Floods and droughts are
    a. strictly natural disasters.
    * b. influenced by human activities.
    c. decreased by increases in human population.
    d. independent of human activity.
    e. totally dependent on human activity.

38. Humans influence the frequency and severity of floods and droughts through
    a. severe loss of vegetation from deforestation.
    b. overgrazing of grasslands.
    c. plowing of prairies.
    d. irrigating of fields.
    * e. all of the above.

39. Which of the following human activities contribute to flooding?
    a. cultivation of land
    b. deforestation
    c. overgrazing
    d. mining
    * e. all of the above.

40 Conflicts over water supplies are most serious in
    * a. the Middle East.
    b. the southwestern United States.
    c. California.
    d. southeast Asia.
    e. eastern Africa.

41. In the western United States, as compared to the eastern United States, the *major* water problem(s) is(are)
    a. flooding.
    b. insufficient water for some urban areas.
    * c. chronic drought and insufficient runoff.
    d. pollution of rivers, lakes, and groundwater.
    e. all of the above.

42. Of the following uses for water, the one that is *least* significant in the eastern United States is
    * a. irrigation.
    b. energy production.
    c. cooling.
    d. manufacturing.

43. In the western United States, the *largest* use for water is
    * a. irrigation.
    b. energy production.
    c. cooling.
    d. manufacturing.

44. The *least* important water concern in the East is
    a. occasional urban shortages.
    b. water pollution.
    * c. shortage of runoff caused by low precipitation.
    d. flooding.

45. The most serious water problem in the West is
    a. occasional urban shortages.
    b. water pollution.
    * c. shortage of runoff caused by low precipitation.
    d. flooding.

11-4 SOLUTIONS: SUPPLYING MORE WATER

46. Water supply can be increased by
    a. water diversion projects.
    b. construction of dams and reservoirs.
    c. desalination.
    d. tapping groundwater.
    * e. all of the above.

47. Small-scale dams
    a. cause extensive flooding of land upstream from the dam.
    b. can cause land subsidence and earthquakes.
    c. avoid most of the problems of large dams.
    d. are useful for trapping irrigation water.
  * e. include c and d.
    f. include all of the above.

48. Large dams and reservoirs
    a. reduce danger of flooding upstream.
    b. are inexpensive to build.
    c. cannot be used for outdoor recreation.
  * d. can be used to provide electric power.
    e. provide a controllable supply of water for irrigating above the dam.
    f. include all of the above.

49. Which of the following statements about disadvantages of large dams is *false?*
    a. They cause the water table to rise, increasing chances of waterlogging above the dam.
    b. Reservoir formation displaces people and destroys wildlife habitat.
    c. They can contribute to the incidence of earthquakes.
    d. They last only 40 to 200 years before siltation requires their abandonment.
  * e. They are useful only for flood control.

50. Dams
    a. are expensive to build.
    b. destroy agricultural land and scenic areas.
    c. inhibit migration of fish.
    d. deprive downstream areas of nutrients.
  * e. include all of the above.

51. Water stored in large reservoirs can be used for
    a. irrigating land below the dam.
    b. recreation.
    c. hydroelectric power.
    d. flood control.
    e. providing water through aqueducts to cities.
  * f. all of the above.

52. China's Three Gorges project will
    a. destroy two cities, each with 100,000 people.
    b. displace 1.2 million people.
    c. flood large areas of farmland.
    d. flood 800 factories.
  * e. include all of the above.

53. All of the following are large water distribution projects *except*
    a. the Aral Sea Project.
    b. the James Bay Project in Canada.
  * c. the Boston Harbor Project.
    d. the California Water Project.

54. California's basic water problem stems from the fact that _____% of the population live south of Sacramento, but _____% of the rain falls north of it.
    a. 50 . . . 50
  * b. 75 . . . 75
    c. 50 . . . 75
    d. 75 . . . 50
    e. 90 . . . 50

55. The size of the fourth largest freshwater lake has been decreased by two-thirds to provide water for agriculture. This lake is located in
    a. China.
    b. Bangladesh.
  * c. the states of Uzbekistan and Kazakhstan.
    d. Africa.
    e. Brazil.

56. Which of the following statements about the Aral Sea is *false*?
  * a. Water has been diverted from the Aral Sea and the two rivers that replenish its water for use in manufacturing.
    b. The volume of the Aral Sea has dropped by 69%.
    c. The salinity levels have risen threefold.
    d. All native fish species have disappeared.
    e. Salt rain kills crops, trees, and wildlife.

57. Climate changes resulting from water diversion projects and drought at the Aral Sea include
    a. more rain.
    b. colder summers.
    c. warmer winters.
  * d. longer growing seasons.
    e. fewer storms.

58. Strategies which could improve the situation at the Aral Sea include all of the following *except*
    a. charging farmers more for irrigation water to reduce waste.
  * b. removing trees which block the diversion project.
    c. introducing water-saving technologies.
    d. using groundwater to supplement irrigation water.
    e. developing a regional integrated water management plan.

59. The term *subsidence* refers to
    a. failure of the groundwater supply.
    b. accumulation of silt behind a dam.
  * c. sinking of ground when water has been withdrawn.
    d. intrusion of salt water into a freshwater aquifer.
    e. money paid to farmers who install water-conserving irrigation devices.

60. Overuse of groundwater can lead to
    a. saltwater intrusion.
    b. subsidence.
    c. aquifer depletion.
    d. groundwater contamination.
  * e. all of the above.

61. Currently in the United States, _____% of the groundwater withdrawn is not replenished.
    a. 10
  * b. 25
    c. 38
    d. 50
    e. 75

62. Groundwater depletion can be slowed by
    a. controlling human population growth.
    b. using irrigation water more efficiently.
    c. discontinuing water-thirsty crops in dry areas.
    d. developing crop strains that require less water.
  * e. all of the above.

63. Groundwater depletion can be reduced by
    a. developing plant varieties that use less water.
    b. using plants more resistant to heat stress.
    c. controlling population growth.
    d. refraining from growing water-thirsty crops in dry areas.
  * e. all of the above.

64. Water withdrawn from the Ogallala Aquifer is used to irrigate _____ of all cropland in the United States.
    a. one-sixth
  * b. one-fifth
    c. one-fourth
    d. one-third
    e. one-half

65. The Ogallala Aquifer
    a. is a fossil aquifer and has a slow recharge rate.
    b. is being used eight times faster than it is being recharged.
    c. will be depleted by one-fourth by 2020 if current withdrawal rates continue.
    d. is the largest aquifer in the United States.
  * e. includes all of the above.

66. Which of the following will occur first?
    a. The Ogallala Aquifer will dry up.
    b. We will reach the year 2020.
    c. The Ogallala Aquifer will reverse direction.
  * d. The high cost of pumping water from the dropping water tables will force a change in agriculture.
    e. The recharge rate of the Ogallala Aquifer will increase.

67. Desalination may be accomplished by
    a. distillation.
    b. reverse osmosis.
    c. salt-eating bacteria.
  * d. a and b.
    e. all of the above.

68. Which of the following statements about desalination is *true?*
    a. The common methods of desalination are reverse osmosis and evaporation.
  * b. Desalination is expensive.
    c. The greatest amount of desalination occurs in the United States.
    d. Desalination is the best approach to solving irrigation problems.
    e. Desalination is an energy-efficient process.

69. Desalination
    a. is expensive.
    b. uses vast amounts of energy.
    c. produces large amounts of salt and other minerals.
    d. can threaten food resources in estuaries when brine is dumped in the ocean.
  * e. includes all of the above.

70.   It is most economically and environmentally sound to focus water resource management on
      a.   increasing the water supply.
      b.   controlling the "mining" of groundwater.
   * c.   increasing the efficiency of the way we use water.
      d.   developing desalination plants.
      e.   seeding clouds to increase quantities of rain.

71.   Cloud seeding
      a.   can cause legal disputes.
      b.   is not useful in dry areas.
      c.   introduces large amounts of chemicals into natural ecosystems.
   * d.   includes all of the above.
      e.   includes none of the above.

11-5 SOLUTIONS: USING WATER MORE EFFICIENTLY

72.   According to the World Resources Institute, what percentage of the water that people use throughout the world is unnecessarily wasted?
      a.   25–30%
      b.   35–40%
      c.   45–50%
      d.   55–60%
   * e.   65–70%

73.   Water waste is due to
      a.   artificially low water prices.
      b.   laws that encourage water use whether or not it is needed.
      c.   the large number of governmental units controlling water use.
      d.   lack of incentive to conserve water on the part of multiple users of a common-property aquifer.
   * e.   all of the above.

74.   Riparian rights
      a.   allow a person to use the groundwater on the land he or she owns.
      b.   is the basic philosophy behind water rights in the West.
   * c.   allow a person who owns land adjoining a flowing stream to use the stream.
      d.   is the only method that can be used to apportion water rights.
      e.   include all of the above.

75.   Which of the following statements is *false?*
      a.   Lasers could be used to aid in contouring agricultural fields to increase even distribution of water.
      b.   Seepage from irrigation canals can be reduced by use of plastic and tile liners.
   * c.   Clearing land will increase its water retention.
      d.   Small check dams of earth and stone can be used in LDCs to retain more water.
      e.   Holding ponds can be used to store rainfall for recycling of crops.

76.   The trickle and drip irrigation systems were developed in
      a.   Egypt.
   * b.   Israel.
      c.   China.
      d.   Japan.
      e.   the United States.

77. Which of the following offers the greatest conservation of water?
    a. center-pivot sprinkler systems
    b. low-energy precision-application (LEPA) sprinkler systems
 * c. trickle or drip irrigation
    d. gravity-flow canal systems
    e. holistic hose system

78. Which of the following statements is *false?*
    a. Ninety-five percent of the residences in New York City do not have water meters.
    b. Leaks in water systems account for 20 to 35 percent of water withdrawn from public supplies.
    c. Watering lawns in arid regions of the United States and Australia uses more water than other household uses in the same area.
 * d. It is not possible with present technology to recycle the water from a single household and reuse it.
    e. Residents of Tucson, which has water-conserving city ordinances, use half the volume of water per capita as residents of Las Vegas.

11-6 SOLUTIONS: SAVING THE EVERGLADES

79. Since the Civil War, Americans have converted the Everglades into
    a. farms.
    b. cities.
    c. pastures.
    d. fruit groves.
 * e. all of the above.

80. Development in northern Florida caused all of the following effects in southern Florida *except*
    a. loss of marshlands.
    b. decline of wading bird populations.
 * c. decreasing salinity.
    d. decreasing productivity of Florida's fisheries.
    e. threatening Miami's water supply.

81. Restoration of the Kissimmee River includes all of the following *except*
    a. filling in half of the canal.
    b. constructing artificial marshes to filter out agricultural pollutants.
 * c. straightening the river to ensure direct flow of water to the south.
    d. altering highway construction to help the Florida panther.
    e. reclaiming wetland areas above and below Lake Okeechobee.

11-7 POLLUTION OF STREAMS, LAKES, AND GROUNDWATER

82. A good indicator of the quality of water for drinking or swimming is the number of colonies of
 * a. coliform bacteria.
    b. algae.
    c. dinoflagellates.
    d. manatees.
    e. seabirds.

83. A body of water can be depleted of its oxygen by
    a. viruses and parasitic worms.
 * b. organic wastes.
    c. sediments and suspended matter.
    d. organic compounds such as oil, plastics, solvents, and detergents.
    e. radioactive isotopes.

Water

84. Which of the following statements is *false*?
    a. Heat can lower dissolved oxygen and make fish vulnerable to disease.
    b. Organic wastes reduce the amount of oxygen in the water supply.
    c. Radioactive wastes and toxins can be concentrated by biological amplification.
  * d. Inorganic nutrients such as fertilizers have no adverse effects on aquatic ecosystems.
    e. Nitrates in drinking water can kill unborn children.

85. Waste heat can
    a. cause algae blooms.
    b. deplete water of oxygen.
    c. kill fish.
    d. decrease water quality.
  * e. do all of the above.

86. Nitrates and phosphates are examples of
    a. disease-causing agents.
    b. oxygen-demanding wastes.
    c. organic plant nutrients.
  * d. inorganic plant nutrients.
    e. radioactive substances.

87. Acids, salts, and metals are examples of
    a. oxygen-demanding wastes.
    b. organic plant nutrients.
    c. inorganic plant nutrients.
  * d. water-soluble inorganic chemicals.
    e. radioactive substances.

88. Heat, organic wastes, and inorganic plant nutrients may all deplete dissolved _____ from water.
    a. nitrogen
  * b. oxygen
    c. particulate matter
    d. minerals
    e. decomposing bacteria

89. One class of pollutants that can cause a population explosion of aerobic bacteria is
    a. disease-causing agents.
  * b. oxygen-demanding wastes.
    c. inorganic chemicals.
    d. organic chemicals.
    e. sediments.

90. One class of pollutants that can cause excessive growth of algae is
    a. radioactive substances.
    b. oxygen-demanding wastes.
  * c. inorganic plant nutrients.
    d. organic chemicals.
    e. sediments.

91. The greatest source of water pollution in terms of total mass is
    a. fertilizers.
  * b. sediments.
    c. oxygen-demanding wastes.
    d. water-soluble inorganic chemicals.
    e. organic chemicals.

92. Which of the following decrease(s) photosynthesis in bodies of water?
    a. disease-causing organisms
    b. inorganic plant nutrients
    * c. sediment or suspended matter
    d. heat
    e. radioactive substances

93. All of the following are nonpoint sources of water pollution *except*
    * a. offshore oil wells.
    b. livestock feedlots.
    c. urban lands.
    d. croplands.
    e. construction areas.

94. Which of the following is a nonpoint source of water pollution?
    a. sewage treatment plant
    b. electric power plant
    c. active and inactive coal mines
    * d. logged forest
    e. offshore oil well

95. Which of the following statements is *false*?
    a. Because of their flow, dilution, and bacterial decay, most streams recover rapidly from pollution by heat and biodegradable waste.
    b. In rapidly flowing rivers, dissolved oxygen is replaced quickly.
    c. The amount of oxygen in rivers declines in dry seasons.
    * d. The amount of oxygen in rivers increases as the water's temperature rises.
    e. Nondegradable pollutants are not eliminated by natural dilution and degradation processes.

96. A stream's recovery from pollutants can be affected by
    a. pollutant overload.
    b. drought.
    c. a dam.
    d. water diversion for irrigation.
    * e. all of the above.

97. Oxygen sag curves
    a. may occur during summer droughts.
    b. may occur when oxygen-demanding wastes are added to the water.
    c. may develop in slow-moving rivers.
    d. may occur downstream from a sewage treatment plant.
    * e. include all of the above.

98. Which of the following statements is *false*?
    a. Requiring cities to withdraw water downstream of the city would reduce pollution.
    * b. Slow-flowing rivers are less susceptible to pollutants than fast-flowing streams.
    c. The width and depth of the oxygen sag curve depend on water volume and flow rate.
    d. Streams can recover from degradable pollutants as long as they are not overloaded.
    e. Temperature and pH can affect the oxygen sag curve.

99. The water pollution and control laws enacted in the 1970s have done all but which one of the following?
    a. reduced or eliminated point-source pollution on rivers
    b. increased the number and quality of wastewater treatment plants
    c. held the line against disease-causing agents and oxygen-demanding wastes
    * d. forced municipalities to take their water supply from the downstream side of the city
    e. none of the above; all have been done

100. Water quality assessment could be improved by
   a. increasing the number of sites where water quality is measured.
   b. locating assessment stations in appropriate sites.
   c. measuring toxics.
   d. using ecological indicators of water quality.
 * e. doing all of the above.

101. Which of the following statements about lake stratification is *true?*
   a. Stratified layers of lakes are characterized by vertical mixing.
   b. Stratification increases levels of dissolved oxygen, especially in the bottom layer.
 * c. Lakes are more vulnerable than streams to contamination by plant nutrients, oil, pesticides, and toxic substances that can destroy bottom life.
   d. Lakes have more flushing than streams.
   e. Changing of water in lakes may take up to 500 years.

102. Of the following chemicals, which is *least* likely to be biologically magnified as it moves through food webs?
   a. radioactive isotopes
   b. mercury compounds
 * c. nitrates
   d. PCBs
   e. DDT

103. Which of the following statements is *false?*
 * a. Rivers are more vulnerable than lakes to contamination by plant nutrients, oil, toxins, and pesticides.
   b. Acid deposition and fallout represent a more serious hazard to lakes than rivers.
   c. Eutrophication is a natural process and can occur without the influence of humans.
   d. Human activities can induce cultural eutrophication.
   e. Eutrophication is caused by inputs of silt and nutrients.

104. Cultural eutrophication is caused by
   a. accelerated erosion of nutrient-rich topsoil.
   b. effluents from sewage treatment plants.
   c. fertilizer runoff.
   d. animal waste runoff.
 * e. all of the above.

105. Which of the following developments of cultural eutrophication would occur last?
   a. fish kills
   b. blooms of algae
   c. dense growth of rooted plants along the shore
   d. increase in aerobic bacteria
 * e. increase in anaerobic bacteria

106. In cultural eutrophication, game fish die from
   a. acid deposition.
 * b. suffocation from lack of oxygen.
   c. toxic substances in the water.
   d. salt.
   e. parasites.

107. All of the following strategies would help prevent cultural eutrophication *except*
   a. banning the use of phosphate detergents.
   b. preventing the runoff of fertilizer from agricultural fields.
   c. advance treatment of municipal sewage.
   d. planting trees along streams to reduce erosion.
 * e. stopping release of toxic heavy-metal pollution.

108. About ____ of the 100,000 medium-to-large lakes in the United States suffer from some degree of cultural eutrophication.
   a. one-fifth
   b. one-fourth
 * c. one-third
   d. one-half
   e. three-fourths

109. Which of the following would *not* reduce cultural eutrophication?
   a. Dredge lake bottoms.
   b. Pump oxygen into lakes.
   c. Control land use to prevent nutrient runoff.
   d. Use herbicides and algicides.
 * e. Prevent as much outflow or drainage as possible from the lake.

110. The Great Lakes possess ____% of all the surface fresh water in the United States.
   a. 25
   b. 50
   c. 66
   d. 75
 * e. 95

111. Less than ____% of the water entering the Great Lakes flows out of the St. Lawrence River each year.
 * a. 1
   b. 8
   c. 16
   d. 25
   e. 33

112. Since 1972, the massive pollution control program for the Great Lakes has reduced
 * a. the coliform level.
   b. the dissolved oxygen level.
   c. the amount of fishing (sport and commercial).
   d. swimming.
   e. none of the above.

113. Currently, the greatest problem facing the Great Lakes is
   a. point-source emission of toxins.
   b. phosphates in detergents.
 * c. toxins found in runoff water as well as acid deposition.
   d. oil spills from tankers using the St. Lawrence Seaway.
   e. sewage.

114. The majority of the toxic chemicals entering Lake Superior come from
   a. chemical plants.
 * b. atmospheric deposition.
   c. runoff from cities.
   d. the adjacent bodies of water.
   e. mining operations.

115. The most-polluted of the Great Lakes is Lake
   a. Huron.
   b. Superior.
 * c. Erie.
   d. Ontario.
   e. Michigan.

116. Groundwater
    a. has turbulent flows that dilute pollutants.
    b. has large populations of decomposing bacteria that break down degradable wastes.
  * c. is cold, which slows down decomposition rates.
    d. may take 5 to 10 years to cleanse itself of wastes.
    e. includes all of the above.

117. Which of the following statements about underground contaminants is *false?*
    a. Degradable organic wastes do not decompose as rapidly underground as they do on the surface.
    b. There is little dissolved oxygen to aid in degradation of wastes.
  * c. Waste products are diluted and dispersed quickly in underground aquifers.
    d. It can take hundreds to thousands of years for contaminated groundwater to cleanse itself of degradable wastes.
    e. Underground toxic substances may get introduced into public or private water supplies.

118. An EPA survey in 1984 found that _____ of the rural household wells violated at least one federal health standard for drinking water.
    a. one-fifth
    b. one-fourth
    c. one-third
    d. one-half
  * e. two-thirds

119. The most common pollutants of rural household wells are
    a. nitrates and phosphates.
    b. phosphates and heavy metals.
  * c. nitrates and pesticides.
    d. heavy metals and nitrates.

120. Groundwater can be contaminated by
    a. underground storage tanks.
    b. deep wells used for disposal of hazardous waste.
    c. landfills.
    d. industrial lagoons.
  * e. all of the above.

121. Which of the following statements is *false?*
    a. A survey by the EPA found that one-third of U.S. industrial-waste ponds and lagoons have no liners to prevent toxic waste seepage.
  * b. The EPA has identified only 100 underground storage tanks leaking their contents into groundwater.
    c. Leakage from underground storage tanks is about equal in volume to the *Valdez* oil spill.
    d. Environmentalists believe that all underground tanks should be monitored for leakage.
    e. Environmentalists believe that operators of older underground tanks should be held liable for cleanup and damage costs from leaks.

11-8 OCEAN POLLUTION

122. About _____ of all U.S. municipal sewage ends up untreated in marine waters.
    a. 3/4
    b. 2/3
    c. 1/2
  * d 1/3
    e. 1/4

238

123. Pollution and habitat disruption cause yearly closing to shellfish harvesters in _____ of the contiguous U.S. coastal waters.
    a. 3/4
    b. 2/3
    c. 1/2
    * d 1/3
    e. 1/4

124. The largest estuary in the United States is
    a. Mobile Bay.
    * b. Chesapeake Bay.
    c. San Francisco Bay.
    d. Puget Sound.
    e. the Gulf of Maine.

125. The major source of oysters and blue crabs in the United States is the
    a. Gulf of Mexico.
    b. San Francisco Bay.
    c. Coast of Maine.
    * d. Chesapeake Bay.
    e. Boston Harbor.

126. In Chesapeake Bay, 60% by weight of phosphates comes from _____ and 60% by weight of nitrates comes from _____.
    * a. point sources . . . nonpoint sources
    b. point sources . . . nitrate rocks
    c. nonpoint sources . . . point sources
    d. phosphate rocks . . . nonpoint sources
    e. phosphate rocks . . . point sources

127. Which of the following aquatic ecosystems is *most* capable of diluting, dispersing, and degrading large amounts of sewage, sludge, and oil?
    a. estuary
    b. swiftly flowing stream
    * c. deep-water ocean
    d. coastal parts of the ocean
    e. large rivers

128. Which of the following marine ecosystems is *least* polluted from ocean dumping and coastal development?
    a. mangrove swamps
    * b. deep-ocean trenches
    c. coral reefs
    d. wetlands
    e. estuaries

129. The majority of the oil pollution of the ocean comes from
    a. blowouts (rupture of a borehole of an oil rig in the ocean).
    b. tanker accidents.
    c. environmental terrorism.
    * d. runoff from land.
    e. leakage from underwater sediments.

130. The effects of an oil spill depend on the
    a. time of year.
    b. type of oil (crude or refined).
    c. distance of release from shore.
    d. weather conditions.
  * e. four conditions above.

131. Which of the following is *false?*
    a. Oil evaporates and undergoes decomposition.
  * b. It takes longer for the environment to recover from a spill of crude oil than refined oil.
    c. Recovery from oil spills is faster in warm water than in cold water.
    d. Estuaries and salt marshes suffer the most damage from oil pollution and cannot be
       effectively cleaned up.
    e. Effects of oil spills on gulfs and bays generally last longer than those on the
       open ocean.

132. The oil company responsible for the oil spill of the *Valdez* was
    a. Alyeska.
    b. Gulf.
  * c. Exxon.
    d. Sunoco.
    e. Texaco.

11-9 SOLUTIONS: PREVENTING AND REDUCING WATER POLLUTION

133. The leading nonpoint source of water pollution is
    a. municipal landfills.
    b. runoff from city streets and storm sewers.
  * c. agriculture.
    d. industrial wastes.
    e. cattle stockyards.

134. Farmers can sharply reduce fertilizer runoff by
    a. using prescribed amounts of fertilizer.
    b. fertilizing only flat areas, not sloped areas.
    c. using slow-release fertilizers.
    d. planting nitrogen-fixing plants.
    e. planting buffer zones between cultivated fields and surface water.
  * f. doing all of the above.

135. Farmers can reduce pesticide runoff by
    a. applying pesticides only when needed.
    b. using biological methods of pest control.
    c. using integrated pest management.
  * d. doing all of the above.
    e. none of the above; all of these methods are already being used.

136. Livestock growers can control runoff of animal wastes from feedlots and barnyards by
    a. increasing animal density.
  * b. diverting runoff of animal wastes into detention basins.
    c. removing buffers between stockyards and surface water.
    d. locating feedlots on gently sloping land so rainwater will naturally clean off
       the stockyards.
    e. none of the above.

137. Reforestation
    * a. reduces soil erosion and pollution from sediment.
      b. increases the severity of flooding.
      c. helps accelerate projected global warming.
      d. decreases biodiversity.
      e. does none of the above.

138. Which of the following would *not* reduce nonpoint source pollution?
      a. Require buffer zones of permanent vegetation between cultivated fields and surface water.
      b. Divert runoff of animal wastes into detention basins to be used as fertilizer.
    * c. Establish wastewater lagoons.
      d. Use biotic control or integrated pest management.
      e. Use crop rotation with legumes.

139. Most U.S. cities
      a. use septic tanks.
      b. have separate storm-water and sewer lines.
    * c. have combined storm-water and sewer lines.
      d. use wastewater lagoons.
      e. use advanced water treatment.

140. Which of the following types of sewage treatment are properly matched?
      a. primary—biological process
      b. secondary—mechanical process
    * c. advanced—physical and chemical processes
      d. secondary—chemical process
      e. primary—chemical process

141. Which of the following types of sewage treatment are properly matched?
      a. primary—removal of pollutants particular to a given area
      b. secondary—removal of suspended solids
      c. advanced—removal of oxygen-demanding wastes
      d. all of the above
    * e. none of the above

142. Which of the following substances are removed to the greatest extent by combined primary and secondary wastewater treatment?
      a. organic pesticides
    * b. organic oxygen-demanding wastes
      c. toxic metals and synthetic organic chemicals
      d. radioactive isotopes
      e. viruses

143. Of the following, the most ecologically reasonable way to dispose of sewage sludge is
      a. incineration.
      b. dumping into the deep trenches of the ocean.
      c. conventional landfills.
    * d. treating with heat and using as fertilizer.
      e. all of the above; they are ecologically sound ways to dispose of sewage sludge.

144. All of the following are used to disinfect or to purify water *except*
      a. ozone.
      b. chlorine.
    * c. iodine.
      d. ultraviolet radiation.
      e. none of the above; they are all used for that purpose.

Water   241

145. Coastal waters can be protected from excess water pollution by
    a. eliminating discharge of toxic pollutants.
    b. promoting water conservation in homes and industries.
    c. banning dumping of sewage sludge.
    d. regulating types and density of coastal development.
  * e. doing all of the above.

146. Water pollution from oil can be prevented by
    a. instituting a national energy policy based on decreased reliance on fossil fuels.
    b. reprocessing oils and greases from service stations.
    c. prohibiting oil drilling in ecologically sensitive areas.
    d. requiring double hulls on oil tankers.
    e. increasing financial liability of oil companies for spills.
  * f. doing all of the above.

147. The only effective way to protect groundwater is to
  * a. prevent contamination.
    b. use monitoring wells.
    c. cover all wells carefully.
    d. treat all water from underground sources.
    e. use only water from artesian wells.

148. Groundwater can be protected *least* by
    a. banning disposal of hazardous wastes in sanitary landfills and deep injection wells.
    b. monitoring aquifers near landfills and underground tanks.
  * c. using advanced sewage treatment.
    d. placing stricter controls on application of pesticides and fertilizers.
    e. establishing nationwide standards for groundwater contaminants.

149. Sustainable use of Earth's water resources involves
  * a. an integrated approach to managing water resources and water pollution throughout each watershed.
    b. continued subsidizing of the market price of water so that there is fair distribution of water.
    c. emphasis on waste management over waste prevention.
    d. emphasis on individual and community responsibility rather than cooperation among political entities.
    e. all of the above.
    f. none of the above.

150. Individuals can help protect water resources by
    a. using less-harmful household cleaners.
    b. using biodegradable dishwashing liquids and detergents.
    c. recycling old motor oil.
    d. making compost piles instead of using commercial inorganic fertilizers.
  * e. doing all of the above.

CHAPTER 12
MINERALS AND SOIL

1. The U.S. Mining Law of 1872
    a. requires miners who purchase public lands to pay 10% of their profits to the federal government.
    * b. encourages extraction on public lands almost free of charge.
    c. requires miners to reclaim the land when they close a mining operation.
    d. encourages miners to prevent land degradation.
    e. places pollution limits on acid mine drainage.

2. A change in the U.S. Mining Law of 1872 that environmentalists would *least* like to see would be
    a. making mining companies responsible for restoring the land.
    b. requiring mining companies to pay a 12.5% royalty on extracted hard-rock minerals.
    * c. making it possible to buy cheap public land and sell it for development at a high profit.
    d. making mining companies pay for environmental cleanups resulting from mining activities.
    e. denying authority to the mine when it comes in conflict with wildlife conservation and recreation.

12-1 GEOLOGIC PROCESSES

3. Earth's interior concentric zones include
    a. the shell.
    b. the crust.
    c. the mantle.
    d. the core.
    e. all of the above.
    * f. b, c, and d.

4. The largest zone of Earth's structure is
    a. the lithosphere.
    b. the core.
    c. the crust.
    * d. the mantle.
    e. the asthenosphere.

5. The asthenosphere is
    a. the outer atmosphere.
    b. the inner core of Earth.
    c. a plastic region in the crust.
    * d. a plastic region in the mantle.
    e. an endangered region of the hydrosphere.

6. Which of the following is *false?* Tectonic plates
    a. produce mountains and ocean trenches.
    * b. are composed of crust and core.
    c. move on the asthenosphere.
    d. move about the speed that fingernails grow.
    e. concentrate many of the minerals we extract and use.

7. Tectonic plates move apart in opposite directions at a(n)
    * a. divergent plate boundary.
    b. transform fault.
    c. asthenospheric plate.
    d. subduction zone.
    e. convergent plate boundary.

8. Tectonic plates move in opposite but parallel directions along a fault at a(n)
   a. divergent plate boundary.
 * b. transform fault.
   c. asthenospheric plate.
   d. subduction zone.
   e. convergent plate boundary.

9. Mountain chains, called oceanic ridges, are formed when
   a. severe erosion takes place.
 * b. tectonic plates move apart in opposite directions.
   c. tectonic plates push together.
   d. Earth's core loses homeostatic balance.
   e. tectonic plates slide past each other.

10. Oceanic trenches form at the boundary of
   a. Earth's crust and the mantle.
   b. Earth's mantle and the core.
   c. the asthenosphere and the lithosphere.
 * d. two converging tectonic plates.
   e. two diverging tectonic plates.

11. The energy sources primarily responsible for Earth's external geological processes are
   a. energy from the sun and magnetism.
 * b. energy from the sun and gravity.
   c. energy from the sun and heat from Earth's interior.
   d. gravity and magnetism.
   e. gravity and the heat from Earth's interior.

12. The agent *most* responsible for erosion is
   a. groundwater.
 * b. streams.
   c. glaciers.
   d. wind.
   e. none of the above.

13. The *most* important agent of mechanical weathering is
   a. gravity.
   b. flowing water.
 * c. frost action.
   d. wind.
   e. carbonic acid.

14. Mechanical weathering accelerates chemical weathering because of an increase in
   a. temperature.
   b. pH.
 * c. surface area.
   d. solubility.
   e. precipitation.

15. If you wanted to study chemical weathering in the field, where would you *most* likely go?
 * a. tropics
   b. subtropics
   c. temperate zone
   d. subarctic
   e. poles

16. Chemical weathering is aided by
    a. mechanical weathering.
    b. higher temperatures.
    c. higher precipitation.
 * d. all of the above.
    e. none of the above.

12-2 MINERALS, ROCKS, AND THE ROCK CYCLE

17. A mineral is
 * a. an inorganic, naturally occurring solid.
    b. an inorganic, naturally occurring liquid.
    c. an organic, naturally occurring solid.
    d. an organic, synthetic solid.
    e. an inorganic, synthetic solid.

18. All of the following minerals are composed of a single element *except*
    a. gold.
    b. diamonds.
 * c. quartz.
    d. sulfur.
    e. silver.

19. All of the following are broad classes of rock *except*
    a. sedimentary.
    b. igneous.
    c. metamorphic.
 * d. crystal.
    e. none of the above; all four are broad classes of rock.

20. Granite is an example of ____ rock.
    a. metamorphic
 * b. igneous
    c. sedimentary
    d. plasticized
    e. (none of the above)

21. The most abundant type of rock in the earth's crust is
    a. metamorphic.
 * b. igneous.
    c. sedimentary.
    d. gemstones.
    e. none of the above.

22. The type of rock that covers most of the Earth's land surface is
    a. metamorphic.
    b. igneous.
 * c. sedimentary.
    d. gemstones.
    e. none of the above.

23 Igneous rocks
    a. are the main source of many nonfuel mineral resources.
    b. often contain gemstones.
    c. are used for monuments.
    d. are used in landscaping.
 * e. include all of the above.

24. Heat and pressure convert
    a. igneous rock into sedimentary rock.
    * b. sedimentary rock into metamorphic rock.
    c. igneous rock into metamorphic rock.
    d. metamorphic rock into sedimentary rock.
    e. the asthenosphere into the lithosphere.

25. Dolomite and limestone are examples of ____ rock.
    a. metamorphic
    b. extrusive igneous
    c. intrusive igneous
    * d. sedimentary
    e. (none of the above)

26. Lignite and bituminous coal are ____ rocks.
    a. metamorphic
    b. igneous
    c. tectonic
    d. ridge
    * e. sedimentary

27. The process *most* directly responsible for the formation of sedimentary rock is
    a. melting.
    b. high temperature and pressure.
    c. metamorphism.
    * d. erosion.
    e. volcanic eruption.

28. Slate, anthracite, and marble are ____ rocks.
    a. primary
    b. secondary
    * c. metamorphic
    d. igneous
    e. sedimentary

29. Metamorphic rock is formed from sedimentary rock when the sedimentary rock is exposed to
    a. high temperature.
    b. high pressure.
    c. chemically active fluids.
    * d. any or all of the above.
    e. none of the above.

30. The change of rocks from one type to another is known as
    a. metamorphism.
    * b. the rock cycle.
    c. petrography.
    d. consolidation.
    e. shifting.

12-3 ENVIRONMENTAL IMPACTS OF EXTRACTING AND USING MINERAL RESOURCES

31. Which of the following categories includes the others?
    a. metallic minerals
    b. fossil fuels
    * c. nonrenewable resources
    d. nonmetallic minerals
    e. none of the above

32. That part of a metal-yielding material that can be economically and legally extracted at a given time is called
     a. a crustal resource.
     b. a mineral.
*   c. an ore.
     d. a rock.
     e. a fossil fuel.

33. Instruments that can detect changes in Earth's magnetic or gravitational fields can
     a. create small artificial earthquakes.
     b. analyze site samples.
     c. detect movement of tectonic plates.
     d. separate aluminum from unwanted compounds.
*   e. detect ore deposits from aircraft and satellites.

34. Subsurface mining _____ than surface mining.
     a. is less expensive
*   b. produces less waste material
     c. is less dangerous
     d. is more disruptive
     e. removes more of the resource

35. Surface mining typically involves the removal of _____ to expose underlying mineral deposits.
     a. quarries
*   b. overburden
     c. highwalls
     d. lowwalls
     e. shafts

36. An example of subsurface mining is
     a. open-pit mining.
     b. strip mining.
*   c. long-wall mining.
     d. dredging.

37. Which of the following is *false*? Strip mining
     a. is a type of surface mining.
     b. is done with bulldozers, power shovels, or wheels that remove strips off Earth's surface.
     c. is used for removing coal.
     d. is used for removing phosphate rock.
*   e. removes deposits of gemstones that are a quarter of a mile deep.

38. Extracting, processing, and using mineral resources
     a. requires enormous amounts of energy.
     b. causes land disturbance and erosion.
     c. causes air pollution.
     d. causes water pollution.
*   e. does all of the above.

39. Mining
     a. of the surface can result in severe erosion by wind and water.
     b. can pollute streams with sediment.
     c. results in toxic wastes percolating down to groundwater.
     d. pollutes the air with dust and toxic substances.
*   e. includes all of the above.

40. Of the following substances, the one *least* likely to run off from acid mine drainage is
    * a. limestone.
    b. radioactive uranium compounds.
    c. lead.
    d. cadmium.
    e. arsenic.

41. Air pollutants resulting from smelting include all of the following *except*
    a. sulfur dioxide.
    b. cadmium.
    c. arsenic.
    * d. methane.
    e. lead.

12-4 WILL THERE BE ENOUGH MINERALS?

42. Reserves are _____ resources.
    * a. identified and profitably extractable
    b. unidentified and profitably extractable
    c. identified and subeconomic
    d. unidentified and subeconomic
    e. all identified

43. When a resource has been economically depleted, we can
    a. recycle or reuse what has already been extracted.
    b. cut down on unnecessary waste of the resource.
    c. find a substitute.
    d. do without.
    * e. do all of the above.

44. Economic depletion of most resources is said to occur when ____% has been extracted.
    a. 50
    b. 60
    c. 70
    * d. 80
    e. 90

45. A depletion curve is typically used to
    a. project remaining undiscovered resources.
    * b. project the depletion time for a resource.
    c. predict when a resource will become a reserve.
    d. project when uneconomic resources will become economically feasible.
    e. predict when an ore will be totally unavailable.

46. Which of the following statements is *false?*
    a. Depletion time is dependent on rate of use and the amount of known reserves.
    b. Usually the first deposits of a mineral to be exploited are high-grade deposits.
    c. All of a resource will not be used because it is not economically feasible to continue to extract the resource.
    d. Finding substitutes for a resource requires a new set of depletion curves.
    * e. Recycling, reusing, and reducing consumption causes a depletion curve to peak sooner and higher.

47. Depletion time can be extended by
    a. recycling and reuse.
    b. discoveries of new resources.
    c. reduced consumption.
    d. higher resource prices.
  * e. all of the above.

48. Minerals essential to the economy of a country are called
    a. strategic minerals.
    b. metamorphic minerals.
  * c. critical minerals.
    d. reserved minerals.
    e. depleted minerals.

49. Minerals essential for national defense are called
  * a. strategic minerals.
    b. metamorphic minerals.
    c. critical minerals.
    d. reserved minerals.
    e. depleted minerals.

50. There are ___ major nonfuel minerals.
    a. 15
  * b. 20
    c. 25
    d. 30
    e. 50

51. All of the following are strategic minerals for which the United States has no reserves and depends on imports *except*
    a. chromium.
    b. platinum.
  * c. titanium.
    d. cobalt.
    e. manganese.

52. Which of the following minerals is *not* matched with the area that has its major supply?
    a. cobalt—Zaire
    b. tin—Malaysia, Bolivia
    c. manganese—South Africa
    d. platinum—South Africa, the Commonwealth of Independent States
  * e. copper—Europe

53. The country that most closely meets all of its mineral resource needs is
  * a. the Commonwealth of Independent States.
    b. Australia.
    c. Germany.
    d. Brazil.
    e. the United States.

54. The United States' stockpile of critical and strategic minerals is supposed to last
    a. 8 weeks.
    b. 6 months.
  * c. 3 years.
    d. 6 years.
    e. 10 years.

55. The United States is heavily dependent on ____ for its supplies of chromium, platinum, manganese, and industrial diamonds.
* a. South Africa
  b. the Commonwealth of Independent States
  c. Canada
  d. Indonesia
  e. Brazil

12-5 INCREASING MINERAL RESOURCE SUPPLIES

56. Higher prices for consumer goods will not necessarily stimulate increased production of nonfuel minerals because
  a. investment capital is usually abundantly available.
  b. all resource supplies are theoretically infinite.
  c. the mining industry is fiercely competitive and poorly regulated.
* d. raw materials typically account for only a small fraction of the price of consumer goods.
  e. of none of the above.

57. Which of the following statements is *false?*
* a. Market prices clearly reflect dwindling mineral supplies.
  b. Consumers have incentives to reduce demands before economic depletion occurs.
  c. Low mineral prices are caused by failure to include the external costs of mining.
  d. Low prices encourage waste and faster depletion.
  e. Low prices encourage more pollution and environmental degradation.

58. Most of the world's easily accessible high-grade mineral deposits
  a. are not worth extracting at today's prices.
  b. cannot be located using known techniques.
  c. lie in unexplored areas of the LDCs.
* d. have already been discovered.
  e. are tightly regulated by international law.

59. According to geological theory, if there were 10,000 sites where a deposit of a particular resource might be found, ____ could probably be (a) producing mine(s).
* a. one
  b. 10
  c. 25
  d. 100
  e. 1,000

60. The example of copper mining illustrates that
* a. it can be profitable to mine low-grade ores.
  b. reserves of minerals decline each year.
  c. economic profits must be balanced with environmental costs.
  d. the reserves of some key metals are renewable.
  e. the cost of energy is not a limiting factor.

61. According to some analysts, mining of low-grade ores may be limited by
  a. energy costs.
  b. availability of fresh water.
  c. environmental impact of mining and processing wastes.
  d. land reclamation costs.
* e. all of the above.

62. All of the following chemicals could be extracted from the ocean with today's technology at a profit *except*
    a. magnesium.
    * b. zinc.
    c. bromine.
    d. sodium chloride.
    e. none of the above; all are economically extractable.

63. The major problem associated with extracting minerals from seawater is that
    a. only common table salt can be cheaply extracted.
    * b. the minerals exist in low concentrations.
    c. seawater contains only the most common metals.
    d. seawater is not owned by any one nation.
    e. the technology is still in the research stage.

64. The continental shelf has significant sources of all of the following except
    a. phosphates.
    b. gravel.
    * c. manganese-rich nodules.
    d. sand.
    e. oil and natural gas.

65. Nodules of ____ are found on the floor of the deep ocean.
    a. chromium
    b. boron
    c. platinum
    * d. manganese
    e. cobalt

66. One reason manganese-rich nodules are an attractive resource is
    * a. they contain several other strategic minerals.
    b. mining them entails few political problems.
    c. mining them is ecologically harmless.
    d. valuable deposits are found nearly everywhere.
    e. the technology for extraction is inexpensive.

67. Many economists believe that
    a. resources cannot be exhausted.
    b. as resources become scarcer, demand for them will decline.
    * c. as resources become scarcer, satisfactory substitutes will be discovered.
    d. there are limits to all resources.
    e. none of the above is true.

68. Because they allow higher temperatures, ____ engines increase fuel efficiency by 30 to 40%.
    a. aluminum
    b. plastic
    * c. ceramic
    d. metal alloy
    e. stainless steel

69. High-strength plastics and composite materials strengthened by carbon and glass fibers are advantageous because they are
    a. stronger than metal.
    b. lighter than metal.
    c. less expensive because they use less energy.
    d. easily molded into any shape and don't need painting.
    * e. all of the above.

70. Which of the following would be easiest to find a substitute for?
    a. phosphates
    b. helium
  * c. steel
    d. copper
    e. manganese

12-6 SOILS: THE BASE OF LIFE

71. The inorganic component of soil is composed of
    a. clay.
    b. sand.
    c. silt.
    d. gravel.
  * e. all of the above.

72. The choice *least* likely to be found in a handful of soil is
    a. decaying organic matter.
    b. billions of living organisms.
    c. air and water.
  * d. heavy metals.
    e. inorganic material.

73. Most of Earth's biodiversity is found in
    a. producers.
    b. herbivores.
    c. carnivores.
    d. omnivores.
  * e. decomposers.

74. A cross-sectional view of the ___ in a soil is properly termed a soil ____.
  * a. horizons . . . profile
    b. horizons . . . sample
    c. surface litter . . . sample
    d. surface litter . . . profile
    e. profile . . . sample

75. The layers that compose a mature soil are known as
    a. strata.
  * b. profiles.
    c. horizons.
    d. laminae.
    e. none of the above.

76. The surface litter horizon is described by the letter ____.
    a. A.
    b. B.
    c. C.
    d. E.
  * e. O.

77. The ____-horizon of a soil contains no organic material.
    a. A
    b. B
  * c. C
    d. E
    e. O

78. The A-horizon of soil is commonly referred to as
   * a. topsoil.
   b. surface litter.
   c. subsoil.
   d. parent rock.
   e. zone of leaching.

79. The topsoil horizon is described by the letter
   * a. A.
   b. B.
   c. C.
   d. E.
   e. O.

80. Topsoil contains all of the following *except*
   a. plant roots.
   b. humus.
   * c. freshly fallen leaves.
   d. some inorganic minerals.
   e. living organisms.

81. Most of nature's recyclers would be found in the
   a. surface-litter layer.
   * b. topsoil.
   c. subsoil.
   d. parent material.
   e. air.

82. Most foodwebs would be found in which layers?
   * a. O and A
   b. A and B
   c. B and C
   d. O and C
   e. O and B

83. The soil layer penetrated only by fractures is the
   a. parent material.
   b. zone of leaching.
   c. subsoil.
   * d. bedrock.
   e. topsoil.

84. The soil layer containing unique colors and often iron, aluminum, humus, and clay leached from higher layers is the
   a. parent material.
   b. zone of leaching.
   * c. subsoil.
   d. topsoil.
   e. bedrock.

85. The soil layer through which dissolved or suspended materials move downward is the
   a. surface litter.
   * b. zone of leaching.
   c. topsoil.
   d. subsoil.
   e. parent material.

86. Freshly fallen leaves, organic debris, and partially decomposed organic matter are indicative of the
   * a. surface litter.
   b. zone of leaching.
   c. parent material.
   d. subsoil.
   e. topsoil.

87. The dissolving of material from the upper layers of the soil and its movement to lower horizons is called
   a. percolation.
   b. weathering.
   c. porosity.
   d. accumulation.
   * e. leaching.

88. Humus is
   a. indicative of poor soils.
   b. light colored or nearly white.
   c. poisonous to soil microorganisms.
   * d. partially decomposed organic matter.
   e. water-soluble and easily washed away.

89. Humus
   * a. coats sand, silt, and clay particles.
   b. creates air spaces between soil particles.
   c. drains water and plant nutrients to the next soil layer.
   d. decreases crop yields.
   e. none of the above.

90. Humus
   a. prevents soil particles from clumping.
   * b. helps topsoil hold water.
   c. has a positive charge.
   d. is the best rock from which to form fertile soil.
   e. is a tasty Arabic food.

91. Topsoil that is _____ in color is the most highly fertile.
   a. gray
   b. red
   * c. dark brown or black
   d. yellow
   e. orange

92. Red and yellow colors in a soil horizon usually indicate a
   a. high percentage of sand.
   b. high percentage of lime and gypsum.
   c. lack of iron oxide.
   * d. low organic matter content.
   e. need for potash fertilizer.

93. Spaces, or pores, in the soil contain
   a. water.
   b. oxygen.
   c. nitrogen.
   * d. all of the above.
   e. none of the above.

94. Decomposers turn organic materials into inorganic ones in a process called
    a. leaching.
    b. infiltration.
    c. weathering.
    d. humifaction.
    * e. mineralization.

95. Leaf mold, a humus-mineral mixture, and silty loam are indicative of
    a. coniferous forest soil.
    * b. deciduous forest soil.
    c. tropical forest soil.
    d. grassland soil.
    e. desert soil.

96. A soil sample that is alkaline, dark, and rich in humus probably came from a
    a. coniferous forest.
    b. deciduous forest.
    c. tropical forest.
    * d. grassland.
    e. desert.

97. A soil sample of closely packed pebbles that is a mixture of minerals and low humus most likely came from a
    a. coniferous forest.
    b. deciduous forest.
    c. tropical forest.
    d. grassland.
    * e. desert.

98. Leaching occurs when
    a. humus is dissolved.
    * b. water removes soluble nutrients.
    c. organic compounds slowly decay.
    d. rock is shattered by frost action.
    e. water is taken up by plant roots.

99. Soil texture helps determine
    * a. porosity.
    b. pH.
    c. color.
    d. nutrient content.
    e. chemical content.

100. Which of the following is *not* a particle size used to determine soil texture?
    a. silt
    * b. loam
    c. clay
    d. sand
    e. none of the above; all are particle sizes

101. Which of the following would be considered the best type of soil in which to grow most crops?
    a. silt
    * b. loam
    c. clay
    d. sand
    e. gravel

102. Which of the following types of soils has the least pore space?
    a. silt
    b. loam
    c. clay
    * d. sand
    e. none of the above; soil has no pore space

103. Which of the following types of soils holds the most water?
    a. silt
    b. loam
    * c. clay
    d. sand
    e. gravel

104. Clay has _____ permeability and _____ porosity.
    a. high . . . high
    b. high . . . low
    * c. low. . . high
    d. low. . . low

105. Sand has _____ permeability and _____ porosity.
    a. high . . . high
    * b. high . . . low
    c. low . . . high
    d. low . . . low

106. Which of the soils would most likely become waterlogged?
    a. silt
    b. loam
    * c. clay
    d. sand
    e. gravel

107. When moistened and rubbed between your fingers, a silt-laden soil feels
    a. gritty.
    * b. smooth (like flour).
    c. sticky or slippery.
    d. crumbly and spongy.
    e. stringy and hard.

108. A dab of soil that is well suited for plant growth will have a _____ texture when moistened and rubbed between your fingers.
    a. stringy and hard
    b. slimy and slick
    * c. crumbly and spongy
    d. sticky or gooey
    e. gritty or sandy

109. A soil that feels _____ has a high sand content.
    a. sticky
    b. smooth (like flour)
    c. stringy and hard
    d. crumbly and spongy
    * e. gritty

110. Alkaline soil can be neutralized or made more acid by adding
    * a. sulfur.
    b. calcium.
    c. phosphates.
    d. sodium.
    e. potassium.

111. An acidic soil is one with a pH of
    a. 10 or more.
    b. exactly 10.
    c. 7 to 10.
    * d. less than 7.
    e. exactly 7.

112. Acidic soil can be partially neutralized by adding
    * a. lime.
    b. water-insoluble compounds.
    c. phosphates.
    d. sulfur.
    e. H^+ ions.

113. The addition of lime
    a. makes soil more acidic.
    * b. causes decomposition of organic material.
    c. increases porosity of the soil.
    d. will change soil texture.
    e. will reduce soil alkalinity.

114. If your farm was downwind from a fossil-fuel-burning power plant, which action would you be least likely to take?
    a. Add lime.
    * b. Add sulfur.
    c. Add manure.
    d. Add fertilizer.
    e. None of the above.

12-7 SOIL EROSION

115. Most soil erosion is caused by
    * a. moving water.
    b. wind.
    c. earthquakes.
    d. volcanoes.
    e. acidification of the soil.

116. Which term includes the others?
    a. sheet erosion
    b. rill erosion
    * c. water erosion
    d. gully erosion
    e. none of the above

117. The greatest source of water pollution is
    * a. sediment from erosion.
    b. runoff of agricultural chemicals.
    c. drainage from urban areas.
    d. industrial point sources.
    e. runoff from stockyards.

118. In tropical and temperate areas, 1 inch of topsoil takes an average of ____ years to form.
   a. 10–50
   b. 100–200
 * c. 200–1,000
   d. 1,000–2,000
   e. 5,000–10,000

119. Which of the following sites would be expected to have the *most* rapid erosion rates?
   a. agricultural land
   b. lumbering sites
   c. rangeland
   d. natural areas
 * e. construction sites

120. Currently, topsoil is eroding faster than it forms on about ____ of the world's croplands.
   a. one-fourth
 * b. one-third
   c. one-half
   d. two-thirds
   e. three-fourths

121. The largest single cause of soil erosion is
 * a. overgrazing.
   b. deforestation.
   c. surface mining without reclamation.
   d. improper irrigation techniques.
   e. cultivation of land with unsuitable soils.

122. The world is losing about ____% of the topsoil from its croplands each decade.
   a. 2
 * b. 7
   c. 11
   d. 17
   e. 22

123. Which of the following practices leads to desertification?
   a. irrigation
   b. overgrazing
   c. soil compaction
   d. deforestation
 * e. all of the above

124. All of the following cattle-producing regions show desertification *except*
   a. the Middle East.
   b. western Asia.
 * c. the eastern half of the United States.
   d. Australia.
   e. sub-Saharan Africa.

125. Desertification is intensified by all of the following *except*
 * a. slow population growth.
   b. high human and livestock population densities.
   c. poor land management.
   d. poverty.
   e. none of the above.

126. Consequences of desertification include
    a   large numbers of environmental refugees.
    b. famine.
    c. declining living standards.
    d. worsening drought.
  * e. all of the above.

127. The cost of a program to prevent desertification and rehabilitate desertified areas could be recouped in
    a. 1 year.
  * b. 3–4 years.
    c. 15–20 years.
    d. 25 years.
    e. 50 years.

128. The United States has already lost about ___ of the original topsoil of the cropland in use today.
    a. one-tenth
    b. one-sixth
    c. one-fifth
  * d. one-third
    e. one-half

129. The major erosion of U.S. cropland soil occurs in
  * a. the Great Plains.
    b. the Southeast.
    c. areas whose crop output is not important.
    d. areas with soil of low to very low fertility.
    e. irrigated lands in the western United States.

130. The Soil Conservation Service
    a. was established in 1955.
    b. is a regional service to the Great Plains.
  * c. provides technical assistance to both farmers and ranchers.
    d. solved the Great Plains' erosion problems.
    e. was established by James Watt.

131. Eighty-six percent of U.S. soil erosion comes from land used
    a. to grow crops for human consumption.
  * b. to graze cattle and raise crops fed to cattle.
    c. for cutting timber.
    d. for mining.
    e. for expansion of urban areas.

12-8 SOLUTIONS: SOIL CONSERVATION

132. Plowing, breaking up, and smoothing soil in fall to plant in the spring is
  * a. conventional tillage.
    b. conservation tillage.
    c. contour farming.
    d. terracing.
    e. strip cropping.

133. Conservation tillage reduces
    a. labor costs.
    b. amount of erosion.
    c. energy consumption.
    d. water loss from the soil.
 * e. all of the above.

134. Which of the following statements is *false?* Conservation tillage
    a. is used in one-third of U.S. cropland.
    b. is not widely used outside the United States.
    c. reduces soil compaction.
 * d. requires increased use of herbicides.
    e. involves using tillers that break up subsurface soil without turning over topsoil.

135. Which of the following practices both reduces erosion and increases soil fertility?
    a. strip cropping
    b. terracing
    c. contour farming
    d. alley cropping
 * e. a and d

136. Contour farming involves
    a. converting a steep slope into a series of terraces.
    b. building a series of small dams.
 * c. plowing at right angles to slopes.
    d. plowing straight downslope or straight upslope.
    e. planting crops in alleys between trees.

137. Planting crops in alternating rows of close-growing plants
    a. creates windbreaks.
 * b. is called strip cropping.
    c. is called crop rotation.
    d. increases erosion rates.
    e. is called alley cropping.

138. An agricultural style that prevents erosion on steep slopes is
    a. conventional-tillage farming.
    b. conservation-tillage farming.
    c. alley cropping.
    d. strip cropping.
 * e. terracing.
    f. contour farming.

139. In alley cropping,
 * a. crops are planted between hedgerows of trees or shrubs that are used for fruits or fuelwood.
    b. terraces are built to prevent swift water runoff.
    c. plowing runs across slopes.
    d. special tillers are used so the topsoil is not disturbed.

140. Which of the following statements about gullies is *false?*
    a. Gullies are created on slopes that are not covered by vegetation.
    b. Small gullies can be seeded with quick-growing plants.
 * c. Once gullies are formed, it is impossible to reclaim the land for agricultural purposes.
    d. Deep gullies can be filled in by building small dams to collect silt.
    e. Channels can be built to divert water from gullies.

141. Which of the following is a true statement about shelterbelts (windbreaks) in the Great Plains?
    a. Most are employed to channel water away from gullies.
    b. They eliminate the need for other soil conservation methods.
    c. They are usually planted in an east-to-west direction.
  * d. Many have been destroyed to make way for farming equipment.
    e. They are planted next to irrigation systems.

142. Which of the following is *not* one of the three major types of organic fertilizer?
    a. green manure
  * b. sewage sludge
    c. compost
    d. animal manure

143. Animal manure is normally not used as a crop fertilizer on large farms because it
    a. is low in fertility.
    b. lacks soil microorganisms.
    c. degrades soil structure.
  * d. is often unavailable or too costly to transport.
    e. encourages pests.

144. Which of the following statements is *false?* Application of animal manure
    a. improves soil structure.
    b. stimulates the growth of soil bacteria and fungi.
  * c. has increased since farms have specialized in animal- or crop-farming operations.
    d. is expensive when manure must be transported from feedlots near urban areas to rural crop-growing areas.
    e. includes none of the above; all are true.

145. Inorganic fertilizer is classified based upon all of the following elements *except*
  * a. calcium.
    b. phosphorus.
    c. potassium.
    d. nitrogen.

146. Commercial inorganic fertilizers commonly contain all of the following *except*
  * a. organic nitrogen.
    b. phosphate.
    c. nitrate.
    d. potassium.

147. Commercially available inorganic fertilizers
    a. lack trace elements.
    b. decrease soil porosity.
    c. decrease soil water-holding capacity.
    d. lower oxygen content of soil.
  * e. have all of the above characteristics.

148. The most cost-effective means of dealing with soil erosion is to
    a. plant crops only on absolutely flat land.
  * b. maintain adequate vegetative cover on soils.
    c. abandon croplands after they lose their topsoil.
    d. use large amounts of inorganic fertilizers.
    e. use organic fertilizers.

149. Inorganic fertilizers
    a. increase soil porosity.
  * b. reduce the soil's ability to hold water.
    c. usually have trace minerals present in sufficient amounts to foster plant growth.
    d. decrease soil compaction.
    e. have all of the above characteristics.

150. Salinization may result from
    a. deforestation.
    b. erosion.
  * c. irrigation.
    d. overgrazing.
    e. use of organic fertilizers.

151. Salt buildup may
    a. stunt crop growth.
    b. decrease yields.
    c. eventually kill crop plants.
    d. eventually make the land unproductive.
  * e. do all of the above.

152. Salinization is often accompanied by
  * a. waterlogging.
    b. high pH.
    c. low pH.
    d. high nutrient levels.
    e. high humus levels.

CHAPTER 13
WASTES: REDUCTION AND PREVENTION

1.   Love Canal is located in
      a.   Ohio.
   * b.   New York.
      c.   California.
      d.   Panama.
      e.   Minnesota.

2.   The company responsible for the dumping of toxic and cancer-causing wastes into an old
      canal excavation called the Love Canal was
      a.   DuPont.
      b.   Monsanto.
   * c.   Hooker Chemical and Plastics Corporation.
      d.   the 3M Company.
      e.   Exxon.

3.   After Love Canal was abandoned, it was sold and used for
   * a.   an elementary school and housing project.
      b.   a shopping mall.
      c.   an amusement park.
      d.   an industrial park.
      e.   a municipal dump site.

4.   Love Canal residents received payments from an out-of-court settlement from
      a.   Occidental Chemical Corporation.
      b.   the city of Niagara Falls.
      c.   the Niagara Falls school board.
      d.   none of the above; the case went to court.
   * e.   all of the above.

5.   The Love Canal incident demonstrates that
   * a.   preventing pollution is safer and cheaper than cleaning it up.
      b.   political officials are alert and sympathetic to their constituents.
      c.   pollutants can be stored safely underground for a long time.
      d.   polluting companies can escape from the costs and responsibility of their actions.
      e.   individuals are powerless to challenge government institutions and industries.

13-1 WASTING RESOURCES: SOLID WASTE AND THE THROWAWAY APPROACH

6.   The most wasteful people on the planet live in
   * a.   the United States.
      b.   the Commonwealth of Independent States.
      c.   Brazil.
      d.   Ethiopia.
      e.   Canada.

7.   Garbage produced directly by households and businesses accounts for _____% of the solid
      waste produced in the United States.
   * a.   less than 2
      b.   5
      c.   10
      d.   15
      e.   20

8. The single largest category of U.S. solid waste is
    *   a. mining waste.
        b. agricultural wastes.
        c. industrial wastes.
        d. municipal wastes.
        e. wastes from oil and natural gas production.

9. The amount of solid waste produced in the United States in 1991 would fill a convoy of garbage trucks stretching around the world almost
        a. two times.
        b. four times.
    *   c. six times.
        d. eight times.
        e. 10 times.

10. Industrial solid wastes include all of the following *except*
        a. sludge from industrial waste treatment plants.
        b. scrap metal.
        c. plastics.
    *   d. garbage.
        e. fly ash.

11. Balloons are least dangerous to
        a. turtles.
        b. fish.
    *   c. zooplankton.
        d. whales.
        e. seals.

12. U.S. consumers throw away enough aluminum to rebuild the country's commercial airline fleet every
        a. three weeks.
    *   b. three months.
        c. six months.
        d. year.
        e. three years.

13. The recycling/composting rate of municipal solid waste produced in the United States is about
        a. 2%.
    *   b. 17%.
        c. 19%.
        d. 25%.
        e. 33%.

14. _____percent of the solid waste produced in the United States is buried in landfills.
        a. Ninety-five
    *   b. Sixty-six
        c. Fifty-five
        d. Forty-four
        e. Thirty-three

15. _____percent of the solid waste produced in the United States is burned.
        a. Seven
    *   b. Seventeen
        c. Twenty-seven
        d. Thirty-seven
        e. Forty-seven

16. Which of the following statements is *true?*
    a. Solid wastes are actually wasted solids.
    b. Our economic and political system rewards waste production rather than waste conservation and consumption.
    c. A low-waste approach is pollution prevention.
    d. A high-waste approach focuses on management of wastes.
 * e. All of the above are true.

13-2 SOLUTIONS: REDUCING AND REUSING SOLID WASTE

17. Which of the following strategies should be given top priority?
    a. incinerate
    b. reuse
 * c. reduce
    d. bury
    e. recycle

18. Which of the following strategies should be given lowest priority?
    a. incinerate
    b. reuse
    c. reduce
 * d. bury
    e. recycle

19. A low-waste approach
    a. eliminates unnecessary packaging.
    b. makes a product last longer.
    c. uses less material (such as in making cars lighter).
    d. eliminates style differences and built-in obsolescence.
 * e. does all of the above.

20 All of the following reflect a low-waste approach *except*
    a. extending the useful lifetime for a product.
    b. substituting less bulky products, such as a transistor for a vacuum tube.
 * c. built-in obsolescence.
    d. modular construction for repair.
    e. better remanufacturing industries.

21. Which is the most advanced approach?
    a. recycling materials
    b. using biodegradable material
    c. creating more durable products
 * d. reducing the amount of materials used
    e. adopting reusable products

22. To minimize passing our environmental problems on to future generations, citizens should
    a. focus on pollution prevention.
    b. reduce waste production.
    c. reuse materials.
    d. recycle wastes.
 * e. do all of the above.

23. In 1991, _____ enacted the world's toughest packaging law.
    a. Japan
  * b. Germany
    c. the United States
    d. Brazil
    e. Australia

24. Environmentalists say that the best way to handle soft-drink and beer containers is to
    a. use landfills.
    b. recycle aluminum cans.
    c. use stainless steel cans.
  * d. use reusable glass bottles.
    e. use recyclable plastic containers.

25. _____ has a beverage container deposit fee that is 50% higher than the cost of the drink, to encourage use of refillable bottles.
    a. Italy
  * b. Ecuador
    c. Germany
    d. Canada
    e. the Philippines

26. The most energy-efficient beverage container on the market is
  * a. refillable glass.
    b. recyclable aluminum.
    c. stainless steel.
    d. degradable plastic.
    e. recyclable plastic.

27. Which statement is *false*? Disposable diapers _____ than cloth diapers.
    a. use up more landfill space
    b. use more trees and plastic resources
    c. produce more air and water pollution
  * d. consume more water
    e. use more energy

28. At the checkout counter, an environmentalist is most likely to
    a. say "plastic please."
    b. say "paper please."
  * c. say "I brought my own bag."
    d. walk out of the store.

29. The *least* desirable final resting place for a used tire is
    a. in the foundation of a low-cost passive solar home.
    b. in asphalt pavement.
    c. in an incinerator to produce electricity.
  * d. in the landfill.
    e. as an artificial reef.

13-3 SOLUTIONS: RECYCLING SOLID WASTE

30. Compost
    a. is a method to decompose many organic wastes.
    b. is used as fertilizer and soil conditioner.
    c. is rich in organic matter and soil nutrients.
    d. can be produced from biodegradable solid waste in large plants, bagged, and sold.
  * e. includes all of the above.

31. Compost is most completely described as
    a. manure.
    b. landfill by-products.
    c. pure garbage.
  * d. soil conditioner and organic fertilizer.
    e. inorganic fertilizer.

32. Compost can be applied to
    a. golf courses.
    b. forests.
    c. roadway medians.
    d. parks.
  * e. all of the above.

33. Which of the following problems with composting can be solved with a Dutch tunnel?
    a. siting problems
    b. contamination by toxic materials
  * c. odor problems
    d. color problems
    e. none of the above

34. Which of the following statements is *false?*
    a. It is more economical to have consumers separate trash before pickup than to use high-technology recovery plants.
    b. Glass, iron, and aluminum can be recovered from solid wastes.
    c. Low-tech resource recovery involves consumers separating trash into four collections: glass, paper, plastics, and metal.
  * d. High-technology recycling provides many more jobs than low-technology recycling.
    e. Recycling and reuse of solid wastes is called resource recovery.

35. Source separation differs from high-technology recycling plants in all but which of the following?
    a. It is cheaper.
    b. It provides greater income for unskilled labor and volunteer organizations.
    c. It produces less air and water pollution.
  * d. It is the most common type of recycling strategy.
    e. It uses less energy.

36. Which of the following substances can be recovered from solid waste?
    a. metals
    b. plastic
    c. paper
    d. glass
  * e. all of the above

37. The most desirable type of recycling is
  * a. primary, or closed-loop recycling.
    b. secondary, or open-loop recycling.
    c. tertiary, or figure-eight recycling.
    d. high-tech resource recovery.
    e. the most cost-effective.

38. Which of the following is *false?* Recycling aluminum
    a. produces 95% less air pollution than using virgin ore.
    b. produces 97% less water pollution than using virgin ore.
  * c. requires 50% less energy than using virgin ore.
    d. is done at a 30% rate in the United States.
    e. cans in the United States would save more aluminum than most countries use.

39. Plastics
   * a. account for about 60% of the debris found on U.S. beaches.
   b. in landfills can release cadmium and plutonium from binders.
   c. break down quickly in landfills.
   d. can economically be converted to fuel oil.
   e. do none of the above.

40. Partially biodegradable plastics need _____ to be broken down.
   a. light
   * b. oxygen and moisture
   c. anaerobic conditions
   d. cool conditions
   e. none of the above

41. Obstacles to recycling in the United States include
   a. throwaway lifestyle attitudes.
   b. lack of inclusion of environmental and health costs in market prices.
   c. tax breaks for mining virgin materials.
   d. lack of large, steady markets for recycled materials.
   * e. all of the above.

42. All of the following are obstacles to recycling *except*
   a. lack of incentives.
   b. lack of markets.
   * c. lack of materials.
   d. lack of appropriate attitudes.
   e. market prices that exclude social and environmental costs.

13-4 LAST RESORTS: BURNING OR BURYING SOLID WASTE

43. Incinerators
   a. are not held liable once the wastes are burned.
   b. reduce garbage volume by 60%.
   c. increase toxicity of waste.
   d. swap waste for air pollution.
   * e. include all of the above.

44. Incinerators are less _____ than recycling.
   a. expensive
   * b. politically acceptable
   c. hazardous
   d. (all of the above)
   e. (none of the above)

45. Which of the following statements is *false?*
   a. The ash from incinerators may contain toxins such as dioxin and heavy metals.
   b. Fly ash is more toxic than bottom ash.
   * c. The ash from incinerators is safe.
   d. Toxic metals can cause cancer and nervous disorders.
   e. The ash from incinerators is currently disposed of in landfills.

46. Which of the following statements about incinerators is *true?*
   a. They are less expensive to operate than landfills.
   b. Incinerators tend to encourage recycling efforts.
   * c. Many incinerators have long-term contracts with cities to supply them with a certain volume of trash.
   d. Incinerators are becoming more efficient and economical.
   e. Incinerators do not release toxic waste products.

47. In contrast to U.S. incineration strategies, in Japan
    a. hazardous wastes are not removed from solid wastes before burning.
    b. workers shovel out bottom ash and dump it into landfills.
  * c. violations of air standards are punishable by large fines.
    d. incinerator workers must have an associate's degree.
    e. none of the above.

48. Because of threats to the health and safety of its citizens, the first state to ban incineration in 1992 was
    a. California.
    b. New York.
    c. Maine.
  * d. Rhode Island.
    e. Wisconsin.

49. Sanitary landfills typically have problems with
    a. rodents and insects.
    b. odor.
    c. open, uncovered garbage.
  * d. traffic, noise, and dust.
    e. all of the above.

50. Underground anaerobic decomposition in a landfill produces
    a. volatile organic compounds.
    b. methane.
    c. hydrogen sulfide.
  * d. all of the above.
    e. none of the above.

51. Which of the following statements about landfill leaching is *false?*
    a. Rain filtering through landfills leaches toxic materials.
    b. Contaminated leachate can seep from the bottom of landfills.
    c. Contaminated groundwater is a problem at some landfills.
    d. Older, unlined landfills may have particularly bad water pollution problems.
  * e. None of the above is false.

52. The *Mobro* garbage barge case illustrates
    a. the first law of thermodynamics.
    b. the second law of thermodynamics.
  * c. there is no "away."
    d. the importance of garbage management approaches.
    e. that land has a cheaper value in LDCs.

13-5 HAZARDOUS WASTE: TYPE AND PRODUCTION

53. A waste is considered hazardous if it possesses one of four properties. Which of the following is *not* one of those properties?
    a. flammable
    b. unstable
  * c. soluble
    d. corrosive
    e. carcinogenic, mutagenic, or teratogenic beyond set limits

54. If a small business produces less than ___ pounds of a waste per month, the EPA does not consider the material hazardous.
   a. 50
 * b. 220
   c. 550
   d. 1,500
   e. 2,000

55. Which of the following substances would be considered hazardous waste?
   a. radioactive material
 * b. cadmium and other heavy metals
   c. materials of any type discarded from households
   d. mining wastes
   e. oil and gas drilling wastes

56. Using conservative estimates, approximately ____ ton(s) of hazardous wastes are produced for each person in the United States per year.
   a. one
   b. seven
   c. 13
 * d. 23
   e. 50

57. _____ percent of the U.S. hazardous waste is not regulated by hazardous waste laws.
   a. Fifty-four
   b. Sixty-four
   c. Seventy-four
   d. Eighty-four
 * e. Ninety-four

58. A serious pollutant that accumulates because it does not degrade is
   a. phosphate.
   b. nitrate.
 * c. lead.
   d. oxygen-demanding waste.
   e. oil.

59. Lead
   a. accumulates in bones.
   b. may be passed from pregnant women to their unborn children.
   c. can damage the central nervous system.
   d. may cause palsy, partial paralysis, blindness, and mental retardation.
 * e. has all of the above characteristics.

60. According to a 1986 EPA study, one in _____ of all children in the United States under age 6 has unsafe blood levels of lead that may retard their mental, physical, and emotional development.
   a. two
   b. four
 * c. six
   d. eight
   e. 10

61. In the United States, lead comes from
    a. paint used in homes between 1978 and 1988.
 * b. drinking water that runs through pipes held together with lead solder.
    c. today's ceramics industry.
    d. burning pine logs in fireplaces.
    e. most vegetables and fruits.

62. Of the following sources of lead in the United States, the one that probably causes the *least* problems is
 * a. chewing on lead pencils.
    b. atmospheric lead that settles on the ground.
    c. pre–1978 paint.
    d. lead solder from seamed food cans.
    e. drinking water contaminated by plumbing containing lead solder.

63. Which of the following would *least* protect children from lead poisoning?
    a. eliminating leaded paint and contaminated dust in housing
 * b. switching from wooden to mechanical pencils
    c. testing all community sources of drinking water for lead contamination
    d. making sure children wash their hands thoroughly before eating
    e. banning incineration of municipal solid waste and hazardous waste

64. Dioxins form during
    a. manufacture of some plastics.
    b. burning of chlorine-containing wastes in municipal and hazardous waste incinerators and cement kilns.
    c. manufacture of some herbicides.
    d. chlorine-bleaching of pulp and paper.
 * e. all of the above.

65. Dioxin can
    a. cause immunological effects.
    b. cause developmental effects.
    c. imitate sex and growth hormones.
    d. cause neurological effects.
    e. a and b.
 * f. do all of the above.

66. Dioxins could *most* directly be reduced by
    a. banning landfills.
 * b. phasing out chlorine in manufacturing and bleaching wood pulp.
    c. recycling.
    d. planting trees.
    e. developing an integrated energy policy.

13-6 SOLUTIONS: DEALING WITH HAZARDOUS WASTE

67. Of the following methods of reducing hazardous wastes, the *most* desirable is
 * a. waste prevention.
    b. conversion to less hazardous materials.
    c. perpetual storage.
    d. deposit in ocean trenches.
    e. use of only biodegradable materials.

68. Of the following methods of reducing hazardous wastes, the *most* desirable is
    a. recycling and reusing hazardous wastes.
    * b. substitution of safer products that don't produce hazardous wastes.
    c. conversion into less hazardous and nonhazardous materials.
    d. incineration.
    e. burial in landfills.

69. In 1992, 13 European nations agreed in principle to eliminate all discharges and emission of chemicals that are
    a. toxic.
    b. likely to bioaccumulate in food webs.
    c. persistent.
    * d. all of the above.
    e. none of the above.

70. Assuming wastes are potentially harmful unless proven otherwise is the
    a. waste management approach.
    b. Peter principle.
    c. entropy principle.
    * d. precautionary principle.
    e. zero discharge principle.

71. Persons living by the precautionary principle are most likely to approach hazardous waste by
    * a. pollution prevention.
    b. detoxification.
    c. deep-well disposal.
    d. recycling.
    e. reuse.

72. The EPA allots about _____% of its waste management budget to encouraging prevention, recycling, and reuse of industrial hazardous waste.
    * a. 1
    b. 10
    c. 20
    d. 30
    e. 40

73. In an EPA study of 28 firms, it was found that investments in waste reduction were in the vast majority of cases returned within
    a. three months.
    * b. three years.
    c. five years.
    d. seven years.
    e. 10 years.

74. The amount most companies spend for waste management is _____% of the total value of the products they sell.
    a. 0.01
    * b. 0.1
    c. 1
    d. 5
    e. 10

75. Hazardous wastes can be converted to less hazardous or nonhazardous materials by
    a. breakdown by bioengineered bacteria.
    b. incineration.
    c. chemical methods.
    d. physical methods.
    * e. all of the above.

76. The *most* comprehensive and effective hazardous-waste detoxification program is in
    a. France.
    * b. Denmark.
    c. Norway.
    d. Yugoslavia.
    e. Sweden.

77. Bioremediation
    a. involves training bacteria to eat new foods.
    b. results in the production of low-level hazardous wastes.
    * c. may be used at considerably less expense than landfills and incineration.
    d. is widely accepted as the best way to cut hazardous wastes.
    e. would decrease global warming.
    f. includes all of the above.

78. Incineration
    * a. can legally be used on all hazardous wastes.
    b. is very expensive.
    c. holds the operators legally responsible for air pollutants resulting from incineration.
    d. results in complete combustion of all hazardous wastes to carbon dioxide and water vapor.
    e. is generally considered to be the safest alternative for managing hazardous wastes.

79. Hazardous-waste expert William Sanjour claims that EPA incinerator regulations won't work because
    a. records can easily be falsified by incinerator operators.
    b. no monitoring of outside air is required in the vicinity of incinerators.
    c. government inspectors are poorly trained.
    d. enforcement officials tend to view the incinerator operator as the client.
    e. there is no incentive for inspectors to find serious violations.
    * f. of all of the above.

80. Most hazardous wastes produced in the United States are disposed of by
    a. landfills.
    * b. deep-well injection.
    c. incineration.
    d. storage in vats, barrels, and similar containers.
    e. salt formations.

81. Hazardous waste deposited in ponds or lagoons
    a. may overflow during hurricanes.
    b. may evaporate into the atmosphere.
    c. may enter groundwater when there are no liners or when liners leak.
    * d. may do all of the above.
    e. may do none of the above.

82. Deep-well disposal of liquid hazardous wastes is
    a. a complex process.
    * b. less visible than other waste-disposal methods.
    c. more carefully regulated than other waste-disposal methods.
    d. the most expensive waste-disposal method.
    e. all of the above.

83. Deep-well injection of hazardous wastes can result in contamination of groundwater by
    a. surface wastes leaching down.
    b. corrosion of well-pipe casings.
    c. leaking seals.
    d. earthquake fractures.
*   e. all of the above.

84. Landfill liners will eventually leak because liners can be
    a. ripped by burrowing animals.
    b. punctured during installation.
    c. dissolved by chemical solvents.
*   d. subject to all of the above.
    e. subject to none of the above.

85. The real cost of dumping hazardous wastes is borne by
    a. the producer of the waste.
    b. the disposer of the waste.
    c. the people whose health is affected by waste disposal.
    d. the taxpayers who pay to clean up disposal messes.
*   e. c and d.

86. In the past decade, there has been an average of approximately ____ accidents in the shipment of toxic wastes each year.
    a. 10
    b. 50
    c. 300
*   d. 1,000
    e. 5,000

87. Which of the following statements about disposal of hazardous wastes is *false?*
    a. Waste disposal firms in the United States and other industrialized nations have shipped hazardous wastes to other countries.
*   b. Most legal U.S. exports of hazardous waste go to Mexico and Panama.
    c. Hazardous wastes have been mixed with wood chips and shipped illegally as burnable material.
    d. Officials of poor LDCs sometimes accept bribes to allow wastes into their countries.
    e. Some Americans call for a ban on U.S. exports of hazardous waste, pesticides, and drugs not approved for use in the United States.
    f. None of the above; all are true.

13-7 HAZARDOUS-WASTE REGULATION IN THE UNITED STATES

88. The Resource Conservation and Recovery Act of 1976
    a. requires the EPA to identify hazardous wastes.
    b. requires the EPA to set standards for hazardous-waste management.
    c. requires all firms that handle more than 100 kilograms of hazardous waste to have a permit stating how such wastes are to be managed.
    d. provides guidelines and financial aid to establish state waste management programs.
*   e. does all of the above.

89. Operators of EPA-licensed hazardous-waste landfills
    a. must allow only specified amounts of landfill leakage.
    b. must monitor the quality of surface water near the landfill.
    c. must report any contamination to Congress.
*   d. must cover the landfills that reach capacity with a leakproof cap.
    e. must be financially responsible for cleanup and damages from leaks up to 50 years after closing.

90. The Superfund program pays
    * a. to clean up inactive or abandoned hazardous-waste dump sites.
    b. to monitor hazardous wastes.
    c. for testing for lead in paint, water, and air samples.
    d. the doctors' bills and lawyers' fees for pollution events.
    e. for all of the above.

91. Which of the following statements about hazardous waste cleanup is *false*?
    a. Cleaning up toxic military dumps is estimated to cost $100-200 billion over 30 years.
    b. The Office of Technology Assessment reports that about 75% of the Superfund cleanups are unlikely to work over the long term.
    * c. It costs 10 to 100 times more to prevent hazardous waste releases than it does to clean them up.
    d. Cleaning up contaminated Department of Energy sites used to make nuclear weapons will run between $200-400 billion over 30 to 50 years.
    e. Out of 1,250 sites on the National Priority List for Superfund cleanup, only 180 sites have been declared clean or stabilized.

92. Individuals can reduce inputs of hazardous waste by all of the following *except*
    a. using rechargeable batteries.
    b. using biodegradable cleaning products.
    * c. changing automobile oil and dumping it carefully in marked storm drains.
    d. consulting local environmental agencies for safe disposal methods for hazardous chemicals.
    e. supporting legislation to put a 10-year moratorium on hazardous-waste incinerators.
    f. none of the above; all methods reduce hazardous-waste inputs.

93. Which of the following questions is *least* likely to be asked by an environmentalist?
    * a. Where do we put all this waste?
    b. Why can't industry change the way it makes things?
    c. Why do we produce so much waste?
    d. Why do we have to produce products that have so many toxic by-products?
    e. Why do we use so much packaging?

94. Toxic racism
    a. is most rampant in South Africa.
    * b. refers to location of landfills and hazardous-waste incinerators in poor areas.
    c. occurs where land is cheap.
    d. occurs only in highly populated areas.
    e. refers to poor people dumping their wastes on rich people's doorsteps.

95. The environmental justice movement attempts to dismantle
    a. exclusionary zoning ordinances.
    b. differential enforcement of environmental regulations.
    c. disparate siting of risky technologies.
    d. dumping of toxic waste on the poor and people of color in the United States and in LDCs.
    * e. all of the above.

96. All of the following are called for by the grass-roots movement for environmental justice *except*
    * a. NIMBY (Not in My Backyard).
    b. holding polluters and elected officials who support them personally accountable.
    c. opposing hazardous-waste landfills, deep-disposal wells, and incinerators.
    d. getting serious about pollution prevention.
    e. banning release of any toxic chemical that bioaccumulates or has a half-life greater than eight weeks.

97. Lois Marie Gibbs feels we should analyze tough issues by
   a. cost-benefit analysis.
   b. asking if the economy will grow.
   c. asking if more jobs will be created.
 * d. asking who will benefit and who will pay the cost.
   e. determining cost effectiveness.

PART IV. BIODIVERSITY: LIVING RESOURCES

CHAPTER 14
FOOD RESOURCES

1. All of the following are perennials being considered as crops *except*
    a. Maximian sunflower.
    b. giant wild rye.
    * c. alpha dandelion.
    d. Illinois bundleflower.
    e. eastern gamma grass.

14-1 HOW FOOD IS PRODUCED

2. To feed the projected 8.6 billion humans by 2015, we must produce as much food in the next 30 years as has been produced in the last _____ years.
    a. 50
    b. 100
    c. 1,000
    d. 5,000
    * e. 10,000

3. Approximately _____ food crops feed the world.
    a. 15
    * b. 30
    c. 45
    d. 60
    e. 75

4. All of the following are among the world's major food crops *except*
    a. wheat.
    * b. soybean.
    c. potato.
    d. rice.
    e. corn.

5. Meat and animal products for humans come mainly from _____ domesticated animal species.
    * a. eight
    b. 15
    c. 21
    d. 30
    e. 50

6. Which of the following statements is *false?*
    a. As a person's income increases, the person eats more grain indirectly in the form of meat and other animal products.
    b. One-third of all the fish that are caught is used as food for livestock.
    * c. People can get more energy per unit of money or labor from meat than from grain.
    d. Almost half of the grain in MDCs is used to feed cattle.
    e. Two-thirds of the world's population have a primarily vegetarian diet.

7. Which of the following types of agriculture is most characteristic of LDCs?
    a. plantation agriculture
    b. traditional intensive agriculture
    * c. traditional subsistence agriculture
    d. industrialized agriculture
    e. none of the above

Food Resources   277

8.	Which of the following types of agriculture is most characteristic of MDCs?
	a.	plantation agriculture
	b.	traditional intensive agriculture
	c.	traditional subsistence agriculture
	*	d.	industrialized agriculture
	e.	none of the above

9.	All of the following crops are commonly grown in plantation agriculture *except*
	*	a.	corn.
	b.	bananas.
	c.	cacao.
	d.	coffee.
	e.	none of the above.

10.	A single type of crop is generally grown in
	a.	plantation agriculture.
	b.	traditional intensive agriculture.
	c.	traditional subsistence agriculture.
	*	d.	industrialized agriculture.
	e.	none of the above.

11.	The type of agriculture that supplements solar energy with human labor and draft animals is
	a.	plantation agriculture.
	b.	traditional intensive agriculture.
	*	c.	traditional subsistence agriculture.
	d.	industrialized agriculture.
	e.	none of the above.

12.	Which of the following is a type of subsistence agriculture?
	*	a.	shifting cultivation on small plots in tropical forests
	b.	intensive crop cultivation plots
	c.	cultivation of large cornfields
	d.	coffee plantations
	e.	none of the above

13.	Industrialized agriculture requires large inputs of
	a.	fossil fuels.
	b.	water.
	c.	inorganic fertilizers.
	d.	pesticides.
	*	e.	all of the above.

14.	Which of the following is characteristic of subsistence farming?
	*	a.	using a diversity of naturally available crop seeds
	b.	planting a single large area in one crop
	c.	using feedlots to raise many animals in a small space
	d.	using massive amounts of fossil fuel during cultivation and harvesting
	e.	using large irrigation systems to increase yields

15.	Since 1950, the majority of the increase in food production is a result of the _____ revolution.
	a.	red
	b.	blue
	*	c.	green
	d.	yellow
	e.	orange

16. Genetic research of the second green revolution produced _____, specially bred for tropical and subtropical climates.
    a. sugarcane and pineapples
    * b. rice and wheat
    c. corn and potatoes
    d. bananas and coconuts
    e. soybeans and peanuts

17. The new crop revolution requires
    a. scientifically bred plant varieties.
    b. irrigation.
    c. pesticides.
    d. fertilizers.
    * e. all of the above

18. The plants of the second green revolution are
    a. dwarf varieties.
    b. fast growing, allowing multiple cropping.
    c. high-yield varieties.
    d. none of the above.
    * e. all of the above.

19. In the 1990s, at least _____% of any increase in world grain output is expected to come from green-revolution techniques.
    a. 50
    b. 60
    c. 70
    * d. 80
    e. 90

20. All of the following factors contributed to a doubling of U.S. food productivity since 1940 *except*
    a. increased use of fossil fuels.
    * b. increased amount of cultivated land.
    c. increased use of pesticides.
    d. increased use of inorganic fertilizers.
    e. favorable climate and soils.

21. In the United States, less than ____% of the people live on farms.
    * a. 2
    b. 4
    c. 6
    d. 8
    e. 10

22. Which of the following industries has the largest total annual sales in the United States?
    a. automotive
    b. housing
    * c. agricultural
    d. steel
    e. high-technology industries

23. Agriculture contributes ____% to the GNP of the United States.
    a. 5
    b. 11
    * c. 18
    d. 30
    e. 45

24. Which of the following requires the greatest input of energy?
    * a. feedlot beef
    b. soybeans
    c. intensive wheat or rice
    d. milk produced by grass-fed cows
    e. apples

25. Which of the following statements is *false?*
    a. Raising animals for food requires more energy than it produces.
    * b. Most plants require more energy from fossil fuels to grow them than they produce.
    c. In the United States, an average of 10 units of nonrenewable fossil-fuel energy is needed to put one unit of energy on the table.
    d. Subsistence farming is more energy-efficient than the U.S. production system.
    e. If all the people in the United States were vegetarians, the country's oil imports could be cut 60%.

26. Intercropping involves growing
    a. several varieties of one crop on a plot of land.
    * b. several different crops on a plot of land.
    c. trees and crops together.
    d. crops in alternate rows.
    e. different crops in alternate years.

27. All of the following are types of interplanting *except*
    a. agroforestry.
    b. polyvarietal cultivation.
    * c. monoculture.
    d. intercropping.
    e. polyculture.

14-2 WORLD FOOD PROBLEMS

28. In describing the years from 1950 to 1984, which of the following statements is *false?*
    a. Grain production almost tripled.
    b. Per capita food production increased.
    c. Food prices, even adjusted for inflation, dropped.
    * d. Most of the world's countries became self-sustaining in food.
    e. Population growth outstripped food production in many areas.

29. Which of the following countries is *not* one of the top five exporters of food?
    a. the United States
    b. Argentina
    * c. Japan
    d. Australia
    e. Canada

30. The term *undernutrition* refers to people who
    * a. eat less than the basic minimum number of daily calories.
    b. eat balanced meals.
    c. eat too much.
    d. suffer from poor food quality.
    e. are anorexic.

31. The term *malnutrition* refers to people who
    a. eat less than the basic minimum number of daily calories.
    b. eat balanced meals.
    c. eat too much.
    * d. suffer from poor food quality.
    e. are anorexic.

32. The term *overnutrition* refers to people who
    a. eat less than the basic minimum number of daily calories.
    b. eat balanced meals.
    * c. eat too much.
    d. suffer from poor food quality.
    e. are anorexic.

33. Overweight people have a greater risk of all of the following *except*
    a. diabetes.
    * b. vitamin deficiencies.
    c. high blood pressure.
    d. stroke.
    e. kidney disease.

34. Overnutrition is characterized by diets
    * a. high in processed foods.
    b. high in fruit and fiber.
    c. high in fresh vegetables.
    d. low in meats.
    e. that are all of the above.

35. The UNICEF program to sharply reduce childhood deaths from improper nutrition would include all of the following *except*
    * a. discouraging breastfeeding.
    b. preventing dehydration by giving infants sugar and salt in water.
    c. giving vitamin A capsules.
    d. providing family planning services.
    e. increasing the education of women.

36. Malnutrition and associated infections could be reduced by
    a. immunization.
    b. teaching women about child care and water sterilization.
    c. spacing births two years apart.
    d. encouraging breastfeeding.
    * e. all of the above.

37. Marasmus is
    a. a weather pattern that has replaced the monsoon in Africa.
    * b. a result of lack of calories and protein.
    c. a vitamin deficiency.
    d. a parasitic infection.
    e. a mineral deficiency.

38. Kwashiorkor
    a. occurs in newborn children.
    * b. is a result of lack of protein.
    c. is a lack of sufficient calories.
    d. is a severe form of diarrhea.
    e. results in blindness in children.

39. Administration of _____ will prevent the majority of blindness in LDCs in the world.
    a. iodine
    b. immunizing shots
    c. sugar and salt in a glass of water
  * d. vitamin A capsules
    e. cataract surgery

40. Iron deficiency anemia in Asia, Africa, and Latin America is *least* likely to occur in
  * a. adult males.
    b. pregnant females.
    c. mature but nonpregnant females.
    d. children.
    e. none of the above.

41. Anemia can be the result of a deficiency in ____.
    a. cobalt.
    b. iodine.
  * c. iron.
    d. calcium.
    e. vitamin A.

42. A 1985 report by a task force of doctors indicated that _____ Americans, mostly children, were hungry.
    a. one in 100
    b. one in 50
    c. one in 23
  * d. one in 11
    e. one in five

43. Which of the following activities causes more pollution and environmental degradation than any other human activity?
    a. transportation
  * b. agriculture
    c. industry
    d. recreation
    e. electric power plants

44. Agriculture can harm the land through
    a. soil erosion.
    b. salinization.
    c. groundwater depletion.
    d. reduction in microorganism diversity.
  * e. all of the above.

45. Industrialized agriculture can affect water resources through
    a. pesticide pollution.
    b. fertilizer enrichment of surface waters.
    c. sediment pollution.
    d. fertilizer seepage into groundwater.
  * e. all of the above.

46. Living organisms can be affected by agriculture through
    a. loss of habitat from clearing land for agriculture.
    b. use of monoculture.
    c. pollution of groundwater with nitrates.
    d. oxygen depletion of surface waters receiving fertilizer runoff.
  * e. all of the above.

47.	Industrial agriculture is more likely to cause _____ than traditional agriculture.
	a.	increased frequency of flooding of lowlands
	b.	sediment pollution in streams and rivers
*	c.	pollution of water by fertilizers and pesticides
	d.	increased threats to human health resulting from flooding
	e.	desertification
	e.	soil erosion

14-3	SOLUTIONS TO WORLD FOOD PROBLEMS

48.	Agricultural experts hope to increase food production by using all of the following plants *except* plants that
	a.	grow in salty soil.
	b.	make their own nitrogen fertilizer.
*	c.	do not require light to produce their food.
	d.	are resistant to drought.
	e.	are more resistant to insects.

49.	Which of the following statements best describes how we hope to improve agricultural yields in the future?
	a.	If the soil is deficient in minerals, we add fertilizer.
	b.	If the land is dry, we add irrigation.
*	c.	If the plants don't fit the environment, we use genetic engineering.
	d.	If the insects produce problems, we add pesticides.
	e.	If we don't have enough muscles, we use fossil fuels.

50.	Which of the following is *not* one of the advances expected through genetic engineering in the next couple of decades?
	a.	developing plants using nitrogen from the atmosphere
	b.	developing plants that can grow in salty soils on less water
*	c.	converting fungi into producers
	d.	increasing the efficiency of photosynthesis
	e.	developing plants resistant to disease

51.	Which of the following statements is *true?*
	a.	Green-revolution plants are less expensive than regular varieties.
	b.	Loss of biodiversity limits green-revolution approaches.
*	c.	Deforestation may limit the potential of future green revolutions.
	d.	The development of monocultures of newly developed plants leads to greater plant diversity.
	e.	The growth of green-revolution plants in response to fertilization and irrigation inputs is expected to exhibit a J-shaped, not an S-shaped curve.

52.	Which of the following plants has been called "supermarket on a stalk"?
	a.	peanut
	b.	soybean
*	c.	winged bean
	d.	rice
	e.	dwarf wheat

53.	Use of perennial crops would reduce
	a.	the amount of fossil fuel used.
	b.	the amount of water used.
	c.	sediment water pollution.
	d.	soil erosion.
*	e.	all of the above.

54. Your text includes all of the following examples of alternative food sources *except*
    a. Amazon River turtles.
    b. babirusa.
  * c. rattlesnakes.
    d. insects.
    e. iguanas.

55. In spite of technological advances, only _____% of Earth's land area is suitable for growing crops.
    a. 5
  * b. 11
    c. 15
    d. 19
    e. 26

56. The decline in the available land for cultivation is the result of
    a. desertification.
    b. urbanization.
    c. waterlogging.
    d. mining.
  * e. all of the above

57. Most of the new cropland in the world is located in two countries, Brazil and _____.
    a. Argentina
    b. India
    c. the Commonwealth of Independent States
  * d. Zaire
    e. Uganda

58. The conversion of grazing land and marginal land to agriculture would result in
    a. increased erosion.
    b. decreased biodiversity.
    c. tremendous capital costs.
    d. enhanced global warming effects.
  * e. all of the above.

59. A large part of Africa cannot be used to grow crops because of
  * a. tsetse flies.
    b. mosquitoes.
    c. screwworm flies.
    d. killer bees.
    e. parasitic nematodes.

60. Tsetse flies spread
    a. malaria.
  * b. sleeping sickness.
    c. elephantiasis.
    d. sylvatic plague.
    e. schistosomiasis.

61. Scientists recommend that plantations are good for growing varieties of all the following species *except*
  * a. winged bean.
    c. rubber trees.
    d. oil palms.
    e. banana trees.

62. Worldwide, humans get an average of _____% of their animal protein from fish and shellfish.
    a. 10
  * b. 20
    c. 30
    d. 40
    e. 50

63. By 2000, the world food catch is expected to return to the 1960 level because of
    a. population growth.
    b. pollution.
    c. overfishing.
  * d. all of the above.
    e. none of the above.

64. Sustainable yield is difficult to estimate because
    a. counting aquatic populations isn't easy.
    b. pollution levels change.
    c. of shifts in climate.
  * d. of all of the above.
    e. of none of the above.

65. Optimum yield is _____ sustainable yield.
    a. greater than
  * b. less than
    c. equal to
    d. greater than or equal to
    e. less than or equal to

66. Drift-net fishing
    a. is used by the United States.
    b. is used primarily in the Atlantic Ocean.
  * c. in one night involves use of enough nets to circle the world.
    d. is the most efficient way to catch a sustained yield of fish.
    e. all of the above.

67. Drift-net fishing can catch and kill
    a. dolphins.
    b. turtles.
    c. seals.
    d. whales and sharks.
  * e. all of the above.

68. Drift-net fishing was reduced when
  * a. consumers boycotted tuna caught by drift nets.
    b. the United States unilaterally decided to stop using this method.
    c. an international treaty caused all countries to stop using this method.
    d. companies manufacturing drift nets ran out of necessary natural resources.
    e. El Niño–Southern Oscillation interfered with radio transmissions.

69. The chief use of the Peruvian anchovy catch was
    a. pizza.
  * b. fish meal for livestock.
    c. food for game fish.
    d. pet food.
    e. fertilizer.

70. El Niño–Southern Oscillation
    a. increases productivity in the Pacific Ocean.
    b. causes the upwelling phenomenon.
    * c. results in a decline in the anchovy population.
    d. brings up toxic wastes from ocean trenches.
    e. does all of the above.

71. El Niño–Southern Oscillation
    a. occurs at predictable intervals.
    b. increases productivity of upwellings off the coast.
    c. cools the normally warm waters of the Humboldt Current.
    d. increases productivity of marine organisms.
    * e. does none of the above.

72. The collapse of the Peruvian anchovy fishery in the 1970s was caused by
    a. biology.
    b. climate and geography.
    c. economics.
    d. politics.
    * e. interactions of all of the above.

73. The consequence of taking a short-term economic view of the anchovy fishery was
    a. loss of income.
    b. loss of jobs.
    c. an increased foreign debt.
    d. a depleted resource.
    * e. all of the above.

74. The Peruvian ____ took over the niche once occupied by the anchovy.
    a. sea urchins
    b. bonita
    c. mackerel
    d. menhaden
    * e. sardine

75. Aquaculture supplies ____% of the world's commercial fish harvest.
    a. 5
    * b. 10
    c. 15
    d. 25
    e. 50

76. Fish ranching is useful for
    a. carp.
    * b. salmon.
    c. tilapia.
    d. lake trout.
    e. swordfish.

77. Fish species cultivated in aquaculture in 71 LDCs include all of the following *except*
    * a. trout.
    b. carp.
    c. clams.
    d. tilapia.
    e. oysters.

78. In MDCs, aquaculture primarily benefits
    a. lower-income people.
    * b. sports fishermen.
    c. bears.
    d. LDCs.
    e. none of the above.

79. Benefits of aquaculture include
    a. high yields per unit area.
    b. use of less fuel than conventional fishing methods.
    c. creation of jobs.
    d. a and b only.
    * e. all of the above.

80. Problems with aquaculture include
    a. destruction of mangrove forests.
    b. loss of fish through pesticide poisoning.
    c. loss of fish through bacterial and viral infections.
    d. waste contamination of nearby surface and groundwater.
    e. b and d only.
    * f. all of the above.

81. Governments often provide assistance to farmers because farmers have little control over
    a. weather.
    b. crop prices.
    c. crop pests.
    d. interest rates and the global market.
    * e. all of the above.

82. Governments can influence the food supply by
    a. keeping food prices artificially low.
    b. giving farmers subsidies.
    c. eliminating price controls.
    d. eliminating subsidies.
    * e. doing all of the above.

83. Governments in many LDCs keep food prices low
    a. to prevent surplus.
    * b. to prevent political unrest in the cities.
    c. to exert political control over the farmers.
    d. to feed their armies.
    e. to punish collectively powerful farmers.

84. Government subsidies may
    a. produce local surpluses.
    b. cause a drop in global food prices.
    c. reduce the incentives for farmers to grow crops.
    d. cause the government to import food.
    * e. do all of the above.

85. The practice of paying farmers not to produce food on one-fourth of the cropland and the purchase of unneeded crops
    a. encourages farmers to produce more.
    b. discourages the use of Earth-sustaining agriculture.
    c. wastes taxpayer dollars.
    d. is a form of welfare for wealthy farmers.
    * e. is all of the above.

86. Between 1945 and 1985, _____ was the largest donor of nonmilitary foreign aid to LDCs.
    a. Japan
    b. Germany
  * c. the United States
    d. Great Britain
    e. the Commonwealth of Independent States

87. Foreign aid
    a. helps the receiving countries.
    b. stimulates economic growth in the donor country.
    c. provides jobs in the donor country.
    d. has been dropping as a percentage of GNP in the United States.
  * e. has all of the above characteristics.

88. The year in which Japan succeeded the United States as the major supplier of international aid was _____.
    a. 1965
    b. 1973
    c. 1981
  * d. 1986
    e. 1990

89. Food relief
    a. prevents people from dying prematurely.
    b. encourages population growth.
    c. ultimately results in more people dying prematurely in the future.
    d. can be thought of in terms of lifeboat ethics.
  * e. all of the above.

90. Garrett Hardin suggests that we
  * a. adopt lifeboat ethics with our food aid.
    b. use food aid as a political weapon.
    c. use food aid as a way to eliminate our surplus crops.
    d. try to support as many people as possible because people are our ultimate resource.
    e. eat lower on the food chain.

91. Large amounts of food aid can
    a. inflate local food prices.
  * b. discourage receiving governments from investing in rural agricultural development to grow sustainable crops.
    c. stimulate mass migration from cities to rural areas.
    d. increase food production.
    e. do all of the above.

92. Some of the food made available to a country does not reach the poor and hungry because of
    a. transportation problems.
    b. robbery, graft, and bribes.
    c. pests and storage problems.
    d. storage problems.
  * e. all of the above.

93. Critics of food relief say that foreign aid should help receiving countries
    a. control population growth.
    b. practice sustainable agriculture.
    c. develop export crops to help pay for food they cannot grow.
    d. with food only when there is a natural disaster.
  * e. in all of the above ways.

94. One of the countries with the most effective land reform program is
    a. India.
    b. Brazil.
    * c. China.
    d. Egypt.
    e. Guatemala.

95. Countries with the greatest need for land reform are found in
    * a. Latin America.
    b. Asia.
    c. Africa.
    d. Europe.
    e. North America.

14-4 SOLUTIONS: SUSTAINABLE FOOD PRODUCTION

96. Sustainable-Earth agriculture is characterized by all of the following *except*
    * a. promoting monoculture.
    b. lack of requirements of massive amounts of fossil fuels.
    c. conserving and building topsoil.
    d. lack of use of many artificial chemicals.
    e. water conservation.

97. All of the following are characteristic of sustainable-Earth agriculture *except*
    a. perennial plants.
    b. nontraditional crops.
    c. diversity of fruits and vegetables.
    * d. large-scale monoculture.
    e. diversity of livestock animals.

98. Sustainable-Earth agriculture is characterized by major use of
    a. massive irrigation projects.
    b. fossil fuels.
    c. pesticides.
    * d. organic fertilizers.
    e. inorganic fertilizers.

99. The sustainable-Earth agriculture philosophy would be supported by
    a. agribusiness companies.
    * b. ecologists.
    c. large farmers.
    d. specialized farmers unprepared for demanding managerial skills.
    e. investors in agricultural products.

100. Sustainable-Earth agriculture
    a. uses pesticides.
    * b. minimizes erosion.
    c. requires high inputs of fossil fuel.
    d. employs massive fertilization programs.
    e. requires exponential population growth.

101. In order to switch to sustainable-Earth agriculture, which of the following practices would
    not be favorable?
    a. Give subsidies and tax breaks to those that use the method.
    b. Establish demonstration programs supporting the sustainable-Earth philosophy.
    c. Increase government support of research.
    * d. Expand the use of the crops of the green revolution.
    e. Set up sustainable-Earth demonstration projects in each county.

102. Sustainable-Earth agriculture
    a. treats agriculture as an industry.
    * b. treats long-term quality of the soil as a top priority.
    c. has a goal of increasing short-term crop production.
    d. depends on genetic engineering of high-yield crops.
    e. has all of the above characteristics.

103. Sustainable-Earth agriculture
    a. emphasizes large-scale farms.
    * b. uses local inputs as much as possible.
    c. maximizes the use of fossil fuels.
    d. promotes subsidies to farmers.
    e. does none of the above.

104. An individual can support the concept of sustainable-Earth agriculture by
    a. eating lower on the food chain.
    b. developing a home garden using appropriate principles.
    c. developing a compost pile.
    d. using biological pest control.
    * e. all of the above.

CHAPTER 15
PROTECTING FOOD RESOURCES: PESTICIDES AND PEST CONTROL

1. Humans have been waging war with insects since
   a. hunting and gathering cultures.
   * b. agriculture began.
   c. the Industrial Revolution began.
   d. World War I.
   e. World War II.

2. Chinese farmers build straw huts in their fields
   a. for storing rice harvests.
   b. for storing farm implements.
   c. for temporary sleeping quarters during harvest.
   * d. for spiders.
   e. for drying tobacco.

3. According to Thomas Eisner, Earth's landlords are
   a. humans.
   b. primates.
   c. whales.
   * d. bugs.
   e. earthworms.

15-1 PESTICIDES: TYPES AND USES

4. A pest is any organism that
   a. spreads disease.
   b. interferes with human activity.
   c. competes with humans.
   d. eats crops.
   * e. does any of the above.

5. In diverse ecosystems, populations of species are *least* likely to be kept in control by
   a. parasites.
   b. disease organisms.
   * c. pesticides.
   d. predators.

6. Which of the following categories includes all of the others?
   a. insecticides
   b. herbicides
   * c. pesticides
   d. fungicides
   e. rodenticides

7. Pesticides kill
   a. rodents.
   b. fungi.
   c. insects.
   d. weeds.
   * e. all of the above.

8.   Which of the following would be used to kill rats and mice?
      a.   herbicides
  *   b.   rodenticides
      c.   fungicides
      d.   insecticides
      e.   none of the above

9.   Which of the following would be used to kill weeds?
  *   a.   herbicides
      b.   rodenticides
      c.   fungicides
      d.   insecticides
      e.   none of the above

10.   First-generation pesticides included
      a.   extracts from insect poisons.
      b.   rotenone.
      c.   inorganic compounds containing toxic metals.
      d.   pyrethrum.
  *   e.   all of the above.

11.   Natural pesticides come from all of the following *except*
  *   a.   inorganic compounds.
      b.   derris plant root.
      c.   chrysanthemum flowers.
      d.   red pepper.
      e.   garlic and lemon oil.

12.   Pyrethrum is a pesticide derived from
      a.   red peppers.
      b.   tomatoes.
      c.   marigolds.
  *   d.   chrysanthemums.
      e.   lichens.

13.   All of the following methods have been recommended to discourage ants *except*
      a.   vinegar.
      b.   boric acid.
  *   c.   populations of banana spiders.
      d.   red pepper.
      e.   crushed mint leaves.

14.   Basil repels
      a.   roaches.
      b.   flies.
      c.   ants.
  *   d.   mosquitoes.
      e.   fleas.

15.   Currently, we use about 1 pound of pesticide for every _____ person(s) on Earth.
  *   a.   1
      b.   10
      c.   100
      d.   1,000
      e.   10,000

16. Since 1964, pesticide use in the United States
    a. has been cut in half.
    b. has stayed the same.
    c. has almost doubled.
  * d. has almost tripled.
    e. has increased tenfold.

17. There are _____ pest-killing ingredients produced in the United States.
  * a. 600
    b. 1,500
    c. 15,000
    d. 55,000
    e. 150,000

18. All of the following are among the crop plants responsible for the great use of pesticides in the United States *except*
    a. corn.
    b. wheat.
  * c. rice.
    d. cotton.
    e. soybeans.

19. The average homeowner in the United States applies about _____ pesticide per unit of land compared to that used by farmers.
    a. half as much
    b. about the same amount of
    c. twice as much
  * d. five times more
    e. ten times more

20. Today, the four major classes of insecticides include all of the following *except*
    a. pyrethroids.
    b. organophosphates.
  * c. bromated hydrocarbons.
    d. carbamates.
    e. chlorinated hydrocarbons.

21. Compared to chlorinated hydrocarbons, organophosphates are _____ persistent and _____ biologically magnified.
  * a. less . . . less
    b. less . . . more
    c. more . . . less
    d. more . . . more
    e. equally . . . equally

22. Major types of herbicides include
    a. contact and systemic.
    b. soil sterilants.
    c. chlorinated hydrocarbons and organophosphates.
    d. carbamates and pyrethroids.
  * e. a and b.
    f. all of the above.

23. Chlorinated hydrocarbons are more likely to _____ than organophosphates and carbamates.
    a. contaminate surface and groundwater
    b. degrade quickly in the environment
    c. be more toxic to organisms other than the targeted pests
    * d. become magnified in the food chain
    e. (all of the above)

24. Which of the following groups of pesticides would persist in an area the longest?
    a. pyrethroids
    * b. chlorinated hydrocarbons
    c. contact insecticides
    d. carbamates
    e. organophosphates

25. Aldrin, dieldrin, endrin, lindane, DDT, and mirex are examples of
    a. pyrethroids.
    * b. chlorinated hydrocarbons.
    c. carbamates.
    d. organophosphates.
    e. systemic insecticides.

26. Rotenone and pyrethrum are examples of
    a. soil sterilants.
    * b. botanicals.
    c. chlorinated hydrocarbons.
    d. carbamates.
    e. organophosphates.

27. Malathion and parathion are examples of
    a. systemic insecticides.
    b. pyrethroids.
    c. chlorinated hydrocarbons.
    d. carbamates.
    * e. organophosphates.

28. Maneb and aldicarb are examples of
    a. contact insecticides.
    b. pyrethroids.
    c. chlorinated hydrocarbons.
    * d. carbamates.
    e. organophosphates.

29. Atrazine and paraquat are
    * a. contact herbicides.
    b. rotenoids.
    c. systemic herbicides.
    d. pyrethroids.
    e. soil sterilants.

15-2 THE CASE FOR PESTICIDES

30. According to pesticide proponents, pesticides
    a. work faster than alternate controls.
    b. increase profit for farmers.
    c. save lives and money.
    d. kill insects that transmit diseases.
    * e. do all of the above.

31. According to proponents of pesticides, which of the following statements is *true?*
    a. Pesticides increase food supplies.
    b. The health risks of pesticides are insignificant compared with their health and other benefits.
    c. Safer and more effective pesticides are constantly being developed.
    d. Insecticides lower food costs.
  * e. All of the above are true.

32. The ideal pesticide
  * a. would kill only the target pest.
    b. would be persistent.
    c. would allow the development of genetic resistance.
    d. would be of equal value to the damage the pest would have caused.
    e. would do all of the above.

33. The ideal pesticide would
    a. be broken down easily.
    b. kill only the target organism.
    c. not allow the development of resistance in the pest.
    d. have no ill effects on nontarget organisms.
  * e. do all of the above.

15-3 THE CASE AGAINST PESTICIDES

34. The most serious drawback to using chemicals to control pests is
  * a. the development of genetic resistance.
    b. the killing of other forms of life.
    c. magnification in the food chain.
    d. their persistence in nature.
    e. the cost.

35. Since 1950, ____ species of insects have developed resistance to one or more insecticides.
    a. 50
    b. 20
  * c. 500
    d. 2,100
    e. 2,500

36. From 1970 to 1988, the incidence of malaria increased ____ times.
    a. two
    b. seven
    c. 15
  * d. 40
    e. 50

37. The primary reason for the increase in malaria is
    a. global warming.
    b. increased spreading of the disease between humans.
    c. environmental protection of swamps where mosquitoes breed.
  * d. genetic resistance to pesticides.
    e. lack of immunization.

38. Broad-spectrum pesticides may increase the number of a pest species through
    a. development of genetic resistance.
    b. killing of predators of the pest species.
    c. killing of parasites that may have kept the population of the pest low.
  * d. all of the above.
    e. none of the above.

39. The pesticide treadmill involves
    a. efforts to overcome genetic resistance.
    b. use of stronger doses of pesticide.
    c. a switch to new chemicals.
    d. use of more frequent doses of pesticide.
  * e. all of the above.

40. David Pimentel's study showed that
    a. since the 1940s, with a 33-fold increase in pesticide use in the United States, crop losses to pests have actually increased to about 37%.
    b. external costs of pesticides in the United States range from $4 billion to $10 billion annually.
    c. use of alternative pest control practices in the United States could cut chemical pesticide use in half without a reduction in crop yields.
    d. cutting pesticide use in the United States in half would cause less than a 1% rise in food prices.
  * e. all of the above are possible.

41. According to the U.S. Department of Agriculture (USDA), no more than _____% of the insecticides applied to crops by aerial spraying reaches the target pests.
  * a. 2
    b. 10
    c. 25
    d. 50
    e. 75

42. According to the USDA, about _____% of herbicides applied to crops reaches the target weeds.
    a. 1
  * b. 5
    c. 25
    d. 50
    e. 75

43. Chlorinated hydrocarbons are
  * a. biologically amplified.
    b. likely to be target-specific.
    c. degraded faster than other organic pesticides.
    d. more toxic than other organic pesticides.
    e. none of the above.

44. Which of the following is more likely to become biologically magnified in a food chain?
  * a. DDT
    b. parathion
    c. carbaryl
    d. malathion
    e. pyrethroid

45. In biological amplification,
    a. organisms in the lower trophic levels accumulate lethal doses of toxins.
    b. the animals at the upper end of a food chain receive lower doses than those below.
  * c. organisms at higher trophic levels have more concentrated levels of toxic substances.
    d. the environment has higher concentrations of toxins than organisms in the food chain.
    e. the toxic substances reach maximum accumulation in one of the intermediate levels of a food chain.

46. Of the following organisms that might occur in a simple aquatic food chain, which organism would have the highest concentration of DDT if it had spilled in the ecosystem several months before being measured?
     a. phytoplankton
     b. large fish
*   c. birds that feed on large fish
     d. minnows
     e. zooplankton

47. In terms of cancer risk, the EPA rated pesticide residues in foods as the ___ most serious environmental health threat.
     a. first
     b. second
*   c. third
     d. seventh
     e. tenth

48. The Delaney clause to protect the public from cancer-causing pesticides is based on
     a. balancing risks and benefits.
     b. balancing costs and benefits.
*   c. no risk.
     d. no unreasonable risk.
     e. standards representing the best available technology.

49. Concerns have been raised about possible links between low levels of pesticides and
     a. birth defects.
     b. nervous system disorders.
     c. endocrine and immune system disorders.
     d. genetic mutations.
*   e. all of the above.

50. According to the Food and Drug Administration, ___% of the food bought in supermarkets has levels of pesticide residues that are above the legal limit.
*   a. 1
     b. 3
     c. 5
     d. 7
     e. 10

51. The world's greatest industrial accident involved
     a. explosives.
*   b. pesticides.
     c. hydrogen.
     d. radioactive material.
     e. fertilizers.

15-4 PESTICIDE REGULATION IN THE UNITED STATES: IS THE PUBLIC ADEQUATELY PROTECTED?

52. Chemicals banned under the Federal Insecticide, Fungicide, and Rodenticide Act (FIFRA) since 1972 include all of the following *except*
*   a. rotenone.
     b. chlorinated hydrocarbon insecticides.
     c. herbicides Silvex and 2,4,5-T.
     d. some organophosphates.
     e. some carbamates.

53. Which of the following statements is *false*?
   a. Pesticide companies can make and export pesticides that have been banned in the United States.
   b. Pesticide companies can use a chemical that has been shown to cause cancer or other harmful effects when the economic benefits exceed the costs.
   c. Pesticides can be left on the market without full health and safety data.
   d. Until 1990, if the EPA banned the use of a chemical on an emergency basis, the EPA had to compensate the pesticide manufacturer.
   * e. The Federal Insecticide, Fungicide, and Rodenticide Act (FIFRA) provides for citizen suits against the EPA.

54. Which of the following statements is *false*?
   a. The EPA is charged with the responsibility of determining if active ingredients in pesticides are carcinogenic.
   b. The National Academy of Sciences says the federal laws regulating use of pesticides in the United States are inadequate and poorly enforced.
   c. The EPA carried out reviews of less than 10% of 600 chemicals it was charged with the responsibility for reevaluating.
   * d. FDA inspectors check half of domestic and imported food for pesticide contamination.
   e. None of the above is false.

55. Which of the following statements is *true*?
   a. The laws governing the regulation of pesticides have numerous weaknesses and loopholes.
   b. Laboratory turnaround time is so long that some contaminated foods are marketed before contamination is detected.
   c. Routine food testing used by the EPA can detect less than half of the pesticides classified by the EPA as moderate to high health hazards.
   d. Little is known about the toxicity of most of the inert ingredients in pesticides.
   * e. All of the above are true.

56. Which of the following statements is *false*?
   a. The Federal Insecticide, Fungicide, and Rodenticide Act of 1972 (FIFRA) required the reevaluation of 600 active ingredients in pesticides.
   b. Some inert or biologically inactive ingredients in a pesticide can cause harm.
   * c. The Federal Insecticide, Fungicide, and Rodenticide Act of 1972 requires the EPA to immediately remove inadequately tested pesticides from the marketplace.
   d. One FIFRA loophole allows sale of up to 15% DDT by weight as an impurity.
   e. Some of the cost of banning pesticides has been passed on to the pesticide companies involved.

57. Each year the Food and Drug Administration tests _____% of domestic and imported food for pesticide contamination.
   * a. less than 1
   b. 5
   c. 10
   d. 15
   e. 20

58. Which of the following statements is *false*?
   a. Pesticide companies can make and export pesticides that have been banned in the United States.
   b. The United States leads the world in the export of pesticides.
   c. The term *circle of poison* refers to the return of banned pesticides to the United States on imported foods.
   d. In 1993, the Clinton Administration recommended banning export of any pesticide denied U.S. registration.
   * e. None of the above is false.

59. Which of the following approaches would be the *least* beneficial in trying to reduce insect damage?
    a. rotating crops
    b. delaying planting
  * c. planting monocultures
    d. planting barrier hedges around agricultural fields
    e. using photodegradable plastic to prevent growth of weeds between rows of crops

60. Using cultivation practices to control pests is discouraged by the need to
    a. avoid bankruptcy.
    b. increase profits.
    c. qualify for government subsidies.
  * d. do all of the above.
    e. do none of the above.

61. Limits to building in resistance to pests include
    a. long development time.
    b. expense.
    c. evolution operating to overcome the resistance.
  * d. all of the above.
    e. none of the above.

62. Which of the following is not considered a biological control organism?
  * a. squirrels
    b. guard dogs
    c. geese, ducks, and chickens
    d. insect-eating birds
    e. spiders

63. Biological control
    a. costs more money than pesticides to use.
    b. is not target-specific.
    c. is easily produced and quickly applied.
    d. has effects only on animals and is almost totally ineffective against weedy plants.
  * e. is often self-perpetuating once established.

64. Which of the following statements is *false?*
  * a. Biological control is a good, quick fix.
    b. Farmers save more money by applying biological controls than by using pesticides.
    c. Biological control is slower acting than pesticides.
    d. Biological control is harder to apply than pesticides.
    e. Biological control may take 10 to 20 years to develop.

65. Biopesticides include
    a. pyrethroids.
    b. Bacillus thuringensis.
    c. carbamates.
    d. organophosphates.
  * e. a and b.
    f. c and d.

66. Insect control by sterilization involves irradiating
    a. eggs.
    * b. males.
    c. females.
    d. larvae.
    e. nymphs.

67. Using insect sterilization as a pest control is difficult because
    a. sterile males may be overwhelmed numerically by nonsterile males.
    b. it depends on knowledge of mating times and behaviors of each target insect.
    c. new nonsterilized males may reinfest after treatment.
    d. of high costs.
    * e. of all of the above.

68. A pheromone is
    a. a new form of chemical insecticide waiting approval by FIFRA.
    b. a strong herbicide.
    * c. a species-specific chemical sex attractant.
    d. a bloodstream chemical that controls an organism's growth and development.
    e. usually the best choice for integrated pest management.

69. Which of the following statements is a weakness of using pheromones?
    * a. They are costly and time-consuming to produce in the laboratory.
    b. They are more effective in the juvenile stage than the adult stage.
    c. Insects develop resistance to pheromones.
    d. They are biologically magnified in nontarget species.
    e. They work only in high concentrations.

70. Pheromones
    a. are species-specific.
    b. have little chance of causing genetic resistance.
    c. are effective in trace amounts.
    d. are not harmful to nontarget species.
    * e. have all of the above characteristics.

71. A hormone is
    a. a new form of chemical insecticide waiting approval by FIFRA.
    b. a strong herbicide.
    c. a species-specific chemical sex attractant.
    * d. a chemical that travels in the bloodstream and controls an organism's growth and development.
    e. usually the best choice for integrated pest management.

72. Which of the following statements is *false?*
    * a. Insect development and metamorphosis are controlled by pheromones.
    b. Each step in the life cycle of a typical insect is controlled by hormones.
    c. Insect hormones can be synthesized in the laboratory.
    d. When applied at certain times during the life cycle, insect hormones cause developmental abnormalities.
    e. None of the above is false.

73. Which of the following statements is *false?*
    * a. Exposure of material to low doses of ionizing radiation makes it radioactive.
    b. The FDA has approved the use of radiation for certain foods.
    c. Radiation extends the shelf life of some perishable foods.
    d. Radiation may destroy some of the food's vitamins and minerals.
    e. New York, New Jersey, and Maine have prohibited the sale and distribution of irradiated foods.

74. Integrated pest management is a
    a. chemical program.
    b. ecological program.
    c. agricultural program.
    d. biological program.
  * e. program that interrelates all of the above.

75. Integrated pest management _____ than pesticides.
  * a. requires more expert knowledge about individual pest-crop situations
    b. is faster acting
    c. requires more fertilizer and irrigation
    d. is more expensive
    e. is less complicated

76. Which of the following statements is *false?*
  * a. The goal of integrated pest control is the complete eradication of the pest.
    b. Integrated pest management is a multidimensional control program.
    c. Integrated pest management involves much sampling to keep up with the pest population.
    d. In integrated pest management, small amounts of pesticides are used at critical times to control pest populations.
    e. Integrated pest management can increase crop yields and reduce crop production costs.

77. All of the following countries are leading in the development of integrated pest management *except*
    a. the United States.
    b. Brazil.
    c. China.
  * d. France.
    e. Indonesia.

78. An integrated pest management program can
    a. reduce inputs of fertilizer and irrigation water.
    b. reduce preharvest pest-induced crop losses by 50 %.
    c. reduce pesticide use.
    d. increase yields and reduce costs.
  * e. do all of the above.

79. Switching to integrated pest management
    a. will be hard to do because it requires a break from tradition.
    b. is strongly opposed by politically and economically powerful agricultural chemical companies.
    c. is difficult because farmers get most of their information from pesticide salespeople or USDA county agents who have supported pesticide use in the past.
    d. is strongly supported by environmentalists.
  * e. has all of the above characteristics.

80. To help protect the food supply and fight pests responsibly, individuals can
    a. write to elected officials to strengthen the Federal Insecticide, Fungicide, and Rodenticide Act.
    b. use minimal amounts of pesticides when necessary and properly dispose of unused amounts.
    c. fix leaking pipes and faucets that attract insects.
    d. allow native plants to cover most of their property.
    e. use natural alternatives to pesticides.
  * f. do all of the above.

81. Which of the following would *not* reduce the threat of pesticide in the food you eat?
   * a. Use imported food whenever possible.
     b. Scrub all food in soapy water.
     c. Grow your own fruits and vegetables using organic gardening methods.
     d. Purchase organically grown foods.
     e. Buy during the growing season and select the less-perfect fruits.

CHAPTER 16
BIODIVERSITY: SUSTAINING ECOSYSTEMS

1. The relationship between the black howler monkeys and the peasant farmers is *best* described as
    a. monkeys win/peasants lose.
    b. monkeys lose/peasants win.
    *   c. monkeys win/peasants win.
    d. monkeys lose/peasants lose.
    e. nonexistent.

16-1 THE IMPORTANCE OF ECOLOGICAL DIVERSITY

2. Which of the following terms includes the others?
    a. ecological diversity
    b. species diversity
    *   c. biodiversity
    d. genetic diversity
    e. none of the above

3. Most wildlife biologists believe that the best way to protect biodiversity is to
    a. pass endangered species laws.
    *   b. protect earth's ecosystems.
    c. use captive breeding programs.
    d. use genetic engineering.
    e. develop programs to protect individual species.

4. The nation which has set aside the most land for public use, enjoyment, and wildlife is
    *   a. the United States.
    b. Sweden.
    c. Japan.
    d. Costa Rica.
    e. Kenya.

5. In the United States, about ____% of the land is public land.
    a. 17
    b. 25
    c. 33
    *   d. 42
    e. 52

6. About three-quarters of federally managed public lands are in
    a. New York State.
    b. Colorado.
    *   c. Alaska.
    d. Arizona.
    e. California.

7. Federally administered public lands contain a large portion of the country's
    a. energy resources.
    b. commercial timber.
    c. grazing land.
    d. minerals.
    *   e. resources, including all of the above.

8. The national forests are managed on
   * a. a sustainable-yield multiple-use basis.
   b. a restricted-use basis.
   c. a maximum timber production schedule.
   d. the basis of political expediency.
   e. an individual basis, depending on local goals and directives.

9. Which one of the following is the most numerous?
   a. national parks
   b. national wildlife refuges
   c. national resource lands
   * d. national wilderness preservation sites
   e. national forests

10. Which one of the following is managed on a multiple-use basis?
    a. national wildlife refuges
    b. national wilderness areas
    c. national parks
    * d. national resource lands
    e. all of the above

11. Which one of the following is managed on a restricted-use basis?
    * a. national parks
    b. national forests
    c. national resource lands
    d. national wildlife refuges
    e. none of the above

12. At the turn of the century, the preservationist and wise-use movements agreed that public lands
    a. should be set aside as wilderness.
    b. should be used for economic gain.
    c. should be in the hands of the states and individuals.
    * d. should be used for the public good and not go into the hands of a few for profit.
    e. should be turned into national parks.

13. Of the following, the *least* supportive of transferring public lands to states and individuals was
    a. Herbert Hoover.
    b. the Sagebrush Rebellion.
    * c. the preservationists.
    d. the current wise-use movement.
    e. Ronald Reagan.

14. Of the following, who is *least* likely to be a member of the current wise-use movement?
    a. a rancher
    b. a coal miner
    * c. a member of the Audubon Society
    d. a logger
    e. a president of an oil company

16-2 SUSTAINING AND MANAGING FORESTS

15. Potentially renewable forests cover about ___ of the land surface of Earth.
    a. one-fourth
    * b. one-third
    c. one-half
    d. two-thirds
    e. three-fourths

16. Most temperate forests are _____ forests; most tropical forests are _____ forests.
    a. primary . . . primary
    b. primary . . . secondary
    c. secondary . . . secondary
 * d. secondary . . . primary
    e. (none of the above)

17. Old-growth forests
    a. result from primary succession.
    b. developed after the abandonment of agricultural lands.
    c. are the predominant forest form in the United States.
 * d. are the predominant forest form in the tropics.
    e. are described by a and d.

18. Secondary forests
    a. result from primary succession.
    b. developed after the abandonment of agricultural lands.
    c. are the predominant forest form in the United States.
    d. are the predominant forest form in the tropics.
    e. are just starting to evolve.
 * f. are described by b and c.

19. Old-growth forests often contain _____ trees such as _____.
    a. massive . . . white pine
 * b. massive . . . coastal redwoods
    c. small . . . redbud
    d. small . . . western hemlock
    e. moderate . . . eastern maple

20. Utilitarian functions of forests include
    a. lumber for housing.
    b. fuelwood.
    c. paper.
    d. medicines.
 * e. all of the above.

21. Many forestlands may be used for forest products and _____
    a. recreation.
    b. mining.
    c. grazing.
 * d. all of the above.
    e. none of the above.

22. The countries responsible for supplying over half of the world's commercial timber include all of the following *except*
    a. Canada.
 * b. Australia.
    c. the former Soviet Union.
    d. the United States.

23. Forests
    a. control erosion and reduce sedimentation.
    b. regulate surface runoff.
    c. aid recharging of aquifers.
    d. reduce severity of flooding.
 * e. do all of the above.

24. Which of the following statements is *false?*
    a. Forests affect humidity, which affects the climate of an area.
    * b. Forests play a minor role in biodiversity.
    c. Through photosynthesis, forests play an important role in the carbon cycle and global warming.
    d. Forests absorb some air pollutants.
    e. Forests reduce noise pollution.
    f. Forests provide an environment of solitude and beauty.

25. If the tropical forests are cut, the climate conditions become
    a. hotter and wetter.
    * b. hotter and drier.
    c. cooler and wetter.
    d. cooler and drier.
    e. stable.

26. Which of the following provides habitats for the largest number of wildlife species?
    a. deserts
    b. oceans
    c. tundra
    d. grasslands
    * e. forests

27. The timber value of a mature tree is approximately $_____, while its ecological value is approximately $_____.
    a. 600 . . . 60,000
    * b. 600 . . . 200,000
    c. 2,000 . . . 60,000
    d. 60,000 . . . 60,000
    e. 200,000 . . . 600

28. The goal of even-aged management of a forest is
    a. sustenance of maximum biological diversity.
    b. production of high-quality timber.
    c. a long-term ecologically oriented approach.
    * d. production of maximum return on a short-term basis.
    e. multiple use of a forest stand.

29. A tree grower using uneven-aged management most likely has the goal of
    a. sustaining biological diversity.
    b. maintaining long-term production of high-quality timber.
    c. producing reasonable economic return.
    d. promoting multiple use of a forest stand.
    * e. all of the above.

30. Selective cutting
    a. encourages crowding of trees.
    b. encourages growth of more mature trees.
    * c. maintains an uneven-aged stand of trees of different species, ages, and sizes.
    d. requires a special seed-distribution plan.
    e. increases fire hazard.

31. Which of the following types of cutting involves at least three cuttings?
    a. selective cutting
    b. seed-tree cutting
    c. clear-cutting
    * d. shelterwood cutting
    e. none of the above

32. Which of the following cutting methods leaves only a few mature seed-producing, wind-resistant trees for regeneration of the forest?
     a. selective cutting
   * b. seed-tree cutting
     c. clear-cutting
     d. shelterwood cutting
     e. even-aged cutting

33. The removal of all trees from a given area in a single cutting to establish a new even-aged stand is called
     a. selective cutting.
     b. seed-tree cutting.
   * c. clear-cutting.
     d. shelterwood cutting.
     e. uneven-aged cutting.

34. Which of the following statements is *false?*
     a. Clear-cutting increases the amount of timber produced per acre.
     b. Lumber companies prefer clear-cutting because it takes less skill and planning than other methods.
     c. Clear-cutting shortens rotation time.
   * d. Clear-cutting requires more roads.
     e. Clear-cutting permits reforesting with genetically improved stock.

35. Clear-cutting on a large scale leads to
     a. erosion.
     b. flooding.
     c. sediment water pollution.
     d. landslides.
   * e. all of the above.

36. Clear-cutting
     a. leaves ugly, unattractive and unnatural forest openings.
     b. reduces biodiversity.
     c. makes trees bordering the clear-cut area more vulnerable to being blown down by windstorms.
     d. increases the size of a harvest.
   * e. does all of the above.

37. Clear-cutting, if done properly, can be useful for some
     a. shade-tolerant species.
   * b. shade-intolerant species.
     c. species that prefer acid soils.
     d. species that prefer alkaline soils.
     e. species that require deep soils.

38. Which of the following statements is the best and cheapest way to protect trees from insects and disease?
   * a. Maintain the biological diversity of the forest.
     b. Use antibiotics on only the infected trees.
     c. Clear-cut and burn all infected areas.
     d. Develop disease-resistant tree species.
     e. Use integrated pest management.

39. Which of the following is the most dangerous type of fire?
    a. surface
    b. prescribed
  * c. crown
    d. ground
    e. none of the above; they are equally dangerous

40. Which of the following statements is *false?*
    a. Wildlife and mature trees are relatively unharmed by surface fires.
  * b. Frequent surface fires increase the chance of a severe crown fire.
    c. Surface fires release minerals that are locked up in vegetation.
    d. Surface fires may actually benefit some plant species and some game animals.
    e. Surface fires increase the activity of nitrogen-fixing bacteria.

41. Which of the following statements is *false?*
    a. Prescribed burning is a controlled surface fire.
    b. Prescribed burning is done at specific times to reduce the amount of air pollution.
  * c. Surface fires kill much wildlife, and such burning should be prevented.
    d. Some wildlife species depend on surface fires to maintain their habitats.
    e. Surface fires help control pathogens and insects.

42. Which of the following statements is *false?*
    a. Fire may help to release minerals locked up inside organisms.
  * b. Fire is universally a dangerous and destructive force that reduces productivity.
    c. Fire may occur periodically without the interference of humans.
    d. Surface fires may be prescribed to reduce the chance of serious fires.
    e. Litter buildup can increase the chance of more serious fires.

43. Which of the following animals do not require periodic fires to maintain their habitats?
    a. quail
    b. deer
    c. muskrats
  * d. squirrels
    e. moose

44. Which of the following statements is *false?*
    a. Conifers are more susceptible to air pollution than hardwoods.
    b. Pollution makes trees more vulnerable to disease and insects.
  * c. Forests at low elevations are exposed to more pollutants.
    d. The only solution to pollution is to reduce it drastically.
    e. Forests downwind from industrial regions are exposed to more pollutants.

45. Exposure of trees to multiple air pollutants makes them more susceptible to
    a. disease.
    b. drought.
    c. insects.
    d. mosses.
  * e. all of the above.

46. In the next few decades, the greatest threat to temperate and boreal forests is expected to be
    a. clear-cutting to create croplands.
    b. insects.
  * c. change in climate.
    d. air pollution.
    e. urbanization.

47. A biologically diverse ecosystem is the best protection against
    a. soil erosion and flooding.
    b. loss of biodiversity.
    c. sediment water pollution.
    d. tree loss from fire, wind, pests, and disease.
    * e. all of the above.

48. Sustainable-Earth forestry includes all of the following *except*
    * a. growth and harvesting of monocultures of high-quality timber.
    b. increased paper recycling.
    c. growth of timber on long rotations.
    d. selective cutting of individual trees or small groups of tree species.
    e. methods of road building that minimize erosion.

49. Sustainable-Earth forestry includes all of the following *except*
    a. use of lightweight equipment for logging.
    * b. clear-cutting only on steep slopes.
    c. leaving dead trees to enhance wildlife habitat and nutrient recycling.
    d. relying on natural controls of pests.
    e. none of the above; all are sustainable methods.

16-3 FOREST MANAGEMENT AND CONSERVATION IN THE UNITED STATES

50. Since 1940, the U.S. demand for wood has been met
    * a. by importing wood to meet increasing demand.
    b. by importing wood to meet decreasing demand.
    c. by depleting commercial forestlands.
    d. by controlling population growth.
    e. by decreasing per capita consumption.

51. U.S. national forestlands
    a. are used for mining operations.
    b. are used for grazing.
    c. receive more visitors than the national parks.
    d. are used for logging.
    * e. are characterized by all of the above.

52. Which of the following statements is *false?*
    a. The U.S. Forest Service has proposed building roads in the national forests that are equivalent to six times the length of the interstate highway system.
    b. The costs of building roads in the national forests was less than the profits available from the sale of timber from those areas.
    * c. All timber sales in national forests are based on fair market value of the timber.
    d. Conservationists propose that money from timber sales not be used to supplement the Forest Service budget.
    e. None of the above is false.

53. Which of the following statements is *false?*
    a. Cattle and sheep are allowed to graze in the national forests.
    b. More people visit the national forests than the national parks.
    c. Laws prohibit timbering in the national forests.
    * d. The management policies of the Forest Service are well accepted and noncontroversial.
    e. The Forest Service is required to manage national forests according to the principles of sustained yield and multiple use.

54. The endangered northern spotted owl
   a. has a high reproductive rate.
 * b. is a symbol of the struggle between environmentalists and timber company owners.
   c. has a high survival rate of juveniles.
   d. feeds at the middle trophic levels.
   e. has 10,000 pairs remaining in the wild.

55. A lumber company viewing a forest is *least* likely to see
   a. the economic value of giant living trees.
   b. jobs.
   c. improved economy.
 * d. recreational opportunity.
   e. profit.

56. An environmentalist viewing a forest is *least* likely to see
   a. ecological value.
   b. aesthetic value.
 * c. economic value.
   d. scientific value.
   e. recreational value.

57. Methods that could reduce destruction of the old-growth forests include all of the following *except*
   a. reducing the annual sale and harvest of timber from public lands.
   b. raising the price of timber sold from public lands.
   c. giving logging towns grants to encourage economic diversification.
   d. enforcing the ban on export of unprocessed logs from the Pacific Northwest.
 * e. taxing the conversion of mills to cut secondary-growth logs.

16-4 COMBATING TROPICAL DEFORESTATION AND THE FUELWOOD CRISIS

58. About _____ % of the earth's land area is tropical forest.
   a. 1
 * b. 6
   c. 12
   d. 16
   e. 25

59. About _____ of the world's tropical forests have already been cleared or damaged.
   a. 1/4
   b. 1/3
 * c. 1/2
   d. 2/3
   e. 3/4

60. Overall reforestation in the tropics averages about _____ trees cut for each tree planted, while reforestation in Africa averages about _____ trees cut for each tree planted.
   a. 100 . . . 200
 * b. 10 . . . 30
   c. three . . . 10
   d. one . . . 10
   e. 30 . . . 10

61. Tropical forests provide habitat to about _____ of Earth's species.
    a. one-fifth
    b. one-fourth
    c. one-third
  * d. one-half
    e. two-thirds

62. Tropical forests supply
    a. hardwood.
    b. coffee and spices.
    c. tropical fruits and nuts.
    d. latex rubber and dyes.
  * e. all of the above.

63. Over 50 years, harvesting of nonwood products from tropical forests would generate _____ revenue as clearing for timber.
    a. three times as much
  * b. twice as much
    c. the same
    d. one-half as much
    e. one-third as much

64. Key active ingredients in one-fourth of the world's drugs come from
  * a. tropical rain forests.
    b. coniferous forests.
    c. deciduous forests.
    d. cacti.
    e. lichens.

65. Drugs originating from tropical forests are used to treat
    a. lung cancer.
    b. heart disease.
    c. malaria.
    d. multiple sclerosis.
  * e. all of the above.

66. Which of the following important food crops did *not* have original strains predominantly from the tropical forests?
    a. rice
    b. corn
  * c. potatoes
    d. wheat

67. As a best approximation, less than _____ % of the flowering plant species in the tropical forests have been examined closely for their possible use as human resources.
  * a. 1
    b. 5
    c. 10
    d. 15
    e. 20

68. Which of the following statements is *false?*
    * a. Tropical biologists know more about how to live sustainably in tropical forests than indigenous tribespeople do.
    b. Tropical forests may be inhabited by indigenous tribespeople who live by hunting and gathering.
    c. Tribespeople may be driven from their homes by bulldozing, burning, and flooding.
    d. Modern cultures may expose tribespeople to disease.
    e. Tribespeople may suffer culture shock when exposed to modern cultures.

69. Tropical forests are being destroyed and degraded by
    a. economic growth and development.
    b. international policies.
    c. poverty.
    d. population growth.
    * e. all of the above.

70. Tropical forests are degraded by
    a. unsustainable small-scale farming.
    b. commercial logging.
    c. mining operations.
    d. damming rivers.
    * e. all of the above.

71. A quarter-pound hamburger made with beef originating from the tropics requires the clearing of approximately _____ of tropical forest.
    a. 1 square foot
    b. 1 square yard.
    * c. 5 square meters.
    d. 20 square meters.
    e. 100 square meters.

72. Shifting ranching
    a. is against government policy in LDCs.
    * b. is very destructive to nutrient-poor soils in the tropics.
    c. is a way to use tropical grasslands.
    d. is a form of nomadic herding.
    e. allows cattle to browse in areas that have been only partially cleared.

73. Currently, the greatest user of tropical lumber is
    a. Indonesia.
    * b. Japan.
    c. Canada.
    d. the United States.
    e. Great Britain.

74. The loss of biodiversity resulting from destruction of tropical forests is harmful to
    a. the current generation of humans.
    b. the next generation of humans.
    c. many future generations of humans.
    d. other species.
    * e. all of the above.

75. The first country to be involved in a debt-for-nature swap was
    a. Mexico.
    * b. Bolivia.
    c. Madagascar.
    d. Java.
    e. Brazil.

76. From the first debt-for-nature swap, we learned that
    a. the budget requirements must be established before the swap is made.
    b. the legislative requirements must be established before the swap is made.
    c. there must be some way to monitor the provisions.
    * d. all of the above are required.

77. Economic steps that can help reduce the destruction and degradation of tropical forests include all of the following *except*
    a. providing aid and debt relief to tropical countries who protect their tropical forests.
    * b. providing economic incentives to convert fuelwood plantations to croplands.
    c. phasing out funding for dams and other projects that are not sustainable.
    d. requiring an environmental impact evaluation for development proposals.
    e. boycotting products of companies involved in tropical forest destruction.

78. Political steps that can help reduce the destruction and degradation of tropical forests include all of the following *except*
    a. international banning on imports of products that destroy existing virgin tropical forests.
    b. banning the burning of tropical forests.
    c. including indigenous tribal people in planning and execution of tropical forestry plans.
    d. giving indigenous people title to tropical forest lands.
    * e. pressuring elected officials to approve the international General Agreement on Tariffs and Trade.

79. Steps that are biological in nature that can help reduce the destruction and degradation of tropical forests include all of the following *except*
    a. funding the Rapid Assessment Program.
    b. setting aside large areas of the world's tropical forests as reserves and parks.
    c. using debt-for-nature swaps and conservation easements.
    * d. discouraging family planning programs.
    e. rehabilitating degraded tropical forests.

80. Fuelwood scarcity is associated with
    a. adding burdens on the poor.
    b. deforestation.
    c. accelerated soil erosion.
    d. reduced cropland productivity.
    * e. all of the above.

81. The fuelwood crisis can be reduced by
    a. burning sun-dried roots of gourds and squashes.
    b. using more efficient stoves.
    c. using fast-growing fuelwood trees.
    d. encouraging local fuelwood planting projects.
    * e. all of the above.

82. Leucaenas are
    a. small mammals.
    * b. fast-growing fuelwood trees.
    c. small farms in tropical rain forests.
    d. an endangered species.
    e. parasitic wasps.

83. The most significant awareness arising from India's Chipko movement is
    a. trees prevent erosion.
    b. soil erosion silts up streams.
  * c. individual actions can bring about change to restore Earth.
    d. technology can help create sustainable societies.
    e. laws protect the land.

16-5 MANAGING AND SUSTAINING RANGELANDS

84. Almost _____ of Earth's ice-free land is rangeland.
    a. one-half
  * b. one-third
    c. one-fourth
    d. one-fifth
    e. one-sixth

85. About 40% of the world's rangeland is used for grazing because the rest is
    a. too dry.
    b. too remote from population centers.
    c. too cold.
  * d. all of the above.
    e. none of the above.

86. Which of the following statements about range grass is *false*?
  * a. When the leaf tip is eaten the leaf stops growing.
    b. Growth of grass is from its base, not its tip.
    c. The lower half of a plant is necessary for a grass to be able to grow back.
    d. Multiple branches of the roots make it hard for grasses to be uprooted.
    e. Range grass has deep, complex root systems.

87. The maximum number of ruminants a rangeland will support is called the
    a. optimum population.
    b. range index.
    c. herbivore count.
  * d. carrying capacity.
    e. stocking number.

88. Carrying capacity of grasslands is influenced by
    a. how long animals graze an area.
    b. kinds of grazing animals.
    c. soil type.
    d. past grazing use.
    e. annual climatic conditions.
  * f. all of the above.

89. Most overgrazing is caused by
    a. drought.
    b. climate changes.
    c. large populations of wild herbivores.
  * d. too many grazing animals for too long a time.
    e. periodic fires.

90. Which of the following plants would be an indicator of overgrazing?
  * a. mesquite
    b. palo verde
    c. short grass
    d. tall grass
    e. all of the above

91.   Wild herbivores _____ than cattle.
  *   a. have a more diversified diet
       b. use vegetation less efficiently
       c. are more subject to disease
       d. require more water
       e. are more vulnerable to predation

92.   The major goal of range management is
       a. soil conservation.
  *   b. maximizing livestock productivity without overgrazing.
       c. recreation.
       d. mining of fossil fuel.
       e. crop production.

93.   The term *stocking rate* refers to
       a. the amount of food supplements.
  *   b. the maximum population the range will support.
       c. the number of a particular kind of animal grazing on a unit of land.
       d. the number of wild animals in a given area.
       e. the number of fish put in rangeland ponds.

94.   Controlling the distribution of grazing animals
       a. is the best way to prevent overgrazing or undergrazing.
       b. is accomplished by the distribution of water holes and salt blocks.
       c. may be achieved by use of fencing combined with supplemental feed.
       d. may be achieved by rotating livestock from one grazing area to another.
  *   e. has all of the above characteristics.

95.   The cheapest and most effective way to remove unwanted vegetation is
       a. herbicides.
       b. controlled burning.
  *   c. trampling.
       d. mechanical removal.
       e. none of the above.

96.   Ranchers with range permits benefit from
       a. low grazing fees.
       b. fencing.
       c. predator control.
       d. weed control.
  *   e. all of the above.

97.   Sustainable management of rangelands would include
       a. elimination of riparian areas.
       b. increasing livestock grazing in underutilized areas.
       c. increasing predator control measures.
       d. reducing grazing fees.
  *   e. increasing funds for restoration of degraded rangelands.
       f. all of the above.

98.   Sustainable management of rangelands would include
       a. creating grazing advisory boards.
  *   b. incentives to ranchers who take steps to improve range conditions on public lands.
       c. giving large ranchers grazing fee discounts.
       d. creating a noncompetitive grazing permit system.
       e. encouraging hunting of deer and antelope.

99.   Currently, the greatest danger to the national parks is
 *   a.   popularity.
      b.   uncontrolled populations of prey because of the decline in predators.
      c.   pollution.
      d.   lack of funding.
      e.   none of the above.

100. The National Park Service is part of the
 *   a.   Interior Department.
      b.   Department of Agriculture.
      c.   Department of Conservation.
      d.   Environmental Protection Agency.
      e.   national resource lands.

101. In MDCs, many national parks are threatened by
      a.   air and water pollution.
      b.   invasion of alien species.
      c.   roads and noise.
      d.   industrial development and urban growth.
 *   e.   all of the above.

102. The single biggest problem for U.S. national and state parks today is
 *   a.   increased number of park visitors.
      b.   air and water pollution.
      c.   invasion of alien species.
      d.   rock collectors.
      e.   climate change.

103. Wolves
      a.   culled herds of herbivores.
      b.   protected vegetation.
      c.   protected species dependent upon vegetation.
      d.   strengthened the gene pool of their prey.
 *   e.   did all of the above.

104. Large predators
      a.   have been excessively hunted and poisoned.
      b.   have sharply declining populations.
      c.   have populations that often cannot be sustained by the small size of some parks.
      d.   are no longer keeping prey species in check.
 *   e.   have all of the above characteristics.

105. Alien species which have caused problems in parks include all of the following *except*
      a.   wild boars.
      b.   mountain goats.
      c.   pepper trees.
 *   d.   coyotes.

106. Human activities which threaten nearby parks include
      a.   logging.
      b.   grazing.
      c.   coal-burning power plants.
      d.   mining.
 *   e.   all of the above.

107. Which of the following statements is *false?*
    a. Costa Rica was once almost completely covered with tropical forests.
    b. Soil erosion was rampant in Costa Rica.
  * c. Indigenous rural families currently control land use in Costa Rica.
    d. Costa Rica has done more than any Latin American country to protect its remaining forests.
    e. Costa Rica is a superpower of biological diversity.

108. Costa Rica's long-term plan is to
    a. emphasize economic development and deemphasize conservation.
    b. emphasize conservation and deemphasize economic development.
    c. continue the current policy of land holding by the elite.
  * d. combine sustainable economic development with conservation.
    e. encourage population growth to develop more problem solvers.

109. In the Guanacaste National Park in Costa Rica, efforts are being made to restore
    a. native grasslands.
    b. tropical rain forests.
  * c. tropical deciduous forests.
    d. coastal wetlands.
    e. inland wetlands.

110. David Janzen's effort to restore the degraded ecosystem of Costa Rica
    a. is doomed to failure because of short-term economic interest.
    b. has been blocked by an interfering government.
  * c. has been labeled a biocultural restoration because of the involvement of 40,000 local people.
    d. is of limited interest to research scientists.
    e. includes all of the above.

111. The Guanacaste National Park of Costa Rica is considered primarily
    a. a tourist destination.
    b. an example of park mismanagement.
  * c. an ecological classroom and training facility.
    d. a pilot program in multiple use of resources.
    e. a source of money for multinational companies.

112. The Wilderness Society and the National Parks and Conservation Association suggested all of the following proposals *except*
    a. establishing the National Park Service as an independent agency.
    b. significantly increasing the pay and number of park rangers.
    c. blocking mining and timbering at park boundaries.
  * d. locating most commercial park facilities only along the main road through the park.
    e. educating the public about the concerns of the parks.

16-7 PROTECTING AND MANAGING WILDERNESS

113. The Wilderness Society estimates that a wilderness area should contain at least
    _____ acres.
    a. 10
    b. 100
    c. 1,000
  * d. 1 million
    e. 1 billion

114. Wilderness areas are needed for
    a. aesthetic reasons.
    b. getting away from noise and stress.
    c. recreation.
    d. preserving biological diversity.
  * e. all of the above.

115. Wilderness areas
    a. maintain biological diversity.
    b. provide undisturbed habitats.
    c. provide natural laboratories to enable scientists to discover how organisms live.
    d. preserve wild species' quality of life.
  * e. do all of the above.

116. Environmentalists say that in order to protect biodiversity, _____ % of the globe's land area must be protected.
    a. 1
    b. 5
  * c. 10
    d. 15
    e. 20

117. Which of the following biomes have the greatest amount of wilderness areas?
    a. tropical rain forests
  * b. deserts and tundra
    c. temperate deciduous forests
    d. chaparral and grassland
    e. coniferous forests

118. In 1981, UNESCO proposed that at least one biosphere reserve be set up in each of Earth's _____ geographical zones.
    a. five
    b. 10
    c. 100
    d. 157
  * e. 193

119. The three zones of a biosphere reserve include all of the following *except* the
    a. core area.
  * b. twilight zone.
    c. transition zone.
    d. buffer zone.

120. Conservationist Norman Myers proposes that all key parks in the Northern Hemisphere be shifted
    a. northward and be separated to prevent spread of disease.
    b. southward and be separated to prevent spread of disease.
  * c. northward and be connected by corridors to allow migration.
    d. southward and be connected by corridors to allow migration.
    e. eastward and turned into habitat islands.

121. _____ % of U.S. land is protected as wilderness with most of it in _____.
    a. Four . . . the West.
    b. Four . . . the East.
  * c. Four . . . Alaska.
    d. Ten . . . the West.
    e. Ten . . . the East.
    f. Ten . . . Alaska.

122. Good wilderness management would do all of the following *except*
    a. extensively patrol the accessible, popular areas.
    b. offer permits to selected areas for those who have demonstrated wilderness skills.
    c. leave some areas undisturbed by humans.
 * d. allow citizens to camp anywhere at their own risk.
    e. none of the above; all would be allowed.

123. Activities allowed in the national Wild and Scenic Rivers System include all of the following *except*
    a. camping.
    b. canoeing.
 * c. motorboating.
    d. fishing.
    e. sport hunting.

124. Individuals can help sustain Earth's land ecosystems by all of the following *except*
    a. planting trees.
    b. reducing consumption of paper products.
    c. using recycled products.
 * d. increasing consumption of beef.
    e. rehabilitating a degraded area.

125. Under the National Wild and Scenic Rivers Act, protection can be offered to rivers and river segments with
    a. cultural and historical value.
    b. geological value.
    c. wildlife and scenic value.
    d. recreational value.
 * e. all of the above.

126. An activity which would not be allowed on a river protected by the National Wild and Scenic Rivers Act would be
    a. swimming.
    b. canoeing.
    c. commercial fishing.
 * d. waterskiing.
    e. camping.

127. The National Trails Act was passed in
    a. 1938.
    b. 1958.
 * c. 1968.
    d. 1978.
    e. 1988.

CHAPTER 17
BIODIVERSITY: SUSTAINING WILD SPECIES

1. Alexander Wilson, an ornithologist in the early 1800s, described a flock of _____ that he estimated was 240 miles long and 1 mile wide and consisted of over 2 million birds.
    a. Carolina parakeets
    b. starlings
    c. sparrows
    * d. passenger pigeons
    e. California condors

2. The passenger pigeon became extinct in
    a. 1874.
    b. 1894.
    * c. 1914.
    d. 1934.
    e. 1952.

3. Passenger pigeons became extinct because of
    a. uncontrolled commercial hunting.
    b. loss of food supplies as forests were cleared for farms.
    c. loss of habitat as forests were cleared for cities.
    * d. all of the above.
    e. none of the above.

4. Passenger pigeons
    a. were used as fertilizer.
    b. were good to eat.
    c. had feathers used in pillows.
    d. were suffocated by burning grass or sulfur below their roosts.
    * e. had all of the above characteristics.

5. The case of the passenger pigeon best illustrates
    a. the strength of natural selection.
    b. the theory of evolution.
    c. the concept of coevolution.
    * d. the role humans play in premature extinction of species.
    e. none of the above.

17-1   WHY PRESERVE WILD SPECIES?

6. Wild plant species have economic value as
    a. the basis for new crop strains.
    b. fiber.
    c. dyes.
    d. oils.
    * e. all of the above.

7. Some domestic species depend upon wild species for
    a. genes.
    b. pollination.
    c. nitrogen fixation.
    * d. all of the above.
    e. none of the above.

8.   Less than _____ % of Earth's plant species have been tested for medical uses.
 *   a.   1
      b.   5
      c.   10
      d.   20
      e.   25

9.   About _____ % of the world's population relies on plants or plant extracts for medicines.
      a.   25
      b.   50
 *   c.   75
      d.   85
      e.   95

10.   Alternatives to using living organisms for research include all of the following *except*
      a.   use of cell and tissue cultures.
      b.   use of computer models.
      c.   use of bacterial cultures.
 *   d.   development of new drugs and surgical techniques.
      e.   use of videotapes.

11.   People regard wildflowers as beautiful, and this is evidence that wildflowers have _____ importance.
      a.   economic
 *   b.   aesthetic
      c.   medical
      d.   ecological
      e.   ethical

12.   Ecotourism is a human activity that emphasizes the _____ importance of wild species.
 *   a.   recreational
      b.   scientific
      c.   medical
      d.   ecological
      e.   ethical

13.   A single male lion living to age 7 in Kenya is worth $515,000 as a
      a.   sporting trophy.
 *   b.   tourist attraction.
      c.   functioning component of the ecosystem.
      d.   specimen to be sold alive to a zoo.
      e.   game species.

14.   Ecological functions of wild species include all of the following *except*
 *   a.   ecotourism.
      b.   climate regulation.
      c.   detoxification.
      d.   water supply regulation.
      e.   nutrient recycling.

15. Which of the following statements is *false?*
    a. The current extinction spasm is caused by humans.
    b. The current rate of extinction is in terms of decades and much faster than ever before.
  * c. Extinction is being balanced by the production of new species.
    d. It is difficult to document loss of species.
    e. If habitat destruction continues at the current rate over decades, this period will rival the great natural extinctions of the past.

16. The species extinction rate is apparently
  * a. increasing rapidly.
    b. always equal to the species formation rate.
    c. about one species per year.
    d. lower now than it has ever been.
    e. constant.

17. An endangered species is any species that can
    a. undergo alteration of its genetic traits.
    b. become rare within the next century.
  * c. soon become extinct in all or part of its range.
    d. eventually become threatened or rare.
    e. alter its own habitat.

18. Endangered species include all of the following *except* the
    a. snow leopard.
  * b. bald eagle.
    c. black rhinoceros.
    d. rare swallowtail butterfly.
    e. giant panda.

19. Of the following, the endangered species is (are) the
    a. American alligator.
    b. grizzly bear.
  * c. giant panda.
    d. bald eagle.
    e. all of the above.

20. The California condor is
  * a. an endangered species.
    b. a threatened species.
    c. an imported species.
    d. an overpopulated species.
    e. a pest.

21. The bald eagle is
    a. an endangered species.
  * b. a threatened species.
    c. an imported species.
    d. an overpopulated species.
    e. a pest.

22. The blue whale is extinction-prone for all of the following reasons *except*
    a. low reproduction rate.
  * b. feeding at the top trophic level.
    c. specialized feeding habits.
    d. fixed migratory patterns.
    e. large size.

23. The timber wolf is extinction prone because of its
    a. specialized feeding habits.
    b. large size.
    c. specialized breeding areas.
    d. fixed migratory patterns.
 * e. preying on livestock.

24. The whooping crane is vulnerable to extinction because of
    a. its small size.
    b. its call.
    c. its low reproduction rates.
    d. its fixed migration pattern.
 * e. c and d.
    f. all of the above.

25. Bald eagles feed primarily on
    a. small rodents.
    b. birds.
 * c. fish.
    d. small mammals other than rodents.
    e. insects.

26. The bald eagle population has declined because of
    a. loss of habitat.
    b. illegal hunting.
    c. reproductive failure caused by pesticides in the fish they eat.
 * d. all of the above.

27. You can thank a bat if you use
    a. surgical bandages.
    b. dates and figs.
    c. life preservers.
    d. rope.
 * e. all of the above.

28. If you eat cashews, avocados, or bananas, you can thank a
 * a. bat.
    b. manatee.
    c. bee.
    d. hummingbird.
    e. butterfly.

29. The primary reason for our extinction crisis is that humans are using about _____% of Earth's terrestrial net primary productivity.
    a. 5
    b. 10
    c. 20
 * d. 40
    e. 60

30. The greatest threat to most species is
 * a. destruction of habitats.
    b. water pollution.
    c. parasites.
    d. buildup of toxic substances in food chains.
    e. sport hunting.

31. Habitat fragmentation threatens species by destroying
    a. breeding grounds.
    b. food sources.
    c. migration routes.
    * d. all of the above.
    e. none of the above.

32. The single greatest cause of extinction and decline in global diversity is habitat destruction of
    a. coral reefs.
    b. grasslands.
    * c. forests.
    d. deserts.
    e. lakes.

33. Since Europeans first settled North America, _____% of the tall-grass prairies in the United States have been destroyed.
    a. 10
    b. 25
    c. 50
    d. 75
    * e. 98

34. You are studying species diversity in the Caribbean islands. Which island would you expect to have the highest number of species?
    * a. a large island near the mainland
    b. a large island far removed from other sites
    c. a medium-sized island is the middle of an island chain
    d. a small island far removed from other sites
    e. a small island near the mainland

35. National parks can be viewed as habitat islands surrounded by
    a. logging.
    b. industrial activity.
    c. energy extraction.
    d. mining.
    * e. all of the above.

36. Songbird populations in the United States are declining from
    a. hunting.
    b. commercial sales of feathers.
    c. logging of tropical forests.
    d. fragmentation of North American forests.
    e. a and b.
    * f. c and d.

37. Forest fragmentation creates "edge" habitat for songbird predators such as
    a. opossums.
    b. raccoons.
    c. skunks.
    d. squirrels.
    * e. all of the above.

38. Which of the following is *false?* Habitat islands
    * a. are wildlife preserves set aside to sustain endangered species.
      b. are fragments of wildlife habitats.
      c. are often too small to support the minimum number of individuals required to sustain a population.
      d. make up most national parks and protected areas.
      e. include none of the above; all statements are true.

39. The North American Indians depended upon bison
      a. meat for their diet.
      b. skin for their tepees and clothing.
      c. gut for their bowstrings.
      d. feces for fuel.
    * e. for all of the above.

40. Bison were least likely to be killed
      a. by hunters to provide food for railroad builders.
      b. by farmers to protect their crops.
    * c. by miners because bison disrupted their deposits.
      d. by ranchers who didn't want competition with their cattle.
      e. by the military as a form of biological warfare.

41. Illegal hunting for profit is called
      a. subsistence hunting.
      b. sport hunting.
    * c. poaching.
      d. commercial hunting.
      e. none of the above.

42. Legal commercial hunting and poaching have been major factors in population declines of
      a. mountain gorillas.
      b. Bengal tigers.
      c. ocelots.
    * d. rhinos.
      e. all of the above.

43. Rhinos are hunted for their ____, which are thought to be an aphrodisiac.
      a. livers
    * b. horns
      c. skins
      d. gallbladders
      e. feet

44. The Carolina parakeet was exterminated because it
    * a. fed on fruit crops.
      b. was an alien species.
      c. carried a viral disease that killed domesticated parakeets.
      d. spread histoplasmosis, a fungal disease.
      e. built nests in houses and was therefore considered a pest.

45. Prairie dogs have almost been driven to extinction by
      a. imported black-footed ferrets.
    * b. poison.
      c. coyotes.
      d. shotguns and rifles.
      e. cattle.

46.   Which of the following is almost extinct because its prime source of food has been killed by poison?
        a. wolverines
    *   b. black-footed ferrets
        c. manatees
        d. alligators
        e. muskrats

47.   Even with the best management, wildlife reserves may be depleted in a few decades because of
        a. depletion of fossil fuels.
    *   b. climatic change brought about by projected global warming.
        c. demands for sport trophies.
        d. biological magnification of pesticides.
        e. water pollution.

48.   Which of the following birds is *least* vulnerable to population declines from DDT biological amplification?
        a. falcon
        b. eagle
    *   c. robin
        d. hawk
        e. owl

49.   Which of the following has the greatest potential danger for wildlife?
        a. DDT
        b. poaching
        c. legal hunting
    *   d. climate change
        e. deforestation

50.   Some exotic plant species that may bring $5,000 to $15,000 to collectors are most likely
        a. mushrooms.
    *   b. orchids or cacti.
        c. bromeliads or ferns.
        d. bonsai or dwarf trees.
        e. palm trees.

51.   Alien species may
        a. have no natural predators.
        b. cause population reductions in native species.
        c. have no natural competitors.
        d. dominate their new ecosystem.
    *   e. have all of the above characteristics.

52.   A water hyacinth can double its population in
        a. 2 days.
    *   b. 2 weeks.
        c. 6 months.
        d. 1 year.
        e. 10 years.

53.   Which of the following is *false?* The water hyacinth
        a. is native to Central and South America.
        b. entered Florida via transplanting from an exhibit.
        c. has rapidly displaced native species.
    *   d. is successfully controlled by mechanical harvesting and herbicides.
        e. was unchecked by natural enemies.

54. Water hyacinths have been most effectively controlled by
    * a. manatees.
    b. mechanical harvesting.
    c. plastic or canvas covers that deprive them of light.
    d. herbicides.
    e. none of the above.

55. The manatee population has become threatened from all of the following *except*
    a. being slashed by powerboat propellers.
    b. becoming tangled in fishing gear.
    c. being hit on the head by oars.
    * d. eating too much water hyacinth.
    e. none of the above; all are reasons for the threatened status of manatees.

56. Agents which were introduced to control water hyacinths include all of the following *except*
    a. grass carps.
    b. water snails.
    c. weevils.
    * d. manatees.

57. Water hyacinths can
    a. absorb toxic chemicals in sewage treatment lagoons.
    b. be fermented to form biogas fuel.
    c. be used as a mineral and protein supplement in cattle feed.
    d. be applied to the soil as fertilizer.
    * e. do all of the above.

58. All of the following are organisms imported into the United States that cause damage *except* the
    * a. prairie dog.
    b. sea lamprey.
    c. Japanese beetle.
    d. house sparrow.
    e. chestnut blight.

17-3 SOLUTIONS: PROTECTING WILD SPECIES FROM EXTINCTION

59. The goals of the World Conservation Strategy include
    a. preservation of species diversity and genetic diversity.
    b. assurance that any use of species is sustainable.
    c. inclusion of women and indigenous people in the development of conservation plans.
    d. encouragement of rehabilitation of degraded ecosystems.
    * e. all of the above.

60. Which of the following is *not* one of the three major approaches to protecting wildlife?
    a. the species approach
    b. the ecosystem approach
    c. the wildlife management approach
    * d. the kinder, gentler human approach

61. Preserving balanced populations of species in their native habitats is one example of
    a. a species approach.
    * b. an ecosystem approach.
    c. a human approach.
    d. a wildlife approach.
    e. a reductionistic approach.

62. CITES is
    * a. a treaty controlling the international trade in endangered species.
      b. a set of regulations controlling the introduction of exotic species.
      c. a pact that supports critical ecosystems that support wildlife.
      d. an international organization dedicated to the preservation of endangered species.
      e. an international cartel that commercially exploits endangered species.

63. The Endangered Species Act of 1973
      a. is one of the world's toughest environmental laws.
      b. makes it illegal to trade any product made from an endangered species.
      c. allows the use of endangered species for approved scientific purposes or if the use enhances the survival of the species.
      d. authorizes identification of endangered species solely on a biological basis.
    * e. includes all of the above.

64. Endangered species are identified and listed by
      a. Fish and Wildlife Service.
      b. Environmental Protection Agency.
      c. National Marine Fisheries Service.
      d. Department of the Interior.
    * e. a and c.

65. There are approximately ____ listed endangered species in the United States.
      a. 100
      b. 200
    * c. 750
      d. 1,200
      e. 2,800

66. There are recovery plans in action for about _____% of the endangered species.
      a. 5
      b. 15
    * c. 30
      d. 45
      e. 60

67. Wildlife protection could be made stronger by
      a. giving federal officials deadlines for implementing recovery plans for endangered species.
      b. requiring development of conservation plans for whole ecosystems.
      c. allowing citizens to file lawsuits if an endangered species faces serious harm or extinction.
      d. increasing funding for endangered species programs.
    * e. all of the above.

68. The group of animals most protected by wildlife refuges is
      a. small mammals.
    * b. migratory waterfowl.
      c. large mammals.
      d. songbirds.
      e. amphibians and reptiles.

69. The discovery of oil and gas reserves in the Arctic National Wildlife Refuge would
      a. enhance national security.
      b. increase oil company profits.
      c. reduce our dependence on foreign oil.
      d. provide a better balance of payments and reduce trade deficit.
    * e. all of the above.

70. The *Exxon* oil spill occurred at
    a. Fairbanks.
    b. Prudhoe Bay.
    c. Anchorage.
    d. Seattle.
  * e. Valdez.

71. Prudhoe Bay is a
    a. gem of biodiversity that shelters several habitat islands.
  * b. site of oil drilling in Alaska.
    c. wildlife refuge in Alaska.
    d. site of a disastrous oil spill in the 1970s.
    e. critical point on the migration path of caribou, polar bears, and the Arctic tern.

72. Which of the following statements is *true?*
    a. The first wildlife refuges were created to protect plant species.
  * b. Guidelines for the management of public lands for wildlife have not yet been established.
    c. Hunting, fishing, and mining are banned in refuges.
    d. The Wildlife Federation, the Nature Conservancy, and similar organizations have control over the gas, oil, coal, and mineral resources on the majority of the wildlife refuges.
    e. All of the above are true.

73. Seed gene banks are
  * a. refrigerated environments with low humidity.
    b. refrigerated environments with high humidity.
    c. warm environments with low humidity.
    d. warm environments with high humidity.
    e. places where products of the human genome project are stored.

74. Gene banks have the following drawback(s):
    a. Lack of funding does not provide enough capacity to store all endangered species.
    b. Some species cannot be stored.
    c. They have too little storage capacity.
    d. Seeds in storage do not evolve and therefore become less fit for reintroduction into a changed environment.
  * e. All of the above are drawbacks.

75. Egg pulling refers to
    a. techniques used to extend the breeding span of captured birds.
  * b. collecting eggs from the wild.
    c. using fertility drugs to increase productivity.
    d. production of hybrids in captive breeding programs.
    e. removal of eggs from mammals for fertilization and artificial insemination to save endangered species.

76. Captive breeding programs of American zoos helped save the nearly extinct
  * a. Arabian oryx.
    b. buffalo.
    c. heath hen.
    d. prairie dog.
    e. African wildebeest.

77. Captive breeding programs in zoos
    a. eliminate the need to preserve critical habitats.
    b. can be used for most species except mammals.
    c. increase the genetic variability of species.
    d. are relatively inexpensive to operate.
    * e. require the captive population to number between 100 and 500.

78. The world's zoos contain _____ species with populations of 100 or more individuals.
    * a. 20
    b. 40
    c. 60
    d. 80
    e. 100

79. Since it will not be possible to save all of the candidates for preservation programs, experts use all but which of the following criteria in making their choices?
    a. those that have the best chance for survival
    * b. those that are the most visible or popular to help save the program
    c. those that have the most ecological value to an ecosystem
    d. those that possess the greatest potential to benefit human beings
    e. none of the above; all these criteria are used

17-4 WILDLIFE MANAGEMENT

80. Examples of the species wildlife management approach include all of the following *except*
    a. identifying and giving species legal protection.
    b. preserving and managing habitat critical to an endangered species.
    c. propagating species in captivity.
    * d. establishing harvest quotas.
    e. reintroducing species into suitable habitats.

81. Those using a wildlife management approach would
    a. manage for sustained yield.
    b. enforce hunting regulations.
    c. establish harvest quotas.
    d. use international treaties to protect migrating game species.
    * e. do all of the above.

82. Most wildlife management in the United States is keyed to
    a. protection of endangered species.
    b. control of population sizes and habitats to maintain diversity.
    * c. management of population sizes and habitats to favor game species.
    d. maximizing of bird diversity.
    e. an equal balance of all of the above.

83. Ideally, wildlife management plans are based on principles of
    a. species habitat requirements.
    b. wildlife population dynamics.
    c. ecological succession.
    d. characteristics of the hunting population.
    * e. all of the above.

84. Stages of ecological succession include
    a. early-successional.
    b. mid-successional.
    c. late-successional.
    d. wilderness.
    * e. all of the above.

85. Weedy pioneer species are found primarily in
    * a. early-successional stages.
    b. mid-successional stages.
    c. late-successional stages.
    d. wilderness.
    e. none of the above.

86. Wilderness species would be expected to flourish best in
    a. early-successional stages.
    b. mid-successional stages.
    c. late-successional stages.
    * d. mature, relatively undisturbed habitats.
    e. none of the above.

87. Moderate-size, old-growth forest refuges are required by
    a. early-successional species.
    b. mid-successional species.
    * c. late-successional species.
    d. wilderness.
    e. none of the above.

88. Partially open areas with plenty of edge habitat are required by
    a. early-successional species.
    * b. mid-successional species.
    c. late-successional species.
    d. wilderness.
    e. none of the above.

89. Sport-hunting laws used to manage populations of game animals include
    a. hunting license requirements.
    b. animal hunting seasons.
    c. hunting equipment regulations.
    d. limitations on the size, sex, and numbers of animals that can be killed.
    * e. all of the above.

90. Flyways are
    a. insect infestation routes.
    * b. bird migration routes.
    c. routes connecting islands.
    d. areas that require insecticide treatment.
    e. none of the above.

91. Waterfowl management includes
    a. regulating hunting.
    b. protecting existing habitats.
    c. building artificial nesting sites and developing new habitats.
    * d. all of the above.
    e. none of the above.

92. Money from the sale of duck stamps is used primarily for
    a. hunter safety and education.
    b. wildlife research and surveys.
    * c. buying land and easements to benefit waterfowl.
    d. land acquisition.
    e. patrolling during hunting season.

93. Fishing can be controlled by
    a. setting fishing seasons.
    b. using hatcheries to restock ponds.
    c. regulating fishing equipment.
    d. establishing quotas.
  * e. all of the above

94. Fishing can be managed by all of the following accepted techniques *except*
    a. breeding genetically resistant fish varieties.
  * b. requiring use of small-mesh nets.
    c. using small dams to control water flow.
    d. protecting habitats from buildup of sediment, pollution, and debris.
    e. fertilizing nutrient-poor lakes.

95. Which of the following is an example of the tragedy of the commons?
  * a. exploitation of whales
    b. mismanagement of wildlife refuges
    c. use of gene banks
    d. use of animals for research purposes
    e. all of the above

96. By international law, exclusive economic zones over which individual countries have legal jurisdiction extend _____ nautical miles from the shore.
    a. 10
    b. 100
  * c. 200
    d. 500
    e. 700

97. Which of the following statements is *false*?
    a. Fishery commissions do not have legal authority to compel members to follow their rules.
    b. Dynamiting and poisoning fish are outlawed.
    c. Commissions may introduce species and build artificial reefs.
  * d. Zones within 200 nautical miles of land are called the "high seas."
    e. Foreign fishing vessels can take certain quotas of fish within exclusive economic zones only with government permission.

98. Fishery commissions can
    a. limit fishing seasons.
    b. regulate fishing gear.
    c. set annual quotas.
    d. set size of fish to keep.
  * e. do all of the above.

99. International laws extend coastal countries' control over fishing to _____ nautical miles off their coastline.
    a. 3
    b. 12
    c. 30
  * d. 200
    e. 500

100. How many major whale species have been fished to extinction?
     a. one of 11
     b. three of 11
     c. five of 11
   * d. eight of 11
     e. 10 of 11

101. The world's largest animal is the _____ whale.
   * a. blue
     b. humpback
     c. gray
     d. sperm
     e. killer

102. Blue whales feed on
     a. plankton.
   * b. krill.
     c. small fish.
     d. sharks.
     e. all of the above.

103. Blue whales are
     a. toothed cetaceans.
     b. baleen whales.
     c. extinct.
     d. endangered.
     e. a and c.
   * f. b and d.

104. Blue whales are vulnerable because
     a. they are so large.
     b. they accumulate in large numbers in their feeding zone.
     c. they swim close to shore.
     d. it takes longer for them to reach sexual maturity than humans, and they have only one offspring every two to five years.
   * e. of all of the above.

PART V. ENERGY RESOURCES

CHAPTER 18
SOLUTIONS: ENERGY EFFICIENCY AND RENEWABLE ENERGY

1. Rocky Mountain Institute in Colorado is
    a. superinsulated.
    b. partially earth-sheltered.
    c. passively heated.
    * d. all of the above.
    e. none of the above.

2. The energy-saving options already available include
    a. monitoring microprocessors.
    b. thinner insulation material.
    c. rolls of solar cells to attach to roofs.
    d. superinsulating windows.
    * e. a and d.
    f. b and c.

18-1   EVALUATING ENERGY RESOURCES

3. Our current dependence on fossil fuels is the number one contributor to
    a. air pollution.
    b. land disruption.
    c. water pollution.
    d. greenhouse gases.
    * e. all of the above.

4. Which is our *best* immediate energy option?
    a. Find and burn more forms of oil, natural gas, and coal.
    * b. Cut out unnecessary energy waste by improving energy efficiency.
    c. Build more and better conventional nuclear power plants.
    d. Increase efforts to develop breeder nuclear fission and nuclear fusion.
    e. Convert to windpower to produce electricity.

5. It takes at least ____ years to phase in new energy alternatives.
    a. five
    b. 10
    c. 20
    d. 30
    * e. 50

6. To make decisions about energy alternatives, we need to ask:
    a. How much will be available in the short term, intermediate term, and long term?
    b. How much will it cost to develop, phase in, and use?
    c. What are potential environmental and social impacts?
    d. What is the estimated net useful energy yield?
    * e. All of the above.

18-2   SOLUTIONS: IMPROVING ENERGY EFFICIENCY

7. To reduce energy waste, we must
    a. reduce energy consumption.
    b. improve energy efficiency.
    c. switch from fossil fuels.
    d. use only solar technologies.
    * e. do a and b.
    f. all of the above.

8. _____ % of all commercial energy used in the United States is wasted.
    a. Ninety-four
    *   b. Eighty-four
    c. Seventy-four
    d. Sixty-three
    e. Forty-three

9. _____ % of all commercial energy used in the United States is wasted unnecessarily.
    a. Ninety-four
    b. Eighty-four
    c. Seventy-four
    d. Sixty-three
    *   e. Forty-three

10. Energy consumption can be reduced by
    a. using mass transit instead of automobiles.
    b. walking for short trips.
    c. turning off unneeded lights.
    d. turning the thermostat down in winter.
    *   e. all of the above.

11. Consumers can reduce their energy demand by evaluating the _____ when they are buying electrical appliances.
    a. internal costs.
    b. external costs.
    c. market prices.
    *   d. life-cycle costs.
    e. fixed costs.

12. The space-heating strategy with the best net energy efficiency is
    a. electric from a nuclear power plant.
    b. electric from a coal-fired plant.
    c. electric from an oil-fired plant.
    *   d. passive solar.
    e. wood stove.

13. Reducing energy use and waste does all of the following *except*
    a. extend the length of time fossil fuels will last.
    b. ease international tensions.
    c. improve global military security.
    *   d. cause an economic recession.
    e. buy time to phase in renewable energy resources.

14. Factors that have prevented the United States from maximizing energy efficiency include all of the following *except*
    a. a temporary glut of low-cost fossil fuels.
    *   b. the United States has the strongest global economy, and the prevailing attitude is "if it ain't broke, don't fix it."
    c. sharp cutbacks in federal support for improvements in energy efficiency and development of renewable resources.
    d. failure of prices to include external costs of nonrenewable resources.
    e. political influence of companies controlling nonrenewable resources.

15. Energy consumption can be reduced by all of the following *except*
    a. using mass transit instead of individual automobiles.
    *   b. turning the thermostat up in wintertime.
    c. turning off unused lights.
    d. purchasing only needed products.
    e. riding a bicycle instead of using a car.

Solutions: Energy Efficiency and Renewable Energy

16. Energy efficiency could be improved by all of the following *except*
    a. buying cars with good fuel mileage.
    b. buying energy-efficient appliances.
    c. keeping car engines tuned.
    * d. removing insulation from attics.
    e. switching from incandescent to fluorescent light bulbs.

17. Improving energy efficiency does all of the following *except*
    a. make nonrenewable fossil fuel supplies last longer.
    b. provide a longer time for phasing in renewable energy sources.
    c. improve national security by reducing dependence on oil imports.
    d. save money.
    * e. eliminate excess jobs.

18. Using and wasting energy affects
    a. acid deposition.
    b. water pollution.
    c. projected global warming.
    d. air pollution.
    * e. all of the above.

19. The largest, cheapest untapped supplies of energy in the United States are
    * a. energy-wasting buildings, industries, and vehicles.
    b. oil fields.
    c. natural gas deposits.
    d. uranium deposits.
    e. geothermal sources.

20. Amory Lovins claims that with existing technology to improve energy efficiency,
    a. we could save $1 trillion annually.
    b. 1.3 million jobs would be created.
    c. the U.S economy could be run on about one-fourth as much energy as it now uses.
    * d. all of the above would be possible.
    e. none of the above would be possible.

21. Cogeneration units
    a. produce both heating and cooling.
    b. provide high-temperature heat and electricity.
    c. involve a battery, in case the main system fails.
    * d. recover some of the wasted energy from conventional boilers to produce heat and electricity.
    e. could produce more electricity than all the U.S. nuclear power plants by 2000.

22. Industry can reduce its energy consumption by
    a. switching to high-efficiency lighting.
    b. recovering and reusing waste heat.
    c. increasing recycling and reuse of materials.
    d. using more high-efficiency motors.
    * e. doing all of the above.

23. Between 1981 and 1991, federal support for research and development to improve industrial efficiency was
    a. increased 60%.
    b. increased 20%.
    c. stabilized.
    d. decreased 20%.
    * e. decreased 60%.

24. Which of the following statements is *false?*
   a. Transportation in the United States consumes 65% of all oil used, up from 50% in 1973.
   * b. The United States has encouraged research in developing smaller, more efficient cars through high gasoline taxes.
   c. Within 10 years, U.S. cars could average 35 mpg at an increased cost of $500 per car.
   d. Some foreign prototype cars are now able to get from 67 to 138 mpg.
   e. Air conditioners in cars and light trucks are responsible for 75% of the country's annual ozone-destroying CFC emissions.

25. According to the U.S. Office of Technology Assessment, by 2010 continued improvements in mileage ratings in new cars and light trucks could
   a. eliminate oil imports.
   b. allow buyers of fuel-efficient cars to pay off the extra cost in one year and then enjoy the extra savings.
   c. save more than $50 billion yearly in fuel costs.
   * d. do all of the above.
   e. do none of the above.

26. Improving the fuel efficiency of automobiles and light trucks could
   a. eliminate the need to import any oil.
   b. save consumers more than $50 billion a year in fuel costs.
   c. have additional costs per car paid off by fuel savings in about a year.
   d. raise the fuel efficiency of the whole U.S. fleet to 35 mpg by 2010.
   * e. do all of the above.

27. Since 1985, at least 10 automobile companies have made fuel-efficient cars that
   a. carry four or five passengers.
   b. are peppy.
   c. meet or exceed 1990 safety standards.
   d. meet or exceed 1990 pollution control standards.
   * e. all of the above.

28. Vehicle safety is most dependent upon
   a. size.
   b. weight.
   * c. design.
   d. none of the above.

29. Of the following measures, the one that would *least* improve vehicle safety is
   * a building with new, stronger, heavier materials.
   b. air bags.
   c. collapsible steering columns.
   d. antilock brakes.

30. Electric cars
   a. are noisy.
   b. have been totally ignored by American car companies.
   c. would be especially helpful for long-distance trips.
   * d. require little maintenance.
   e. must be recharged twice a day.

31. Electric car performance could *best* be improved by
   a. building with stronger materials.
   b. improving the steering mechanism.
   c. modifying the transmission to improve efficiency.
   * d. producing a longer lasting battery with a higher charge density.
   e. finding a way that cloud cover could be controlled.

32. The textbook cites the _____ as a monument to energy waste.
    * a. World Trade Center in New York
    b. Sears Tower in Chicago
    c. Georgia Power Company building in Atlanta
    d. Transamerica Building in San Francisco
    e. Superdome in New Orleans

33. The textbook cites the _____ as a highly energy-efficient structure.
    a. World Trade Center in New York
    b. Sears Tower in Chicago
    * c. Georgia Power Company building in Atlanta
    d. Transamerica Building in San Francisco
    e. Superdome in New Orleans

34. About _____ of the heat loss from U.S. buildings is through closed windows.
    a. one-fifth
    b. one-fourth
    * c. one-third
    d. one-half
    e. two-thirds

35. Superinsulation R-8 to R-10 windows will pay for themselves in heat savings in about _____ year(s).
    a. one-half a
    b. one–two
    * c. two–four
    d. six–eight
    e. 10–12

36. All of the following are already commercially available *except*
    * a. microprocessors programmed to do different tasks to heat or cool a home.
    b. superinsulated windows that have insulation values similar to walls so that a homeowner could have as many windows as he or she wants.
    c. small-scale cogeneration units.
    d. socket-type fluorescent light bulbs using one-fourth as much electricity and lasting about 10 times longer than incandescent bulbs.
    e. appliances that work on direct current from solar cells.

37. An average house in Sweden consumes _____ as much energy as a U.S. house of the same size.
    a. one-half
    * b. one-third
    c. one-fourth
    d. one-fifth
    e. one-sixth

38. Lighting requires about _____ of the electricity generated in the United States and other industrial countries.
    a. one-half
    b. one-third
    * c. one-fourth
    d. one-fifth
    e. one-sixth

39. Current incandescent bulbs are ____% efficient.
    * a. 5
      b. 10
      c. 20
      d. 33
      e. 50

40. Which of the following is *not* one of the four recommended guidelines for an individual to save energy?
      a. Insulate, caulk, and weatherstrip your home.
      b. Use energy-efficient windows.
    * c. Use only electricity for heating space and water.
      d. Get as much heating and cooling from natural sources as possible.
      e. Evaluate the cost of consumer goods on the basis of lifetime, not initial, costs.

18-3 DIRECT USE OF SOLAR ENERGY FOR HEAT AND ELECTRICITY

41. Perpetual and renewable energy resources include all of the following *except*
      a. the sun.
      b. the wind.
      c. biomass.
      d. Earth's internal heat.
    * e. natural gas.

42. Development of perpetual and renewable energy resources would
      a. save money and create jobs.
      b. eliminate the need for oil imports.
      c. produce less pollution per unit of energy.
      d. increase national security.
    * e. do all of the above.

43. Windows designed to capture solar energy face
      a. north.
      b. east.
    * c. south.
      d. west.

44. The Enertia solar envelope house is heated and cooled
      a. actively by solar energy and Earth's thermal energy.
      b. actively by solar energy and tree plantings.
    * c. passively by solar energy and Earth's thermal energy.
      d. passively by solar energy with a natural gas backup system.
      e. by a solar power tower.

45. All of the following are features of passive solar design *except*
      a. adobe walls and flagstone floor used for heat storage.
    * b. coniferous trees blocking the sun all year.
      c. windows on the south side of the house.
      d. summer cooling vents in the roof.
      e. earth tubes bringing in cool air during the summer.

46. All of the following can be used for cooling a house in warm weather *except*
      a. cooling vents in the roof.
      b. earth tubes and tanks buried 20 feet underground.
      c. deciduous trees.
      d. window overhangs.
    * e. foil sheets under the floor.

47. The Copper Cricket is a roof-mounted _____ solar water heater that can provide _____ of a typical home's hot water needs.
    a. active . . . about one-half
    b. active . . . all or most
    c. passive . . . about one-half
  * d. passive . . . all or most
    e. None of the above; the Copper Cricket is a new, heavily insulated refrigerator.

48. Which of the following is *true* of passive solar systems compared to nonsolar heating systems?
    a. The technology is well developed and can be installed quickly.
  * b. The system adds 5% to the initial cost but reduces lifetime costs by 30–40%.
    c. They require more materials and more maintenance.
    d. The systems deteriorate more readily and need to be replaced more often.
    e. None of the above is true.

49. Advantages of solar space heating include all of the following *except*
    a. a free energy source.
  * b. low to moderate net useful energy.
    c. well-developed active and passive technologies.
    d. no carbon dioxide additions to the atmosphere.
    e. low air and water pollution.

50. Disadvantages of solar space and water heating systems include all of the following *except*
    a. high initial costs.
    b. solar collectors are not aesthetically pleasing to some people.
    c. passive systems require owners to open and close windows and shades to regulate heat distribution.
    d. owners need laws to prevent blockage of access to light.
  * e. none of the above; they are all disadvantages.

51. Heliostats
    a. can track the sun.
    b. can focus sunlight on a central heat-collection point.
    c. can produce temperatures high enough for making high-pressure steam to run turbines.
    d. can use molten salt to store the sun's heat to produce electricity at night or on cloudy days.
  * e. can do all of the above.

52. The most promising approach to intensifying solar energy is
  * a. nonimaging optics.
    b. convex lenses.
    c. concave lenses.
    d. convex mirrors.
    e. special diffraction gratings.

53. Solar power plants
    a. pollute air and water.
    b. take three to five years to construct.
    c. require more land than coal-burning plants plus coal deposit lands.
  * d. produce electricity almost as cheaply as a new nuclear plant.
    e. have all of the above characteristics.

54. Cells that convert solar energy directly into electricity are called
    a. electrosolar chips.
  * b. photovoltaic cells.
    c. helioelectric units.
    d. photoelectric cells.
    e. heliostatic chips.

55. Solar cells
    a. have moderate net energy yield.
  * b. are ideal for providing electricity to rural areas.
    c. are being installed in Mexico City.
    d. have been incorporated into basement pipes.
    e. can be coated on a regular window pane.

56. Which of the following statements is *false*?
    a. Solar cells are reliable and quiet and have no moving parts.
    b. Solar cells require minimal maintenance.
  * c. During the 1980s, the U.S. government increased its expenditures for solar research.
    d. Solar cells could provide one-half of the electricity needed by the United States by 2050.
    e. In the 1980s, the United States lost a considerable amount of the solar-cell market
        to Japan.

57. Photovoltaic cells
    a. should last 30 years or more if encased in glass or plastic.
    b. can be installed quickly and easily.
    c. can be built in solar-cell packages of varying sizes.
    d. produce minimal air and water pollution during operation.
  * e. have all of the above characteristics.

58. Photovoltaic cells
    a. are very inexpensive to install.
    b. produce massive levels of water pollution during production.
  * c. can produce moderate levels of water pollution during manufacture.
    d. are difficult to install.
    e. contribute significant amounts of greenhouse gases.

18-4 PRODUCING ELECTRICITY FROM MOVING WATER AND FROM HEAT
      STORED IN WATER

59. Hydroelectric power may be
    a. large scale.
    b. small scale.
    c. pumped storage.
  * d. all of the above.

60. Pumped-storage hydropower systems
  * a. are used to produce power during peak periods.
    b. involve the use of large dams.
    c. depend on stream flow to control power generation.
    d. may vary in output during different seasons.
    e. are a relatively inexpensive way to produce electricity.

61. Which of the following countries produces the greatest proportion of its electricity by
    hydroelectric plants?
    a. Austria
    b. Switzerland
  * c. Norway
    d. Italy
    e. France

62. Which of the following continents uses the smallest proportion of the hydroelectric power available to it?
   a. Asia
 * b. Africa
   c. Europe
   d. Australia
   e. South America

63. The world's largest producer of electricity from hydropower is
   a. Australia.
 * b. the United States.
   c. China.
   d. Austria.
   e. Canada.

64. The era of building large dams in the United States is drawing to a close because of
   a. lack of suitable sites.
   b. high construction costs.
   c. opposition from conservationists.
 * d. all of the above.

65. Hydroelectric plants
   a. need to be shut down frequently for maintenance checks.
   b. offer low net useful energy yield.
   c. have relatively high operating and maintenance costs.
 * d. help control flooding and supply a regulated flow of irrigation water to areas below the dam.
   e. produce moderate air pollutants during operation.

66. Which of the following is a disadvantage of hydroelectric plants?
   a. high pollution
 * b. high construction costs
   c. high operation and maintenance costs
   d. low functional life span
   e. low net useful energy yield

67. Tidal power
 * a. has a free energy source.
   b. has high operating costs.
   c. has low net useful energy yield.
   d. has moderate air pollution during operation.
   e. disturbs large amounts of land.

68. Tidal power plants
   a. have high construction costs.
   b. have variable daily electrical output.
   c. can suffer from storm damage.
   d. can suffer corrosion from seawater.
 * e. have all of the above disadvantages.

69. Which of the following statements about tidal power plants is *false*?
   a. Seawater may corrode their metal.
 * b. The electricity produced each day is uniform, and therefore no backup system is needed.
   c. The net useful energy yield is moderate.
   d. There are relatively few appropriate sites for tidal power plants.
   e. The facilities can easily be destroyed by storms.

70. Which of the following statements about wave power is *false?*
    a. Wave power is created primarily by wind.
    b. Net useful energy is moderate.
  * c. Wave power could produce 10% of the world's energy by 2000.
    d. Wave power is constrained to a few coastal areas with the right conditions.
    e. Construction costs are moderate to high.

71. Which of the following statements about ocean thermal energy conversion (OTEC) is *false?*
    a. OTEC depends upon the temperature differential between surface and deep waters of tropical oceans.
    b. The source of energy for OTEC is limitless.
  * c. OTEC requires an extensive backup system.
    d. Land area is not required for OTEC.
    e. Nutrients brought up from the ocean bottom when water is pumped for OTEC power plants may be used to nourish fish and shellfish.

72. An experimental saline solar-pond power plant has been operating
    a. in Egypt.
    b. in the Canary Islands.
    c. in Hawaii.
    d. in Australia.
  * e. on the Israeli side of the Dead Sea.

73. Freshwater solar ponds
  * a. can be used for hot water and space heating.
    b. are located in isolated mountain areas.
    c. are found in salt flats and deserts.
    d. provide both cooling and heating.
    e. have a low net useful energy yield.

18-5 PRODUCING ELECTRICITY FROM WIND

74. The greatest concentration of wind turbines is found in
    a. Kansas.
    b. Texas.
  * c. California.
    d. Florida.
    e. Oklahoma.

75. Which of the following statements about wind power is *false?*
    a. Danish companies have taken over the global market for wind turbines.
    b. By the middle of the next century, wind power could provide 10 to 25% of the electricity used in the United States.
  * c. Land used for wind turbines cannot be used for grazing cattle.
    d. Europeans are spending ten times more for wind power research and development than the United States.
    e. The federal budget for research and development of wind power was cut 90% during the 1980s.

76. Wind power
    a. is an unlimited source of energy at favorable sites.
    b. offers quick construction time.
    c. has a moderate to high net useful energy yield.
    d. emits no air pollution during operation.
  * e. includes all of the above.

77. Wind farms
    a. may be noisy.
    b. may interfere with flight patterns of migratory birds.
    c. require backup energy if the wind is not steady.
    d. may cause visual pollution.
 * e. have all of the above disadvantages.

18-6 PRODUCING ENERGY FROM BIOMASS

78. Biomass includes
    a. wood and wood products.
    b. agricultural wastes.
    c. garbage.
    d. organic plant matter converted to biofuels.
 * e. all of the above.

79. Biomass fuels are
    a. solid.
    b. liquid.
    c. gaseous.
 * d. all of the above.

80. Biomass, mostly from the burning of wood and manure, supplies about _____ of the energy
    used in LDCs.
    a. three-fourths
    b. two-thirds
 * c. one-half
    d. one-third
    e. one-fourth

81. Biomass can be used for
    a. heating space.
    b. heating water.
    c. propelling vehicles.
    d. producing electricity.
 * e. all of the above.

82. Burning of biomass
    a. releases more carbon dioxide per ton burned than does coal.
    b. releases more sulfur dioxide and nitric oxide per unit of energy produced than does
       uncontrolled burning of coal.
    c. requires little land.
 * d. can cause soil erosion, water pollution, and loss of wildlife habitat.

83. A major disadvantage of using biomass for energy is
 * a. large land requirements.
    b. higher nitrous oxide and sulfur dioxide emissions than other sources of energy.
    c. lack of versatility in its use and application.
    d. that it is not renewable.
    e. all of the above.

84. About _____% of the people living in LDCs heat their dwellings and cook their food by
    burning wood or charcoal from wood.
    a. 10
    b. 30
    c. 50
 * d. 70
    e. 95

85. Which country leads the world in using wood as an energy source?
    a. Canada
  * b. Sweden
    c. China
    d. Brazil
    e. Costa Rica

86. Which of the following species is least likely to be found in a biomass plantation?
    a. sycamore
    b. poplar
    c. water hyacinth
  * d. oak
    e. cottonwood

87. Biomass plantations
    a. require minimal use of pesticides.
    b. require minimal use of fertilizers.
    c. generally increase the biodiversity of an area.
  * d. require large areas of land.
    e. might compete with food crops for prime farmland.

88. Wood
    a. has a low net energy yield when burned near the harvest site.
    b. for burning is usually found near urban areas.
  * c. harvesting can result in accidents.
    d. is burned in stoves that require little maintenance.
    e. burning is nonpolluting.

89. Bagasse is a(n)
    a. energy plantation.
  * b. residue after sugarcane has been harvested.
    c. exotic plant that stores hydrocarbons.
    d. gas given off from landfills.
    e. type of catalytic converter.

90. Ecologists say that it makes the *least* sense to use crop residues as
  * a. fuel for energy.
    b. food for animals.
    c. a way to retard erosion.
    d. a fertilizer.

91. Japan, western Europe, and the United States have built incinerators that burn _____ to produce energy.
    a. tires
    b. biomass
    c. uranium
    d. solar cells
  * e. trash

92. Gaseous and liquid biofuels include
    a. biogas.
    b. liquid methanol.
    c. wood alcohol.
    d. liquid ethanol.
  * e. all of the above.

93. Biogas digesters are
     * a. very efficient, slow, and unpredictable.
     b. very efficient, fast, and predictable.
     c. very inefficient, slow, and unpredictable.
     d. very inefficient, fast, and predictable.
     e. moderately efficient, fast, and unpredictable.

94. A usable fuel produced by landfills is
     a. methanol.
     b. gasohol.
     c. ethanol.
     * d. methane.
     e. ethane.

95. Which of the following statements is *false?*
     a. Methane is a more potent greenhouse gas than carbon dioxide.
     b. Anaerobic decomposition of organic matter produces methane.
     c. Methane is produced from manure at feedlots and during sewage treatment.
     * d. Burning methane produces more global warming than simply releasing it into the atmosphere.
     e. None of the above is false.

96. Which of the following substances is not considered to be a source of ethanol?
     a. sorghum
     b. sugar beets
     * c. potatoes
     d. corn
     e. sugarcane

97. Gasohol is gasoline mixed with
     * a. ethyl alcohol.
     b. methane.
     c. methanol.
     d. butane.
     e. all of the above.

98. Which of the following statements is *false?*
     a. Super unleaded gasoline is a form of gasohol.
     b. Ethanol costs about three times more to produce than gasoline.
     c. Gasohol accounts for 8% of the gasoline sales in the United States.
     d. Ethanol has helped Brazil cut oil imports.
     * e. Methanol is cheaper to produce than gasoline.

18-7 THE SOLAR-HYDROGEN AGE

99. Hydrogen gas can be used to
     a. heat buildings.
     b. fuel cars, trucks, and planes.
     c. power factories.
     d. generate electricity.
     * e. do all of the above.

100. Hydrogen gas can be burned in
      a. a simple reaction with oxygen gas.
      b. fuel cells that produce electricity.
      c. specially designed automobile engines.
      d. a furnace.
      * e. all of the above.

101. Which of the following statements is *false?*
    a. When burned, hydrogen produces virtually no air pollutants.
    b. Some metals can store and release hydrogen.
  * c. Fuel tanks of metal-hydrogen compounds would tend to explode in an accident.
    d. Experimental cars have been running on hydrogen for years.
    e. Hydrogen can be collected by passing electrical current through water.

102. Which of the following statements is *false?*
    a. Hydrogen would be a good fuel for airplanes.
    b. Hydrogen gas could be stored at high pressures and distributed by pipeline.
  * c. Burning hydrogen releases low amounts of carbon dioxide.
    d. Hydrogen could be stored in depleted oil wells.
    e. Hydrogen can be phased in by mixing with natural gas.

103. Which of the following statements is *false?*
    a. Only trace amounts of hydrogen occur naturally on Earth.
    b. Hydrogen production will always require more energy than the energy released when hydrogen is burned.
  * c. It will take at least 100 years to make a transition to a solar-hydrogen civilization.
    d. The use of hydrogen as an energy source will always be dependent on other energy sources.
    e. The key to the spread of a solar-hydrogen revolution is the development of ways to use solar energy to produce electricity to form hydrogen gas.

104. All of the following countries are leaders in hydrogen research *except*
  * a. China.
    b. the United States.
    c. Japan.
    d. Germany.

105. A switch to hydrogen as a primary energy source is constrained by
    a. hydrogen production.
    b. economics.
    c. politics.
  * d. all of the above.
    e. none of the above.

106. Large-scale funding of hydrogen research would generally be *least* opposed by
    a. electric utilities.
  * b. sustainable developers.
    c. fossil-fuel companies.
    d. automobile manufacturers.

18-8 GEOTHERMAL ENERGY

107. Geothermal energy is stored in the form of
    a. dry steam.
    b. wet steam.
    c. hot water.
  * d. all of the above.

108. Geothermal energy can be used for all of the following *except*
    a. heating space.
    b. producing electricity.
  * c. transportation fuel.
    d. producing high-temperature heat for industry.
    e. none of the above.

109. The *most* desirable form of geothermal energy is
    * a. dry steam.
      b. wet steam.
      c. hot water.
      d. hot lava.
      e. geopressurized zones.

110. The *most* common type of geothermal energy occurs in the form of
      a. hot air.
    * b. hot water.
      c. wet steam.
      d. dry steam.
      e. geopressurized zones.

111. Which of the following statements about geothermal power is *false?*
      a. It is relatively cheap where available.
      b. It releases air pollution, including radioactive isotopes and sulfur dioxide.
    * c. It releases much carbon dioxide.
      d. It results in moderate to high water pollution.
      e. It sometimes causes noise and odor.

112. Magma is
      a. a deep source of fossil fuel.
    * b. molten rock.
      c. an air pollutant given off by geothermal energy.
      d. a cooled lava flow.
      e. a geyser.

113. An advantage associated with the development and use of geothermal energy systems is that
    * a. little or no carbon dioxide is produced.
      b. geothermal power plants do not require cooling water.
      c. all geothermal energy sources are renewable.
      d. there is no risk of harmful environmental impact.

114. Eco-Lair is
      a. an organization dedicated to saving energy.
    * b. the energy-efficient home of the author of the textbook.
      c. a set of condominiums designed to have a minimal impact on the environment.
      d. a wildlife refuge for migratory birds.
      e. an underground community.

115. Eco-Lair has all of the following features *except*
      a. a solar room.
      b. a tankless instant water heater fueled by LPG.
      c. active solar collectors to provide hot water.
    * d. roof-mounted photovoltaic cells.
      e. earth tubes.

CHAPTER 19
NONRENEWABLE ENERGY RESOURCES

1.   Chernobyl
      a.   was the worst nuclear disaster ever.
      b.   resulted in exposure of 500,000 people to dangerous radioactivity.
      c.   resulted in 31 human deaths from radiation shortly after the accident.
      d.   flung radioactive debris and dust into the atmosphere, where they circled the globe.
  *   e.   included all of the above.

2.   Chernobyl resulted in all of the following *except*
      a.   evacuation of many families from land they had farmed for centuries.
      b.   separation of families from their domestic animals.
  *   c.   nuclear winter.
      d.   acute radiation sickness.
      e.   recognition that major nuclear accidents in one location affect many locations.

19-1 OIL

3.   Petroleum is a gooey liquid consisting primarily of
  *   a.   hydrocarbon compounds.
      b.   nitrogen.
      c.   sulfur.
      d.   oxygen.
      e.   silicon.

4.   Maximum oil recovery would include
      a.   primary oil recovery.
      b.   secondary oil recovery.
      c.   tertiary oil recovery.
      d.   a and b.
  *   e.   a, b, and c.

5.   Primary oil recovery involves
  *   a.   pumping out the oil that flows by gravity into the bottom of the well.
      b.   using steam to soften the heavy oil.
      c.   using carbon dioxide to force heavy oil into the well.
      d.   using water under pressure to remove oil.
      e.   removing from 50 to 80% of the oil in a well.

6.   Secondary oil recovery involves
      a.   pumping out the oil that flows by gravity into the bottom of the well.
      b.   using steam to force oil into the well.
      c.   using carbon dioxide to force heavy oil into the well.
  *   d.   using water under pressure to remove oil.
      e.   removing from 50 to 80% of the oil in a well.

7.   Tertiary oil recovery involves
      a.   pumping out the oil that flows by gravity into the bottom of the well.
      b.   using steam to soften the heavy oil.
      c.   using carbon dioxide to force heavy oil into the well.
      d.   using water under pressure to remove oil.
  *   e.   b and c.

8. Which of the following would be used to extract the greatest amount of oil from a well?
   a. primary oil recovery
   b. secondary oil recovery
 * c. enhanced oil recovery
   d. bubble method of oil recovery
   e. none of the above.

9. Crude oil components are separated by
   a. gravity.
 * b. heat and distillation.
   c. pressure.
   d. filtration.
   e. centrifugal force.

10. Distillation of crude oil can produce
    a. asphalt.
    b. heating oil.
    c. diesel oil.
    d. gasoline.
  * e. all of the above.

11. Petrochemicals are
  * a. used as raw materials for manufacturing.
    b. removed from oil before it is refined.
    c. impurities that must be burned or buried.
    d. additives to bring up the octane level of gasoline.
    e. added to heavy oil so that it can be pumped.

12. You can thank petrochemicals if you use
    a. paints.
    b. medicines.
    c. plastics and synthetic fibers.
    d. inorganic fertilizers.
  * e. all of the above.

13. It is assumed that most of the world's remaining undiscovered deposits of oil are in
    a. South America.
    b. Mexico.
    c. the Commonwealth of Independent States.
  * d. the Middle East.
    e. Africa.

14. The greatest use of oil in the United States is for
  * a. transportation.
    b. generation of electricity.
    c. commercial and residential heating and cooling.
    d. industrial uses.
    e. petrochemicals.

15. Since 1985, oil extraction in the United States has
    a. increased rapidly.
    b. increased slightly.
    c. stayed the same.
  * d. declined.
    e. peaked and decreased exponentially.

16. At present consumption rates, known world crude oil reserves will be depleted in ____ years.
    a. 25
    b. 35
  * c. 42
    d. 60
    e. 100

17. Saudi Arabia could supply the world's oil needs for ____ years if it were the world's only source of oil.
    a. five
  * b. 10
    c. 15
    d. 20
    e. 33

18. All estimated, undiscovered, recoverable deposits of oil in the United States could meet U.S. demand for ____ years.
    a. two
    b. five
  * c. 10
    d. 20
    e. 35

19. If they were the only source available, the crude oil reserves under Alaska's North Slope could supply the world's oil demand for
    a. one week.
  * b. six months.
    c. one year.
    d. five years.
    e. 10 years.

20. Oil is widely used because it
    a. is relatively cheap.
    b. is easily transported.
    c. has a high net useful energy yield.
    d. is very versatile.
  * e. all of the above.

21. A disadvantage of oil is that it
  * a. produces more carbon dioxide than any other fuel.
    b. produces destruction of nature through oil spills.
    c. can contaminate groundwater supplies.
    d. will be commercially depleted within 35 to 80 years.
    e. has all of the above disadvantages.

22. Oil has all of the following disadvantages *except*
  * a. difficulty in transport between countries.
    b. release of carbon dioxide.
    c. release of sulfur oxides.
    d. release of nitrogen oxides.
    e. oil spills.

23. Kerogen is
    a. the active ingredient in kerosene.
    b. a waste product from production of oil from oil shale.
    c. a fuel supplement added to diesel oil.
  * d. the active ingredient in oil shale.
    e. a residual produced during the fractionation and distillation of crude oil.

24. Oil shale and tar sands
    a. are principal sources of conventional crude oil.
  * b. contain large supplies of heavy oils.
    c. constitute a small but cheap supply of crude oil.
    d. are usable only for aviation fuel.
    e. are usable only as solid fuels.

25. Shale oil is
    a. pure in its naturally occurring form.
    b. a light oil found in ancient sand deposits.
  * c. extracted from the kerogen in oil shale.
    d. commercially produced in about 25 nations.
    e. relatively inexpensive to produce and refine.

26. States with potentially recoverable deposits of shale oil include all of the following *except*
    a. Colorado.
  * b. Iowa.
    c. Utah.
    d. Wyoming.
    e. none of the above.

27. Shale oil processing requires large amounts of
  * a. water.
    b. electricity.
    c. zinc.
    d. time.
    e. aluminum.

28. Disadvantages of shale oil as an energy source include all of the following *except*
    a. massive land disruption.
    b. air and water pollution.
    c. considerable energy input for the energy output obtained.
    d. large amounts of water.
  * e. potentially recoverable U.S. deposits are only enough to meet the country's crude oil demand for five years.

29. Shale oil and tar sands _____ than conventional oil deposits.
    a. are easier to extract
  * b. have lower net useful energy yield
    c. have less environmental impact
    d. are sources of more useful products
    e. (all of the above)

30. Bitumen is
    a. a type of coal.
    b. a deep shale oil deposit.
  * c. high-sulfur heavy oil.
    d. an octane-raising gasoline additive.
    e. a type of impurity found in diesel fuel.

31. Which of the following countries has the greatest tar sand deposits?
    a. Saudi Arabia
  * b. Canada
    c. Venezuela
    d. Kuwait
    e. the United States

32. Tar sand normally contains all but which of the following?
    * a. coal
    b. bitumen
    c. sand
    d. water
    e. clay

19-2 NATURAL GAS

33. Natural gas from wells consists of 50 to 90%
    * a. methane.
    b. ethane.
    c. propane.
    d. butane.
    e. LPG.

34. Liquefied petroleum gas consists of
    a. methane.
    * b. butane and propane.
    c. ammonia.
    d. nitrogen oxides.
    e. heavy hydrocarbons.

35. The country with the largest reserve of natural gas is
    a. Canada.
    * b. the Commonwealth of Independent States.
    c. the United States.
    d. India.
    e. Costa Rica.

36. In the United States, natural gas is used primarily
    a. to blow up weather balloons.
    * b. for space heating and industrial purposes.
    c. to produce electricity.
    d. as a fuel for vehicles.
    e. to produce petrochemicals.

37. The world's supplies of conventional natural gas are projected to last
    a. at least 300 years at 1984 usage rates.
    b. about 200 years if annual usage increases at 2% a year.
    * c. 80 years at present consumption rates.
    d. until unconventional natural gas technologies are fully developed.
    e. 10 years at declining consumption rates.

38. Natural gas has a ____ net useful energy yield and is a ____ fuel compared to coal and oil.
    a. high . . . dirty
    * b. high . . . clean-burning
    c. low . . . dirty
    d. low . . . clean-burning
    e. similar . . . comparably dirty

39. Which of the following statements about liquified natural gas (LNG) is *false?*
    a. Natural gas has to be liquefied before it can be transported by ship.
    * b. Shipment of LNG is safe and inexpensive.
    c. LNG requires refrigeration and pressure.
    d. Conversion of natural gas to LNG reduces net energy yield by one-fourth.
    e. None of the above is false.

40.   A key option during the switch from oil to other new energy sources is
      a. coal.
      b. nuclear.
      c. biomass.
  *   d. natural gas.
      e. solar.

19-3 COAL

41.   Because of its high heat content and low sulfur content the most desirable type of
      coal is ___.
      a. bituminous
      b. lignite
  *   c. anthracite
      d. peat
      e. soft

42.   About two-thirds of the world's proven coal reserves are found in
      a. the Middle East.
  *   b. China, the United States, and the Commonwealth of Independent States.
      c. Canada.
      d. Mexico and Central America.
      e. South America.

43.   The world's *most* abundant conventional fossil fuel is
      a. crude oil.
      b. natural gas.
      c. biomass.
  *   d. coal.
      e. LP gas.

44.   Identified coal reserves in the United States should last about ____ years at current usage.
      a. 10
      b. 50
      c. 100
      d. 200
  *   e. 300

45.   Land subsidence can result from
      a. surface mining.
      b. higher ocean levels.
  *   c. underground coal mining.
      d. natural gas extraction.
      e. oil wells.

46.   Sulfur dioxide, nitrogen oxide, carbon dioxide, and particulate matter are all
  *   a. waste products of coal-fired power plants.
      b. additives to make coal burn cleanly.
      c. removed almost entirely when pollution controls are used.
      d. solid wastes that must be disposed of properly.
      e. harmless by-products of burning coal.

47.   ____ is the dirtiest fossil fuel to burn.
      a. Oil
      b. Natural gas
  *   c. Coal
      d. Wood
      e. Butane

48. Using coal
    a. causes severe land disturbance.
    b. can pollute nearby streams with acids and toxic metals.
    c. releases large amounts of sulfur dioxide, nitrogen oxides, and particulate matter.
    d. produces more carbon dioxide per unit of energy than other fossil fuels.
  * e. all of the above.

49. Fluidized-bed combustion is a relatively cheap method of burning coal
  * a. more cleanly and efficiently.
    b. to produce extremely hot ionized gases.
    c. in underground deposits too deep to mine.
    d. without having to crush or powder it.
    e. to close mines.

50. Which of the following statements about synfuels is *false?*
    a. Synfuels can be transported by pipeline.
    b. Synfuels are more expensive than coal.
  * c. Synfuel combustion, by itself, produces more air pollution than coal.
    d. Synfuels are more versatile than coal.
    e. Synfuel production requires large amounts of water.

51. Synfuels produced from coal
    a. are less versatile fuels than solid coal.
    b. are available only in the United States.
  * c. require large amounts of water for production.
    d. are the dirtiest burning of all fossil fuels.
    e. release less carbon dioxide per unit energy than coal.

19-4 CONVENTIONAL NUCLEAR FISSION

52. The Atomic Energy Commission convinced utilities to use nuclear power to generate electricity because
    a. government picked up one-fourth the cost with no cost overruns allowed.
    b. it was predicted that nuclear energy would produce electricity at extremely low costs.
    c. the utilities were protected from liability to the general public.
  * d. of all of the above.
    e. none of the above.

53. Since 1978, _____ new nuclear power plants have been ordered and then built in the United States.
  * a. no
    b. four
    c. 11
    d. 23
    e. 52

54. In 1993, the 108 licensed nuclear power plants in the United States generated about _____ % of the country's electricity.
    a. 1
    b. 5
    c. 10
  * d. 20
    e. 40

55. The nuclear power industry in the United States has declined because of
   a. high construction costs.
   b. the possibility of wastes being used in nuclear weapons.
   c. false assurances and cover-ups by government and industry officials.
   d. concerns about radioactive waste disposal.
   * e. all of the above.

56. Light-water reactors generate about _____% of the electricity generated worldwide by nuclear power plants.
   a. 95
   * b. 85
   c. 75
   d. 65
   e. 55

57. The fissionable fraction of the fuel in a nuclear reactor is
   * a. uranium-235.
   b. uranium-238.
   c. uranium-239.
   d. plutonium-239.
   e. plutonium-235.

58. A light-water reactor includes
   a. the core.
   b. control rods.
   c. moderator.
   d. coolant.
   * e. all of the above.

59. Control rods in a reactor
   a. contain uranium.
   * b. absorb neutrons.
   c. contain plutonium.
   d. reduce heat.
   e. contain coolant.

60. The moderator in a nuclear reactor
   a. releases neutrons.
   b. absorbs neutrons.
   c. reflects neutrons.
   * d. slows down neutrons.
   e. stops neutrons.

61. The *most* common moderator used in nuclear reactors is
   a. graphite.
   b. boron.
   c. argon.
   * d. water.
   e. helium.

62. Coolants which may be used in nuclear power plants include all of the following except
   a. water.
   * b. liquid nitrogen.
   c. argon.
   d. helium.

63. Each year, about ____ of the fuel assemblies in a nuclear reactor must be replaced because of their spent condition.
   a. one-tenth
   b. one-fifth
   c. one-fourth
   * d. one-third
   e. one-half

64. The moderator at Chernobyl was
   a. Gorbachev.
   b. Medvedev.
   c. water.
   * d. graphite.
   e. tritium.

65. After spent fuel rods have cooled a few years, they are transported to
   a. a nuclear waste repository.
   b. a fuel-reprocessing plant.
   c. storage pools away from the reactor.
   * d. all of the above.
   e. none of the above.

66. All of the following countries have nuclear fuel reprocessing plants *except*
   * a. the United States.
   b. France.
   c. Japan.
   d. West Germany.
   e. Great Britain.

67. The nuclear fuel cycle includes all of the following *except*
   a. the uranium mine.
   b. fuel fabrication.
   * c. the cyclotron.
   d. interim underwater storage.
   e. geologic disposal.

68. If the fuel pellets in spent fuel rods are processed to remove plutonium and other very long-lived radioactive isotopes, the remaining radioactive waste should be safely stored on the order of ____ years.
   a. 100
   b. 1,000
   * c. 10,000
   d. 100,000
   e. 1 million

69. Nuclear plants
   a. don't emit air pollutants.
   b. result in low to moderate water pollution and land disruption if the fuel cycle operates normally.
   c. have multiple safety systems with backups that greatly decrease the likelihood of a catastrophic accident.
   d. including their entire fuel cycles contribute about one-sixth the carbon dioxide per unit of electricity as coal plants.
   * e. have all of the above characteristics.
   f. have none of the above characteristics.

70. Which of the following statements about nuclear reactor safety features is *false?*
    a. Concrete and steel shields surround the reactor vessel.
    b. Control rods are automatically inserted into the core to stop fission in emergencies.
    c. Large filter systems and chemical sprayers inside the containment building remove radioactive dust from the air.
    d. An emergency core-cooling system floods the core automatically within one minute to prevent meltdown of the core.
    e. Backup systems automatically replace each major part of the safety system in the event of a failure.
    * f. None of the above is false.

71. Of the following nuclear accident sites, which had the *most* serious accident?
    a. Three Mile Island
    b. Brown's Ferry
    * c. Chernobyl
    d. Liverpool, England
    e. Paris, France

72. A meltdown of the reactor core would occur if
    a. control rods were inserted into the core.
    * b. too much coolant was lost.
    c. the proportion of uranium-238 was too high.
    d. the containment building developed an air leak.
    e. water entered the reactor core.

73. Three Mile Island is located in
    a. New Jersey.
    * b. Pennsylvania.
    c. Massachusetts.
    d. Connecticut.
    e. New York.

74. The accident at the Three Mile Island nuclear power plant in 1979 involved all of the following *except*
    a. a series of mechanical failures.
    b. a loss of reactor core coolant.
    c. operator errors unforeseen in safety studies.
    d. loss of unknown amounts of radioactive material into the atmosphere.
    * e. a complete reactor core meltdown.

75. There is widespread lack of confidence in the Nuclear Regulatory Commission's ability to enforce nuclear safety because of evidence from 1987 Congressional hearings that
    a. high-level NRC staff obstructed investigations of criminal wrongdoing by utilities.
    b. NRC staff suggested ways utilities could evade commission regulations.
    c. NRC staff provided utilities with advance notice of surprise inspections.
    * d. includes all of the above.
    e. includes none of the above.

76. Which of the following proposals for disposing of high-level radioactive wastes is at present a technological impossibility because we don't know how it could be done?
    a. Shoot it into the sun.
    * b. Convert it into harmless isotopes.
    c. Bury it under ice sheets in Antarctica.
    d. Dump it into downward-descending bottom sediments.
    e. Use it in shielded batteries to run small electric generators.

77. Low-level radioactive wastes are produced by
    a. industries.
    b. hospitals.
    c. nuclear power plants.
    d. universities.
  * e. all of the above.

78. The useful operating life of today's nuclear power plants is supposed to be _____ years.
    a. 20
    b. 30
  * c. 40
    d. 50
    e. 60

79. Nuclear power plants wear out when
    a. the reactor's pressure vessel becomes brittle.
    b. tubes in the reactor's steam generator are weakened over decades of pressure and temperature changes.
    c. pipes and valves throughout the system corrode.
  * d. all of the above happen.
    e. none of the above happen.

80. All of the following are methods of decommissioning nuclear power plants *except*
    a. immediate dismantling.
    b. entombment.
  * c. decomposition.
    d. mothballing.
    e. none of the above.

81. Which of the following makes the statement *false?* Nuclear power plant decommissioning costs
    a. may be as low as $225 million for mothballing.
    b. may be as high as $1 billion per reactor.
  * c. are ready to be dispersed for all retiring plants from interest-bearing accounts.
    d. will be passed on to ratepayers.
    e. will be passed on to taxpayers.

82. Today at least _____ countries sell nuclear technology in the international marketplace.
    a. four
  * b. 14
    c. 24
    d. 34
    e. 44

83. To slow the spread of bomb-grade radioactive material,
    a. uranium must be mined more efficiently.
    b. international safeguards currently in place need to be simplified.
    c. civilian reprocessing plants must be built as quickly as possible.
    d. all of the above must be done.
  * e. none of the above must be done.

84. *Forbes* magazine describes the failure of the nuclear power program as the largest _____ disaster in U.S. business history.
  * a. managerial
    b. technological
    c. economic
    d. environmental
    e. political

85. Early in the last decade, the U.S. Council for Energy Awareness was formed to support _____ energy and downgrade other energy sources.
    a. solar
    * b. nuclear
    c. wind
    d. fossil fuel
    e. geothermal

86. Which of the following statements is *true?*
    a. The development of the nuclear industry to produce electricity will significantly reduce U.S. dependence on foreign oil.
    b. Replacing coal-burning power plants with nuclear plants will prevent the greenhouse effect.
    * c. The U.S. Council for Energy Awareness has designed a massive advertising campaign to resell nuclear power to the public.
    d. The cost of constructing nuclear plants is declining.
    e. Seventy-five percent of the uranium used in U.S. nuclear power plants is mined in the United States.

87. _____ of the uranium used for fuel in the United States is imported.
    a. One-tenth
    b. One-third
    * c. One-half
    d. Two-thirds
    e. Three-quarters

88. Newly designed nuclear power plants
    a. will not produce nuclear wastes.
    b. use a completely different technology from nuclear weapons.
    c. produce wastes which cannot be used to make nuclear weapons.
    d. have all of the above characteristics.
    * e. have none of the above characteristics.

89. Breeder nuclear fission reactors convert
    a. fast-moving neutrons into slow-moving ones.
    b. high-level wastes into harmless isotopes.
    * c. uranium-238 into plutonium-239.
    d. uranium-235 into uranium-238.
    e. uranium radioisotopes into sodium.

90. Which of the following statements is *false?*
    a. A breeder reactor generates more fuel than it consumes.
    b. If the breeder reactor's safety system failed, the reactor could lose some of its liquid sodium coolant.
    c. All experimental breeder reactors built so far have caught fire.
    d. Breeder reactors could produce a nuclear explosion.
    * e. Breeder reactors would not extend the supplies of uranium in the world.

91. Since 1966, experimental breeder reactors have been built in
    a. the United Kingdom.
    b. the former Soviet Union.
    c. Germany.
    d. Japan.
    * e. all of the above.

92. A construction project completed in 1986 brought the world's first full-sized commercial breeder reactor into operation in
    a.   the United States.
    b.   the Commonwealth of Independent States.
    c.   Germany.
*   d.   France.
    e.   Great Britain.

93. The coolant employed in a breeder reactor is liquid
*   a.   sodium.
    b.   plutonium.
    c.   nitrogen.
    d.   lithium.
    e.   aluminum.

94. Compared to a D-T nuclear fusion reaction, a D-D nuclear fusion reaction
    a.   is much easier to initiate.
    b.   releases only 1% as much energy.
*   c.   requires a higher ignition temperature.
    d.   uses fuel that is much scarcer.
    e.   has none of the above characteristics.

95. Which of the following poses a serious obstacle to the development of commercial nuclear fusion power plants?
*   a.   exceedingly high ignition temperatures
    b.   scarcity of deuterium
    c.   public concerns about safety
    d.   dwindling supplies of uranium
    e.   disposal requirements of plutonium-239

96. In nuclear fusion experiments, scientists have tried to force deuterium and tritium atoms together with
    a.   electromagnetic reactors.
    b.   high-speed particles.
    c.   laser beams.
*   d.   all of the above.
    e.   none of the above.

19-5 SOLUTIONS: A SUSTAINABLE-EARTH ENERGY STRATEGY

97. Which of the following statements is *false?*
    a.   There is not enough financial capital to develop all energy alternatives.
    b.   We should not depend on only one source of energy but should develop a mix of perpetual and renewable energy resources.
*   c.   Energy production should be centralized as much as possible to increase efficiency.
    d.   Improving energy efficiency is the best option available to produce more energy.
    e.   Individuals, communities, and countries should get more of their heat and electricity from locally available renewable and perpetual energy resources.

98. Which of the following approaches emphasizes today's prices for short-term economic gain and inhibits long-term development of new energy resources?
*   a.   free-market competition
    b.   energy prices kept artificially high
    c.   energy prices kept artificially low
    d.   all of the above
    e.   none of the above

99. Governments use _____ to manipulate the energy playing field.
    a. tax breaks
    b. payment for long-term research and development
    c. price controls
    d. subsidies
    * e. all of the above

100. Keeping energy prices artificially low
    a. encourages waste and rapid depletion of energy resources getting favorable treatment.
    b. protects consumers from sharp price increases.
    c. can help reduce inflation.
    d. discourages the development of energy alternatives not getting favorable treatment.
    * e. does all of the above.

101. In the United States, nuclear power and fossil fuels receive _____ times more government subsidies per unit of energy than energy-efficient and renewable energy resources.
    a. 10
    b. 20
    c. 100
    * d. 200
    e. 1000

102. Keeping energy prices artificially high does all of the following *except* that it
    a. encourages improvements in energy efficiency.
    b. reduces dependence on imported energy.
    c. dampens economic growth.
    d. places a heavy burden on the poor.
    * e. causes high unemployment.

103. If energy prices were kept artificially high,
    a. energy efficiency would be encouraged.
    b. dependence on imported energy sources would be reduced.
    c. the use of energy resources that would be limited in the future would decline.
    d. inflation would be encouraged.
    * e. all of the above would be possible.

104. The estimated net useful energy of oil, natural gas, and coal is
    a. high and increasing.
    * b. high and decreasing.
    c. moderate and stable.
    d. low and increasing.
    e. low and decreasing.

105. The fossil fuel with the *least* environmental impact is
    a. oil.
    * b. natural gas.
    c. coal.
    d. oil shale.
    e. tar sands.

106. The *least* expensive perpetual resource is
    * a. improving energy efficiency.
    b. hydroelectricity.
    c. tidal energy.
    d. photovoltaics.
    e. wind energy.

107. The renewable and perpetual resources with the *best* availability in the short term and the long term are
    a. large- and small-scale dams.
  * b. low temperature solar heating and improved energy efficiency.
    c. photovoltaics and high-temperature solar heating.
    d. wind energy and geothermal energy.
    e. biomass and biofuels.

108. The energy resource that *most* improves the quality of the environment is
  * a. improved energy efficiency.
    b. wind energy.
    c. biomass.
    d. hydrogen gas.
    e. ocean thermal gradients.

109. Individuals can save energy and money by
    a. avoiding electrical space heating.
    b. insulating and caulking buildings.
    c. using energy-efficient appliances.
    d. turning the thermostat down on water heaters.
  * e. doing all of the above.

SECTION THREE

APPENDIXES

APPENDIX A

CONCEPT MAPS

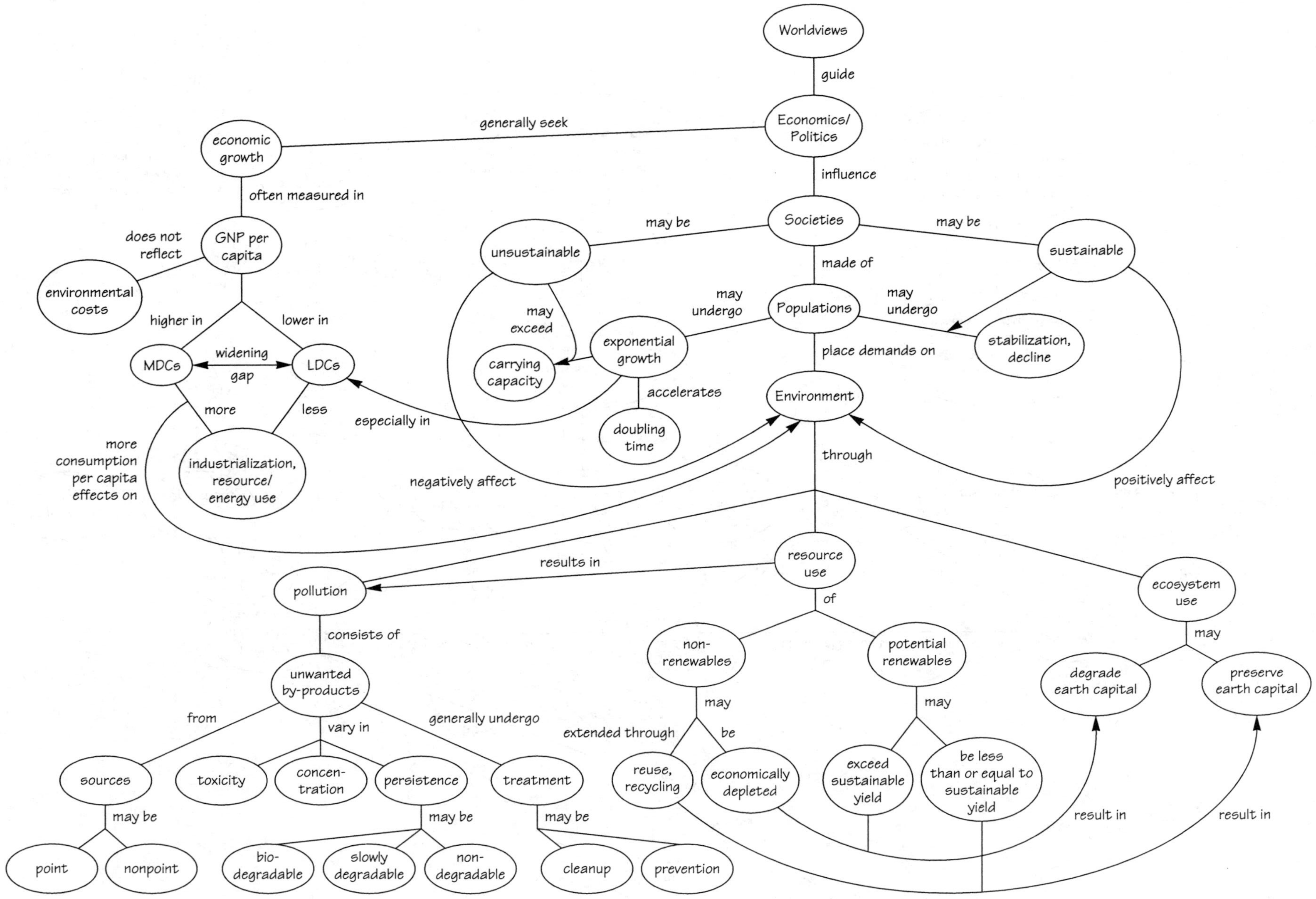

Map 1. Overview

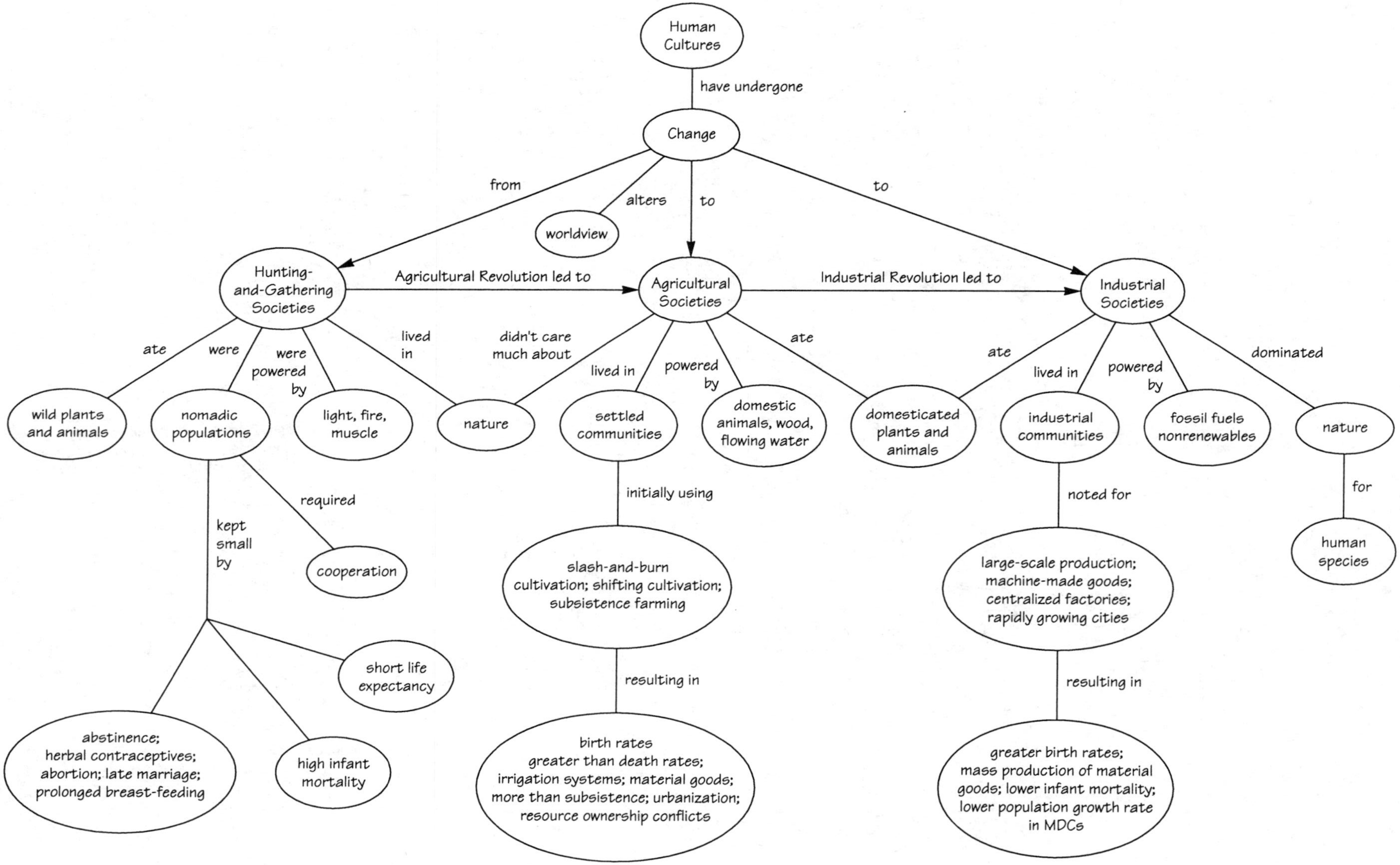

Map 2. Cultural Changes

Human Cultures —have undergone→ Change

Change —alters→ worldview

Change —from→ Hunting-and-Gathering Societies

Change —to→ Agricultural Societies

Change —to→ Industrial Societies

Agricultural Revolution led to

Industrial Revolution led to

Hunting-and-Gathering Societies:
- ate → wild plants and animals
- were powered by → light, fire, muscle
- nomadic populations → required → cooperation
- nomadic populations → kept small by → short life expectancy; high infant mortality; abstinence; herbal contraceptives; abortion; late marriage; prolonged breast-feeding
- lived in → nature
- didn't care much about → nature

Agricultural Societies:
- ate → domesticated plants and animals
- powered by → domestic animals, wood, flowing water
- lived in → settled communities
- settled communities → initially using → slash-and-burn cultivation; shifting cultivation; subsistence farming → resulting in → birth rates greater than death rates; irrigation systems; material goods; more than subsistence; urbanization; resource ownership conflicts

Industrial Societies:
- ate → domesticated plants and animals
- powered by → fossil fuels nonrenewables
- lived in → industrial communities → noted for → large-scale production; machine-made goods; centralized factories; rapidly growing cities → resulting in → greater birth rates; mass production of material goods; lower infant mortality; lower population growth rate in MDCs
- dominated → nature → for → human species

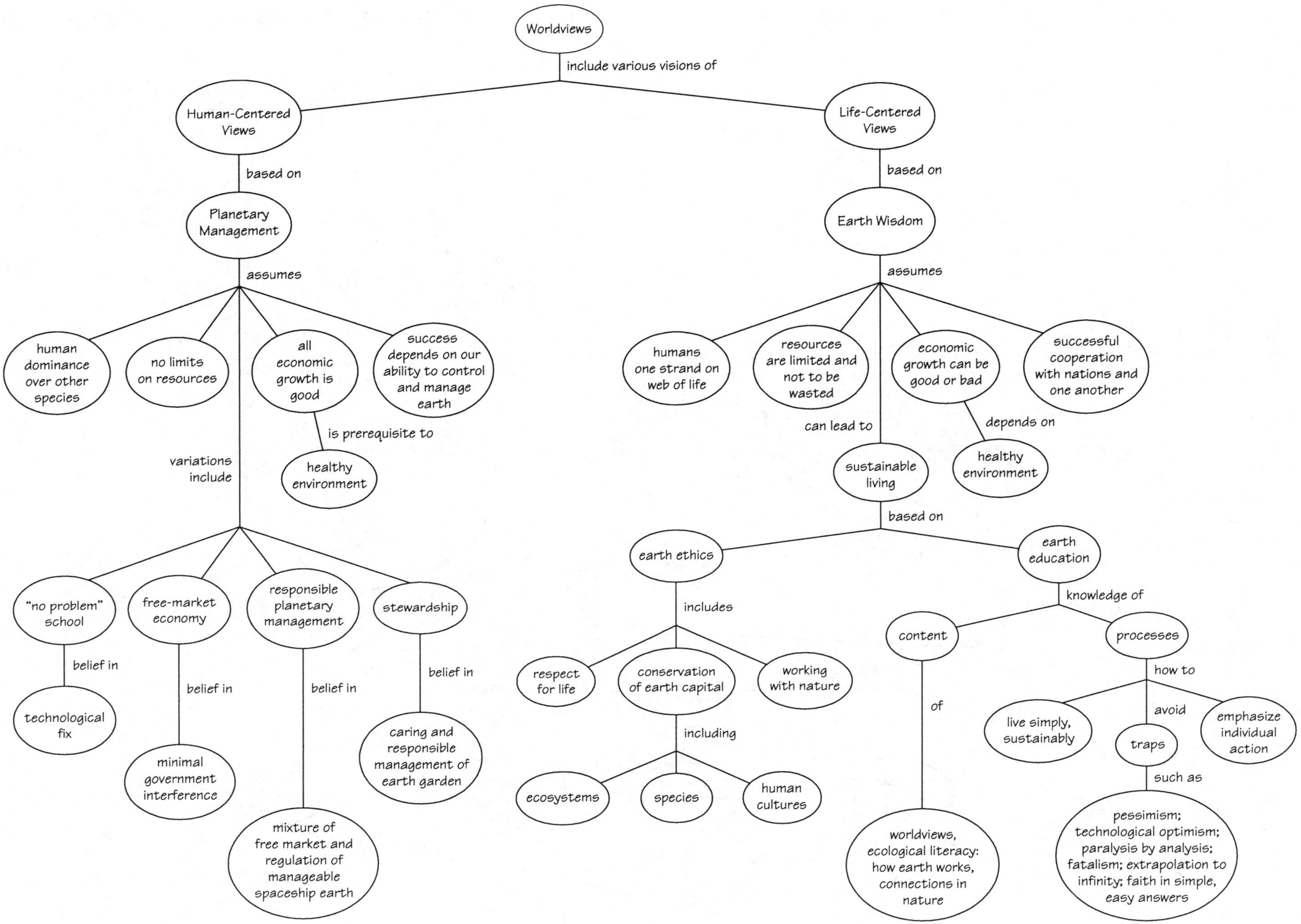

Map 3. Worldviews

Worldviews — include various visions of

Human-Centered Views

based on **Planetary Management**

assumes:
- human dominance over other species
- no limits on resources
- all economic growth is good — is prerequisite to — healthy environment
- success depends on our ability to control and manage earth

variations include:
- "no problem" school — belief in — technological fix
- free-market economy — belief in — minimal government interference
- responsible planetary management — belief in — mixture of free market and regulation of manageable spaceship earth
- stewardship — belief in — caring and responsible management of earth garden

Life-Centered Views

based on **Earth Wisdom**

assumes:
- humans one strand on web of life
- resources are limited and not to be wasted — can lead to — sustainable living
- economic growth can be good or bad — depends on — healthy environment
- successful cooperation with nations and one another

sustainable living based on:

earth ethics

includes:
- respect for life
- working with nature
- conservation of earth capital — including:
  - ecosystems
  - species
  - human cultures

earth education

- content — of — worldviews, ecological literacy; how earth works, connections in nature
- knowledge of processes — how to:
  - live simply, sustainably
  - avoid traps — such as — pessimism; technological optimism; paralysis by analysis; extrapolation to infinity; faith in simple, easy answers
  - emphasize individual action

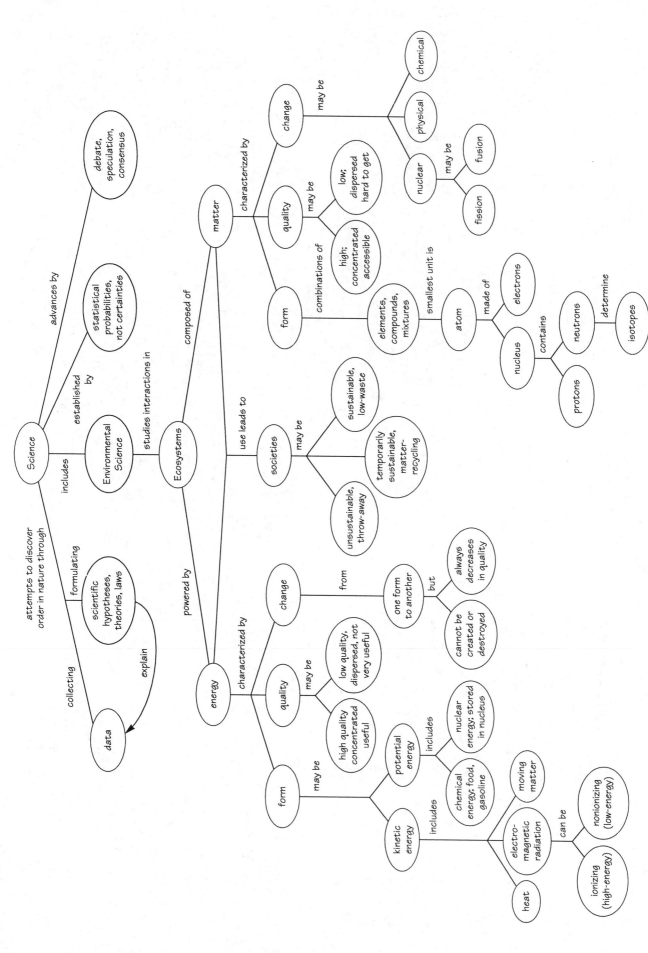

Map 4. Matter and Energy Resources

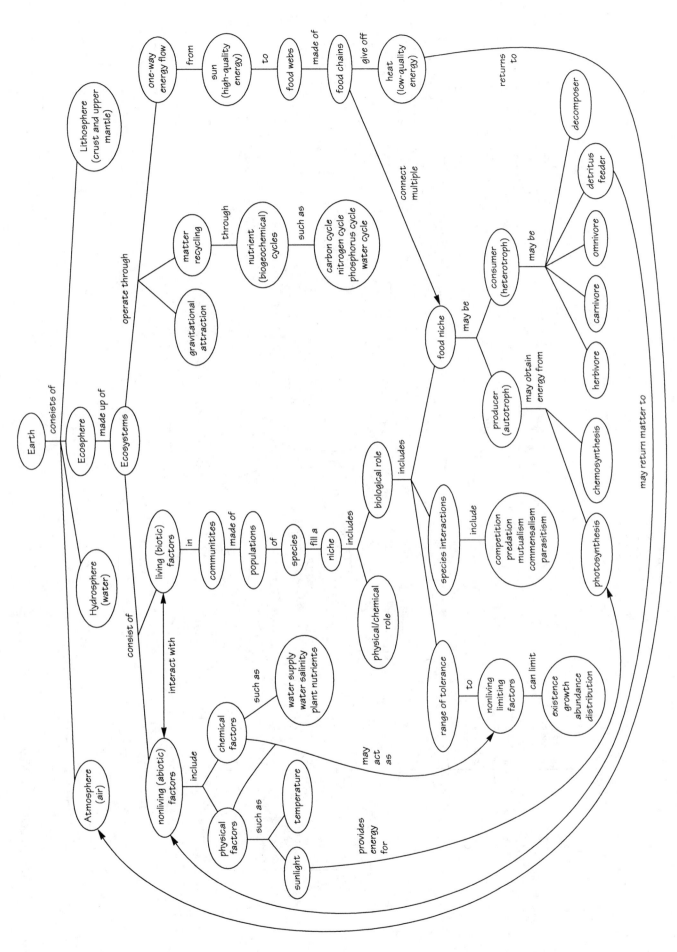

Map 5. How Ecosystems Work

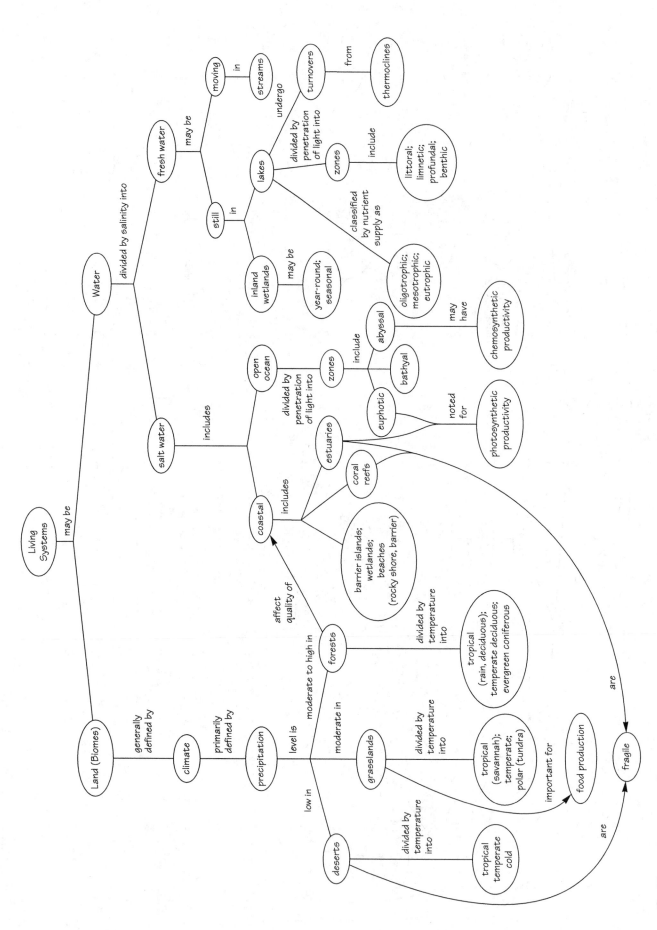

Map 6. Ecosystem Types

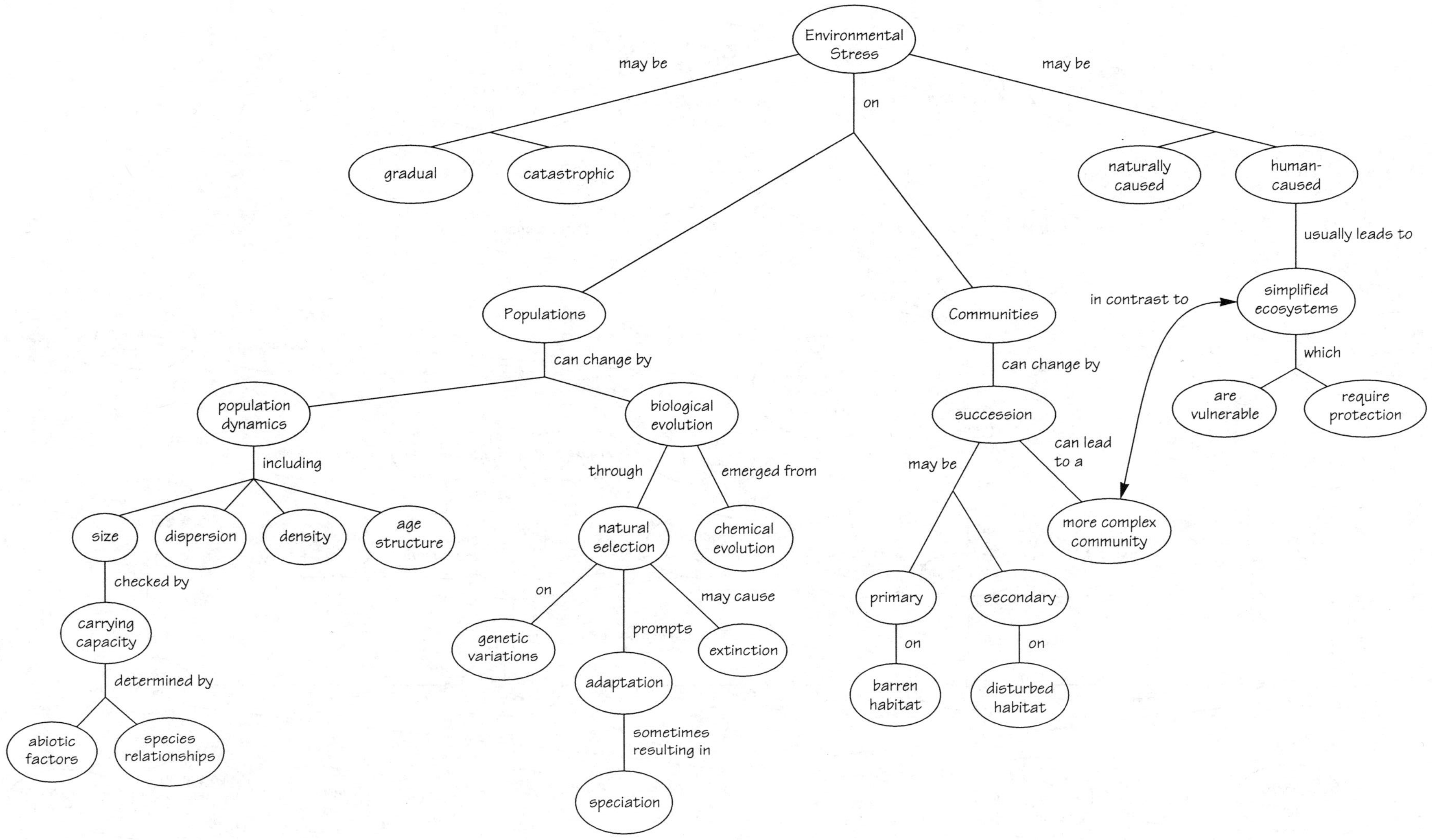

Map 7. Environmental Stress

Environmental Stress
— may be — gradual
— may be — catastrophic
— may be — naturally caused
— may be — human-caused — usually leads to — simplified ecosystems
— on — Communities
— on — Populations

simplified ecosystems
— which — require protection
— which — are vulnerable
— in contrast to — more complex community

Communities — can change by — succession
succession — can lead to a — more complex community
succession — may be — secondary — on — disturbed habitat
succession — may be — primary — on — barren habitat

Populations — can change by — biological evolution
biological evolution — emerged from — chemical evolution
biological evolution — through — natural selection
natural selection — may cause — extinction
natural selection — prompts — adaptation — sometimes resulting in — speciation
natural selection — on — genetic variations

Populations — population dynamics — including — age structure, density, dispersion, size
size — checked by — carrying capacity — determined by — species relationships, abiotic factors

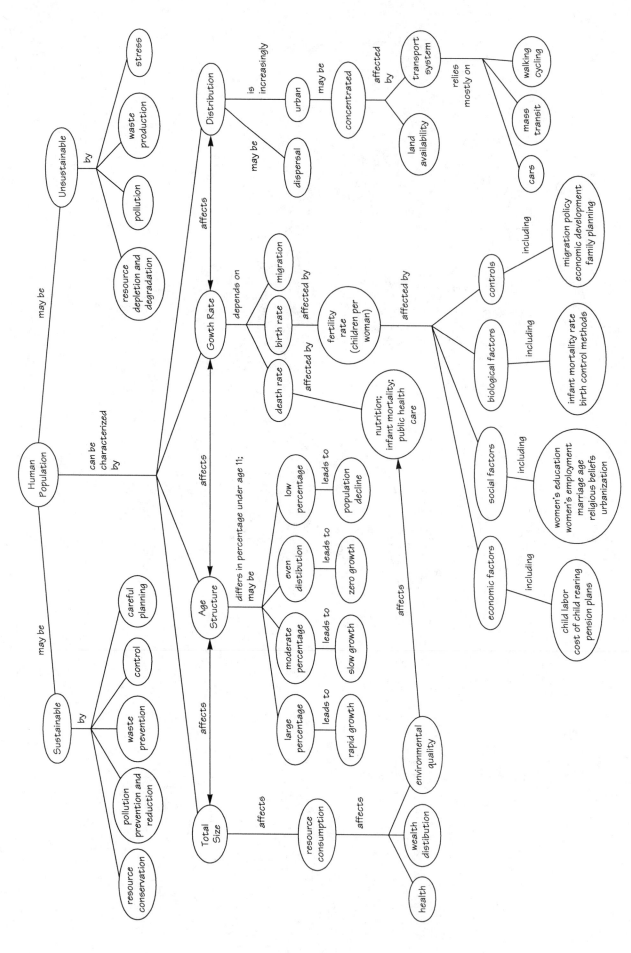

Map 8. Human Population

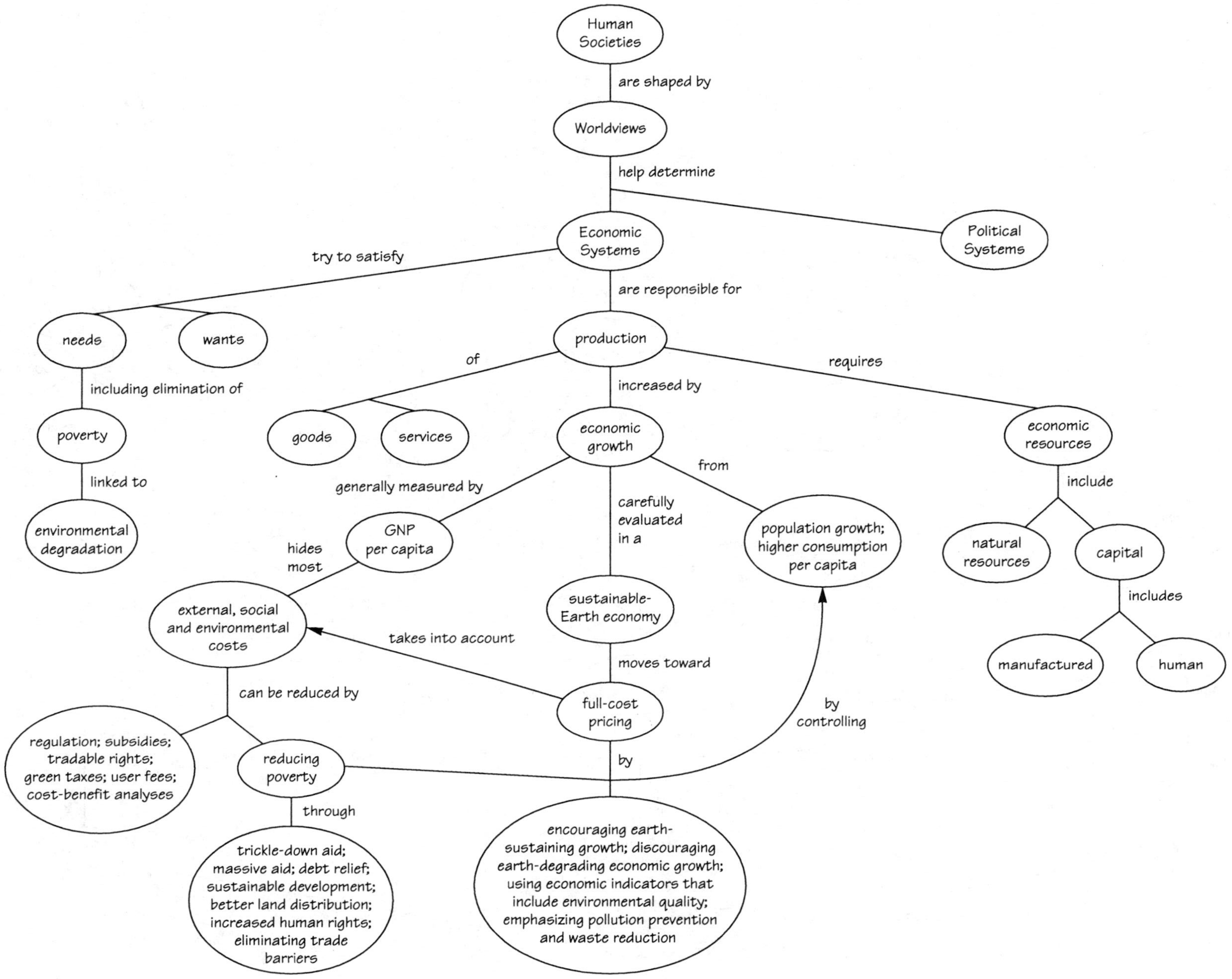

Map 9. Economics

Concept map (reading along the connections):

Human Societies — are shaped by — Worldviews — help determine — Economic Systems

Economic Systems — are responsible for — production — increased by — economic growth

Economic Systems — try to satisfy — needs / wants
- needs — including elimination of — poverty — linked to — environmental degradation

production — of — goods / services

production — requires — economic resources — include — natural resources / capital
- capital — includes — manufactured / human

Worldviews — help determine — Political Systems

economic growth — from — population growth; higher consumption per capita

economic growth — generally measured by — GNP per capita — hides most — external, social and environmental costs

economic growth — carefully evaluated in a — sustainable-Earth economy — moves toward — full-cost pricing

full-cost pricing — takes into account — external, social and environmental costs

external, social and environmental costs — can be reduced by — reducing poverty — through — trickle-down aid; massive aid; debt relief; sustainable development; better land distribution; increased human rights; eliminating trade barriers

full-cost pricing — by — regulation; subsidies; tradable rights; green taxes; user fees; cost-benefit analyses

full-cost pricing — by controlling — population growth; higher consumption per capita

sustainable-Earth economy — encouraging earth-sustaining growth; discouraging earth-degrading economic growth; using economic indicators that include environmental quality; emphasizing pollution prevention and waste reduction

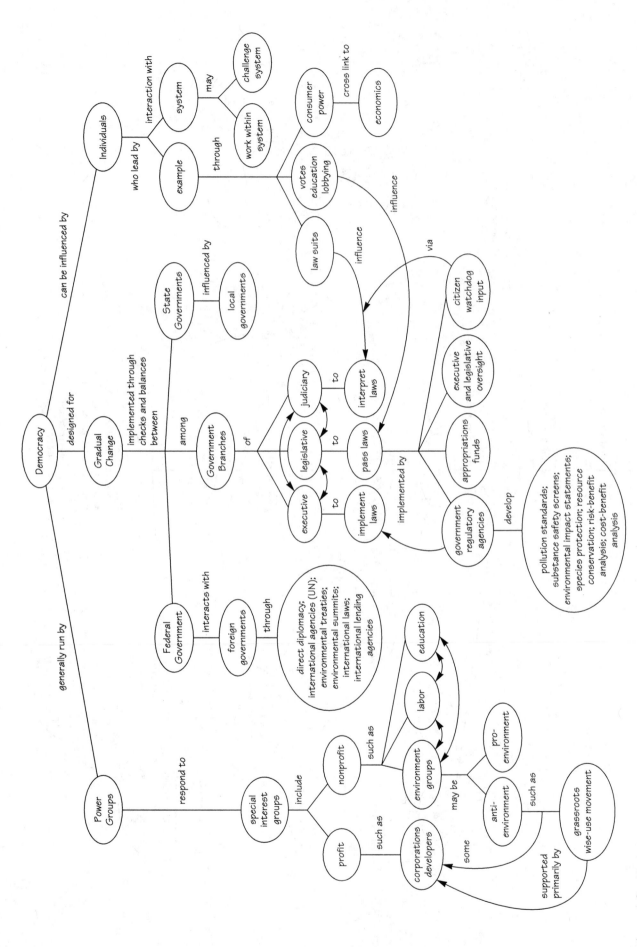

Map 10. Politics

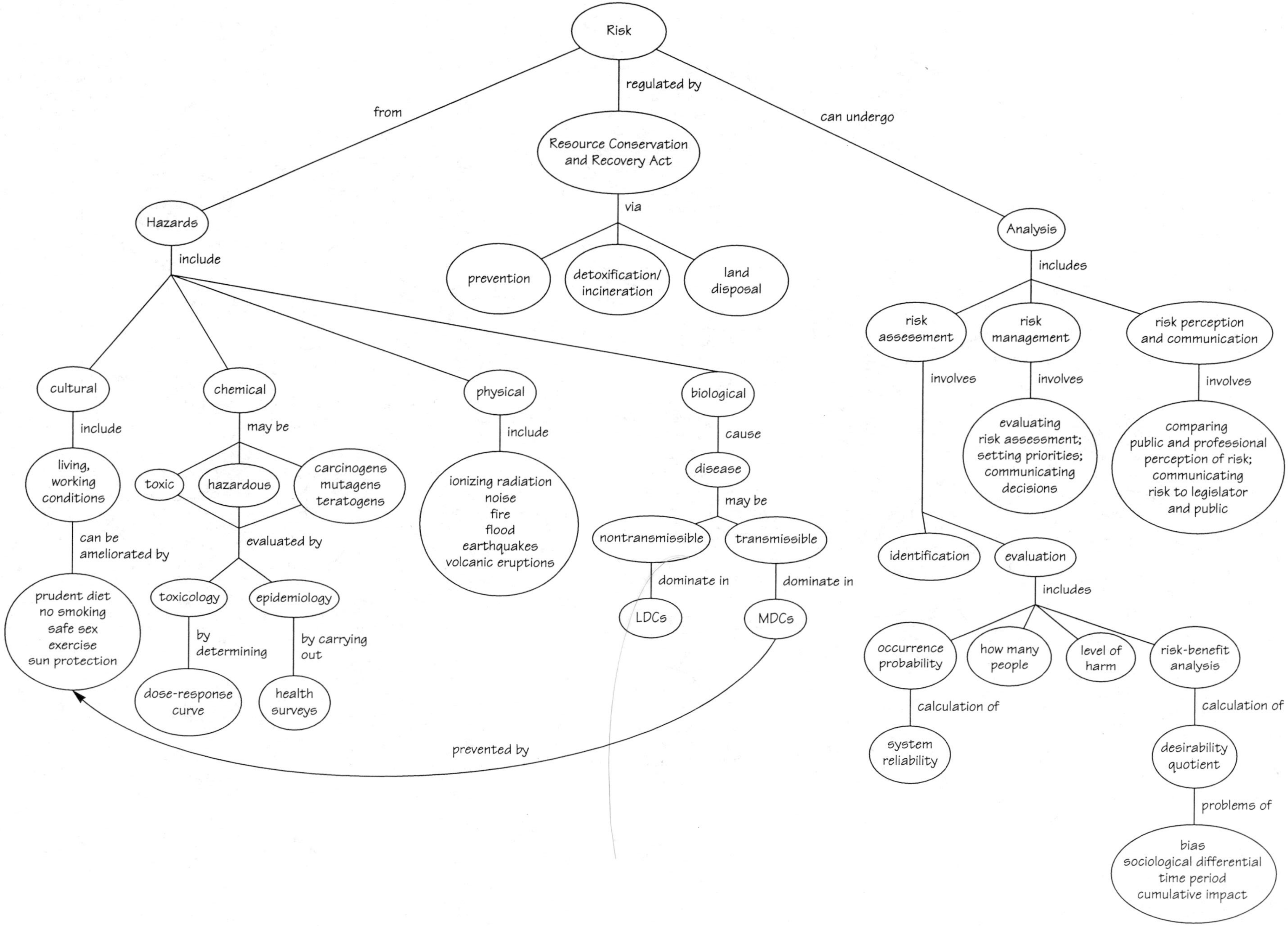

Map 11. Risk

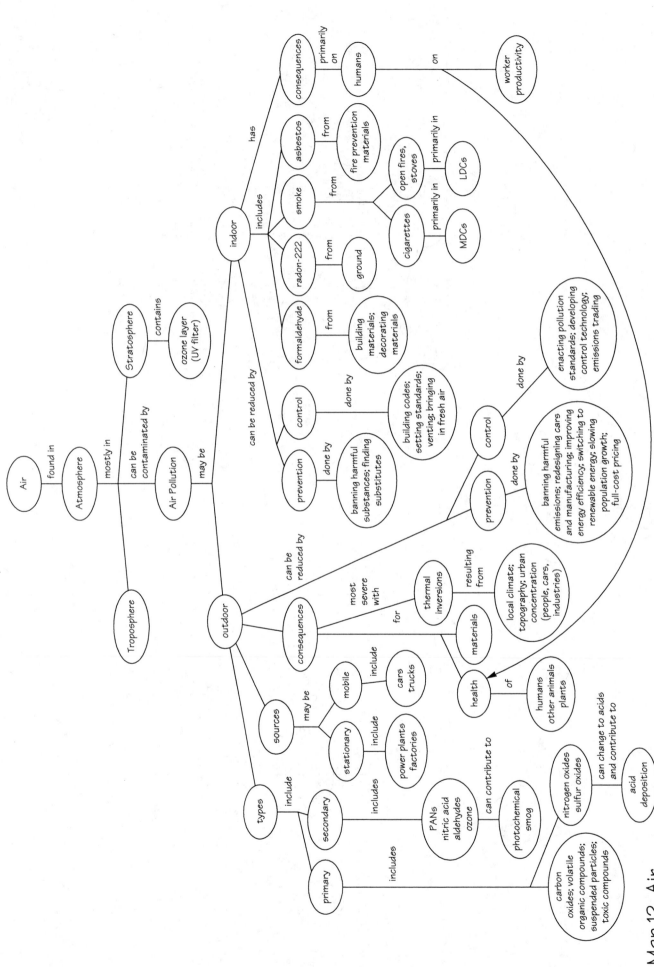

Map 12. Air

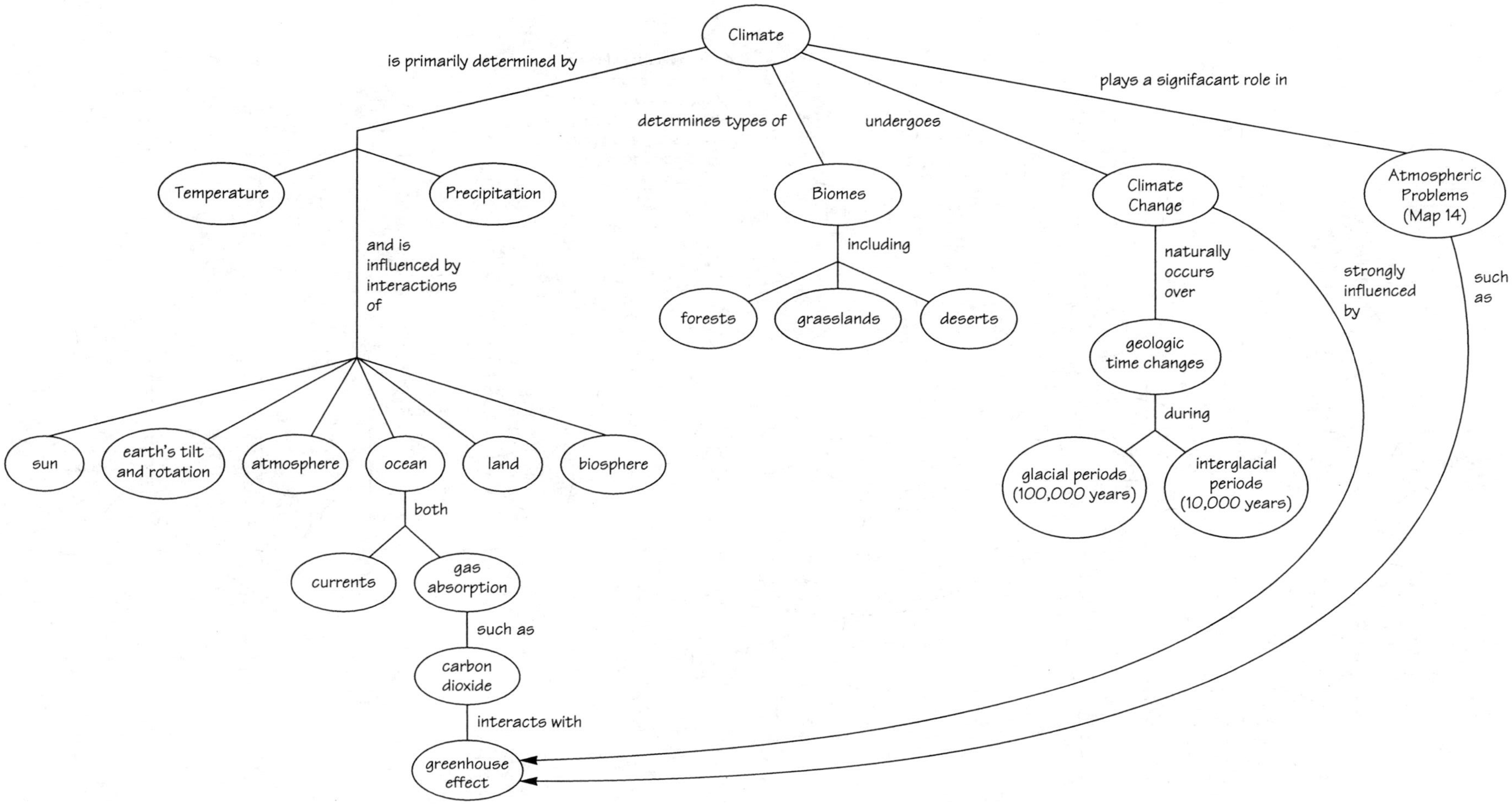

Climate

is primarily determined by

plays a significacant role in

Atmospheric
Problems
(Map 14)

such
as

strongly
influenced
by

determines types of

undergoes

Climate
Change

naturally
occurs
over

geologic
time changes

during

interglacial
periods
(10,000 years)

glacial periods
(100,000 years)

Biomes

including

deserts

grasslands

forests

Temperature

Precipitation

and is
influenced by
interactions
of

sun

earth's tilt
and rotation

atmosphere

ocean

land

biosphere

both

currents

gas
absorption

such as

carbon
dioxide

interacts with

greenhouse
effect

Map 13. Climate

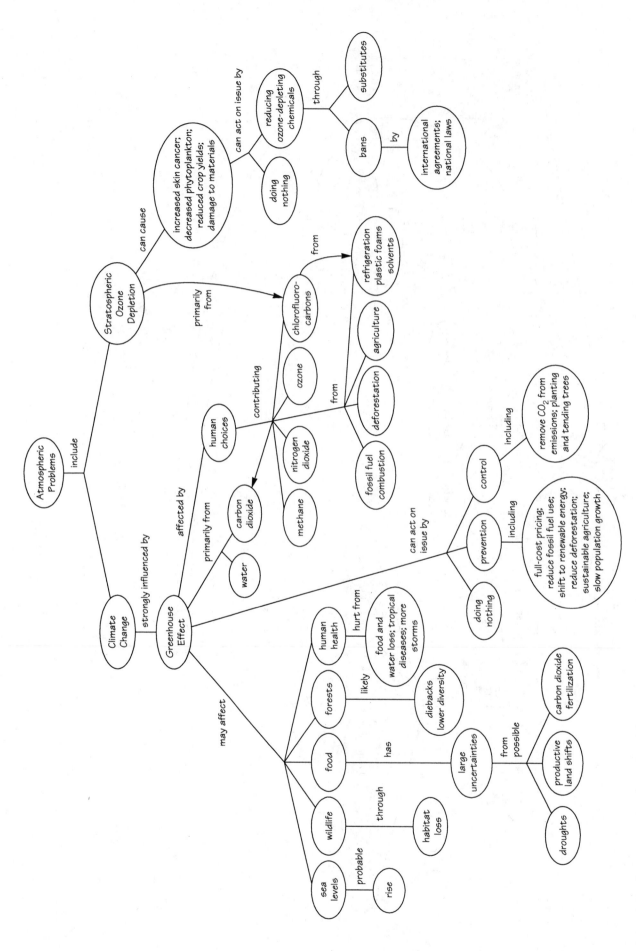

Map 14. Atmospheric Problems

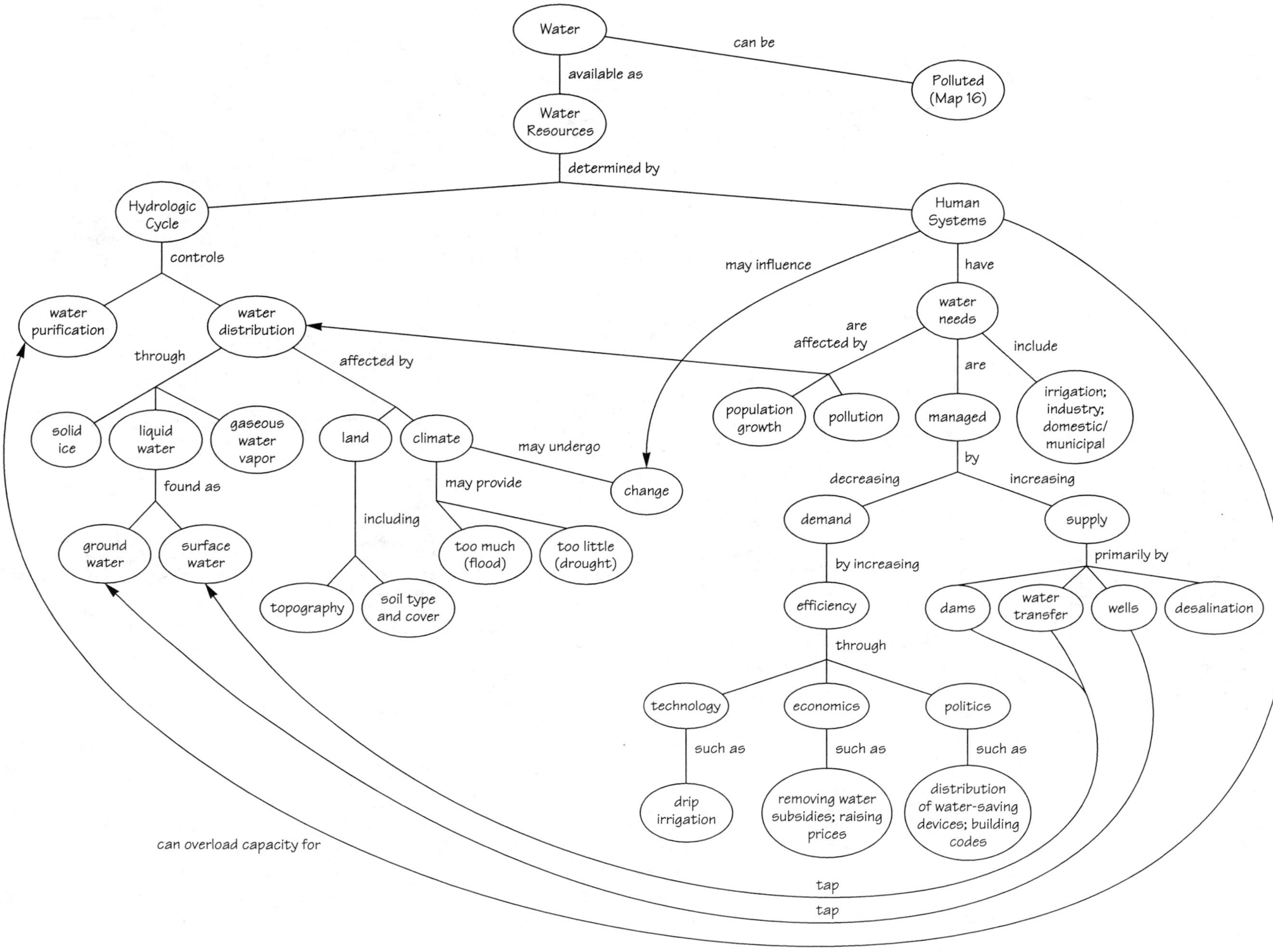

Map 15. Water

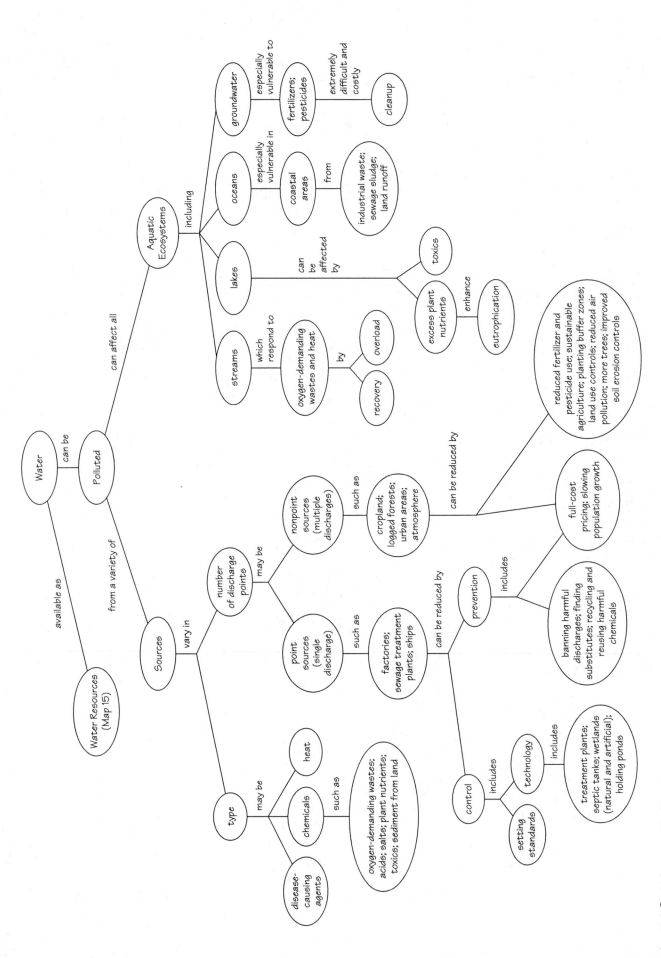

Map 16. Water Pollution

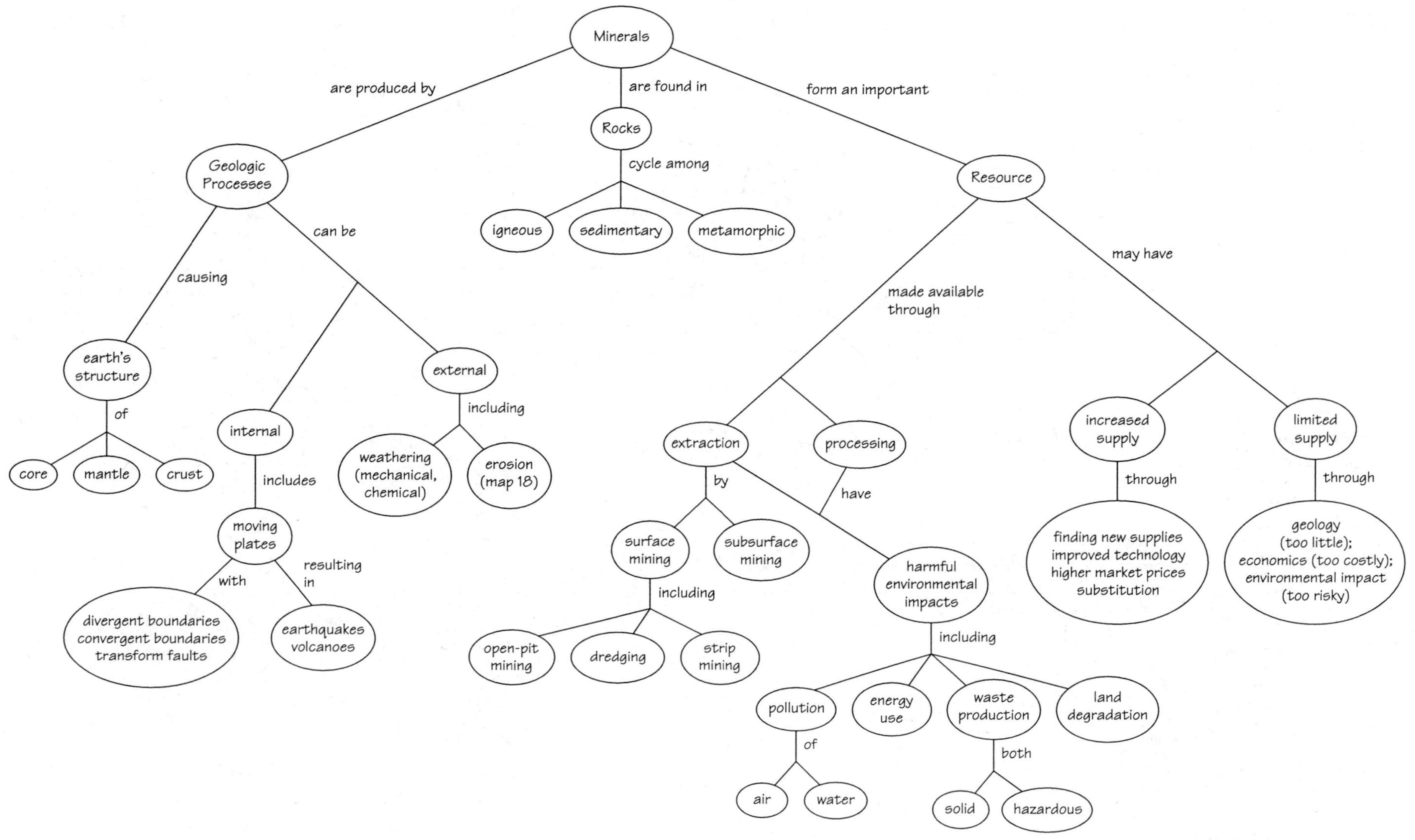

Minerals

Minerals —are found in→ Rocks —cycle among→ igneous, sedimentary, metamorphic

Minerals —are produced by→ Geologic Processes —can be→
- external —including→ weathering (mechanical, chemical), erosion (map 18)
- internal —includes→ moving plates
  - —resulting in→ earthquakes, volcanoes
  - —with→ divergent boundaries, convergent boundaries, transform faults

Geologic Processes —causing→ earth's structure —of→ core, mantle, crust

Minerals —form an important→ Resource

Resource —may have→
- increased supply —through→ finding new supplies, improved technology, higher market prices, substitution
- limited supply —through→ geology (too little); economics (too costly); environmental impact (too risky)

Resource —made available through→ extraction, processing

extraction —by→
- subsurface mining
- surface mining —including→ open-pit mining, dredging, strip mining

extraction, processing —have→ harmful environmental impacts —including→
- pollution —of→ air, water
- energy use
- waste production —both→ solid, hazardous
- land degradation

Map 17. Minerals

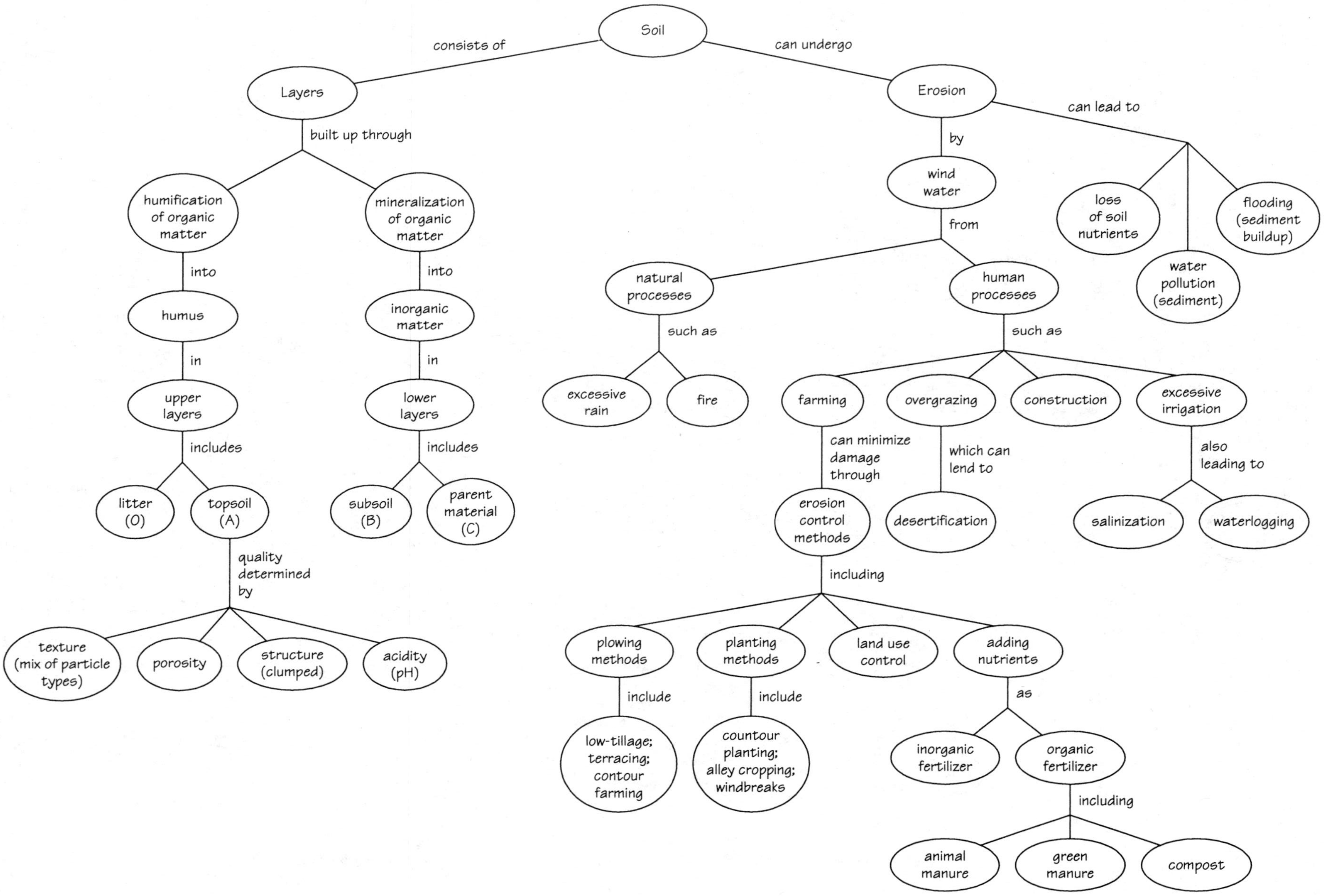

Map 18. Soil

Soil
- **can undergo** → **Erosion**
  - **by** → **wind / water**
    - **can lead to**:
      - loss of soil nutrients
      - flooding (sediment buildup)
      - water pollution (sediment)
    - **from** → **human processes**
      - **such as**:
        - construction
        - excessive irrigation
          - **also leading to**:
            - salinization
            - waterlogging
        - overgrazing
          - **which can lead to** → desertification
        - farming
          - **can minimize damage through** → **erosion control methods**
            - **including**:
              - adding nutrients
                - **as**:
                  - inorganic fertilizer
                  - organic fertilizer
                    - **including**:
                      - animal manure
                      - green manure
                      - compost
              - land use control
              - planting methods
                - **include**: countour planting; alley cropping; windbreaks
              - plowing methods
                - **include**: low-tillage; terracing; contour farming
    - **from** → **natural processes**
      - **such as**:
        - excessive rain
        - fire
- **consists of** → **Layers**
  - **built up through**:
    - mineralization of organic matter
      - **into** → inorganic matter
        - **in** → lower layers
          - **includes**:
            - subsoil (B)
            - parent material (C)
    - humification of organic matter
      - **into** → humus
        - **in** → upper layers
          - **includes**:
            - litter (O)
            - topsoil (A)
              - **quality determined by**:
                - texture (mix of particle types)
                - porosity
                - structure (clumped)
                - acidity (pH)

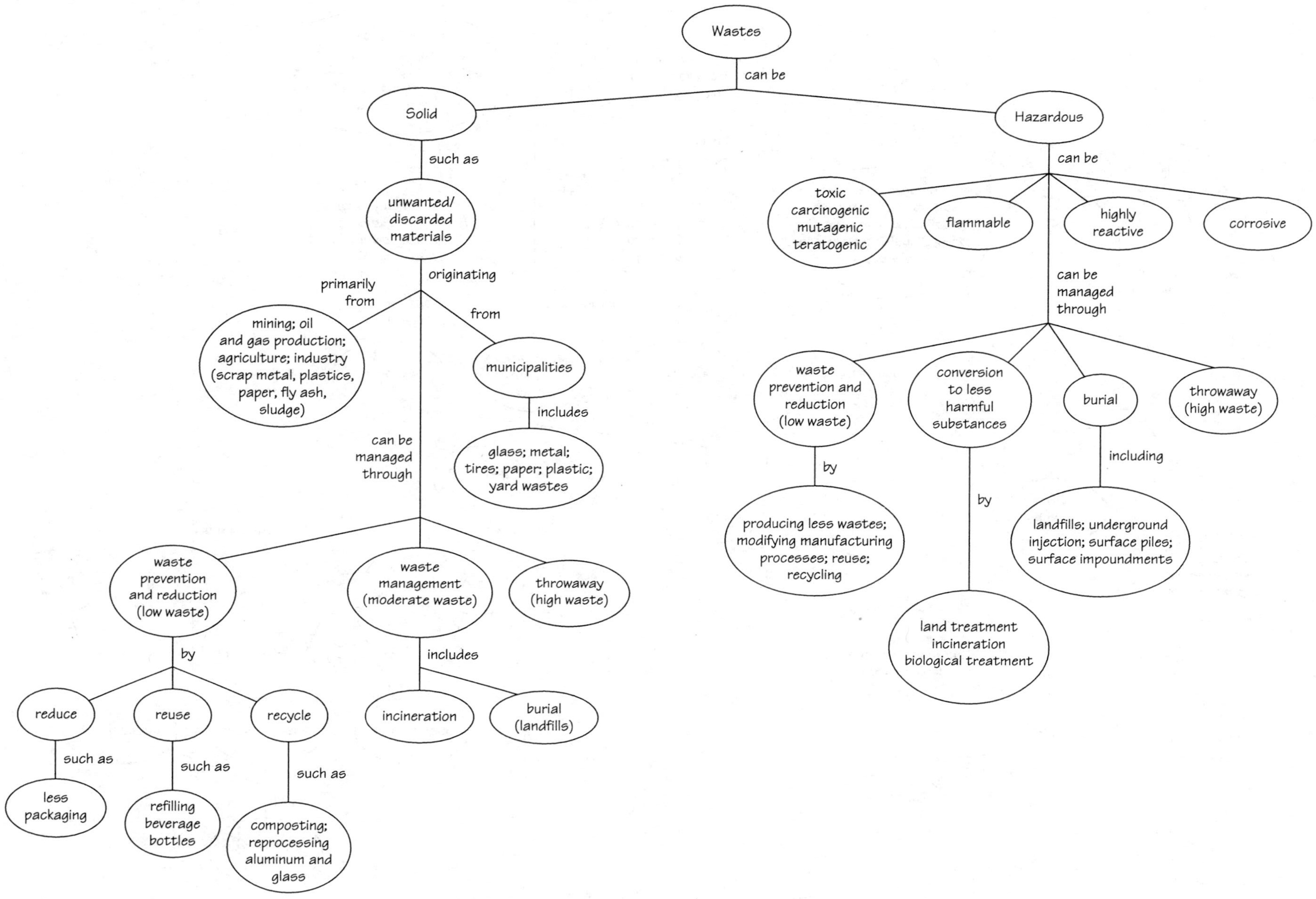

Wastes

can be

Solid

such as → **unwanted/discarded materials**

originating from → **municipalities**
- includes → glass; metal; tires; paper; plastic; yard wastes

primarily from → mining; oil and gas production; agriculture; industry (scrap metal, plastics, paper, fly ash, sludge)

can be managed through →

waste management (moderate waste)
- includes → throwaway (high waste)
- includes → incineration
- includes → burial (landfills)

waste prevention and reduction (low waste)
- by → **reduce** — such as → less packaging
- by → **reuse** — such as → refilling beverage bottles
- by → **recycle** — such as → composting; reprocessing aluminum and glass

Hazardous

can be → corrosive; highly reactive; flammable; toxic, carcinogenic, mutagenic, teratogenic

can be managed through →

burial
- including → landfills; underground injection; surface piles; surface impoundments

conversion to less harmful substances
- by → land treatment; incineration; biological treatment

waste prevention and reduction (low waste)
- by → producing less wastes; modifying manufacturing processes; reuse; recycling

Map 19. Wastes

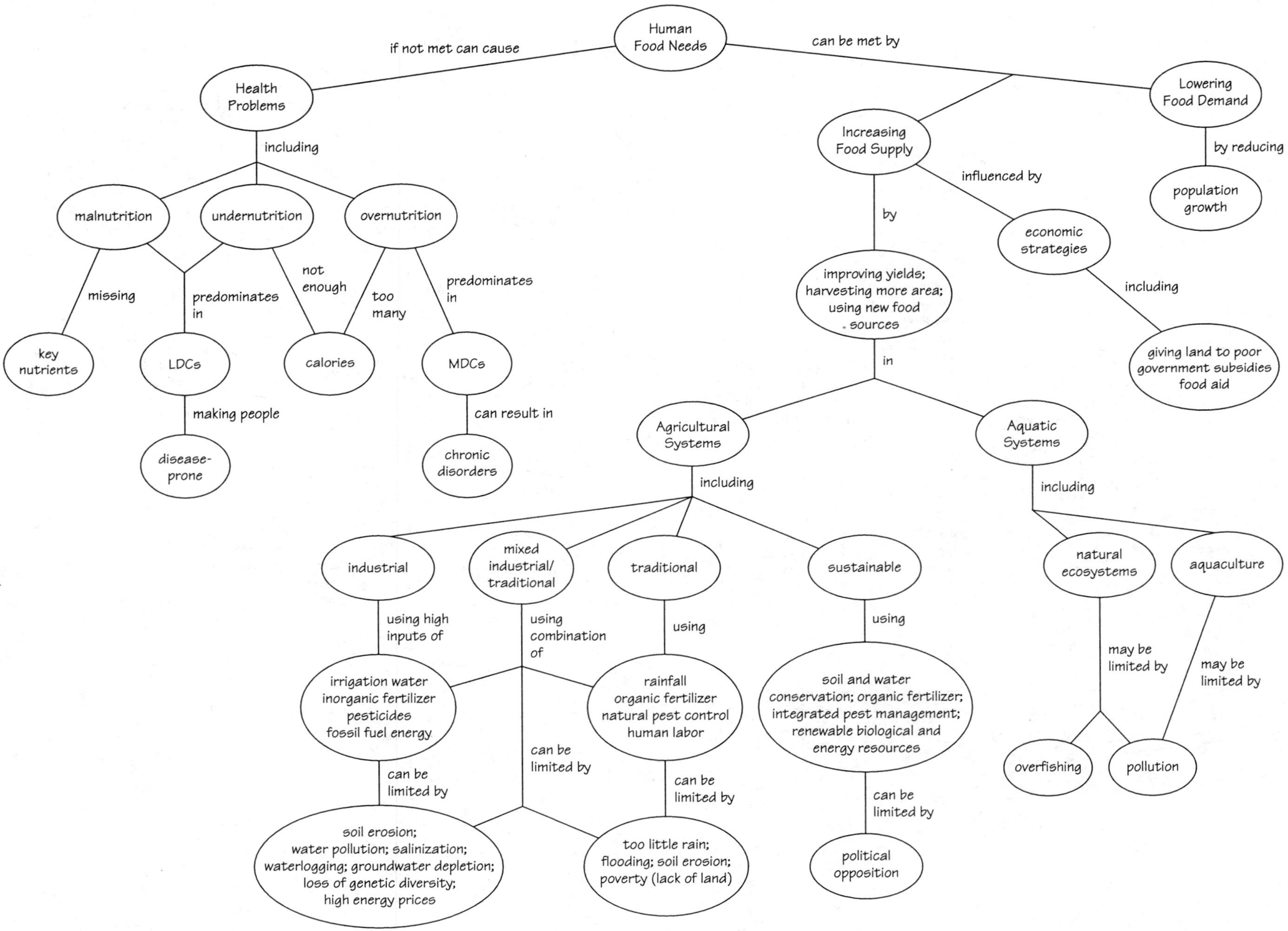

Map 20. Food Resources

Human Food Needs

can be met by → **Increasing Food Supply**
- influenced by → **economic strategies**
  - including → giving land to poor; government subsidies; food aid
- by → **improving yields; harvesting more area; using new food sources**
  - in → **Aquatic Systems**
    - including → **natural ecosystems**
      - may be limited by → **pollution**, **overfishing**
    - **aquaculture**
      - may be limited by → **pollution**
  - in → **Agricultural Systems**
    - including → **sustainable**
      - using → soil and water conservation; organic fertilizer; integrated pest management; renewable biological and energy resources
        - can be limited by → **political opposition**
    - **traditional**
      - using → rainfall; organic fertilizer; natural pest control; human labor
        - can be limited by → too little rain; flooding; soil erosion; poverty (lack of land)
    - **mixed industrial/traditional**
      - using combination of
    - **industrial**
      - using high inputs of → irrigation water; inorganic fertilizer; pesticides; fossil fuel energy
        - can be limited by → soil erosion; water pollution; salinization; waterlogging; groundwater depletion; loss of genetic diversity; high energy prices

can be met by → **Lowering Food Demand**
- by reducing → **population growth**

if not met can cause → **Health Problems**
- including → **overnutrition**
  - too many → **calories**
  - predominates in → **MDCs**
    - can result in → **chronic disorders**
- **undernutrition**
  - not enough → **calories**
  - predominates in → **LDCs**
    - making people → **disease-prone**
- **malnutrition**
  - missing → **key nutrients**

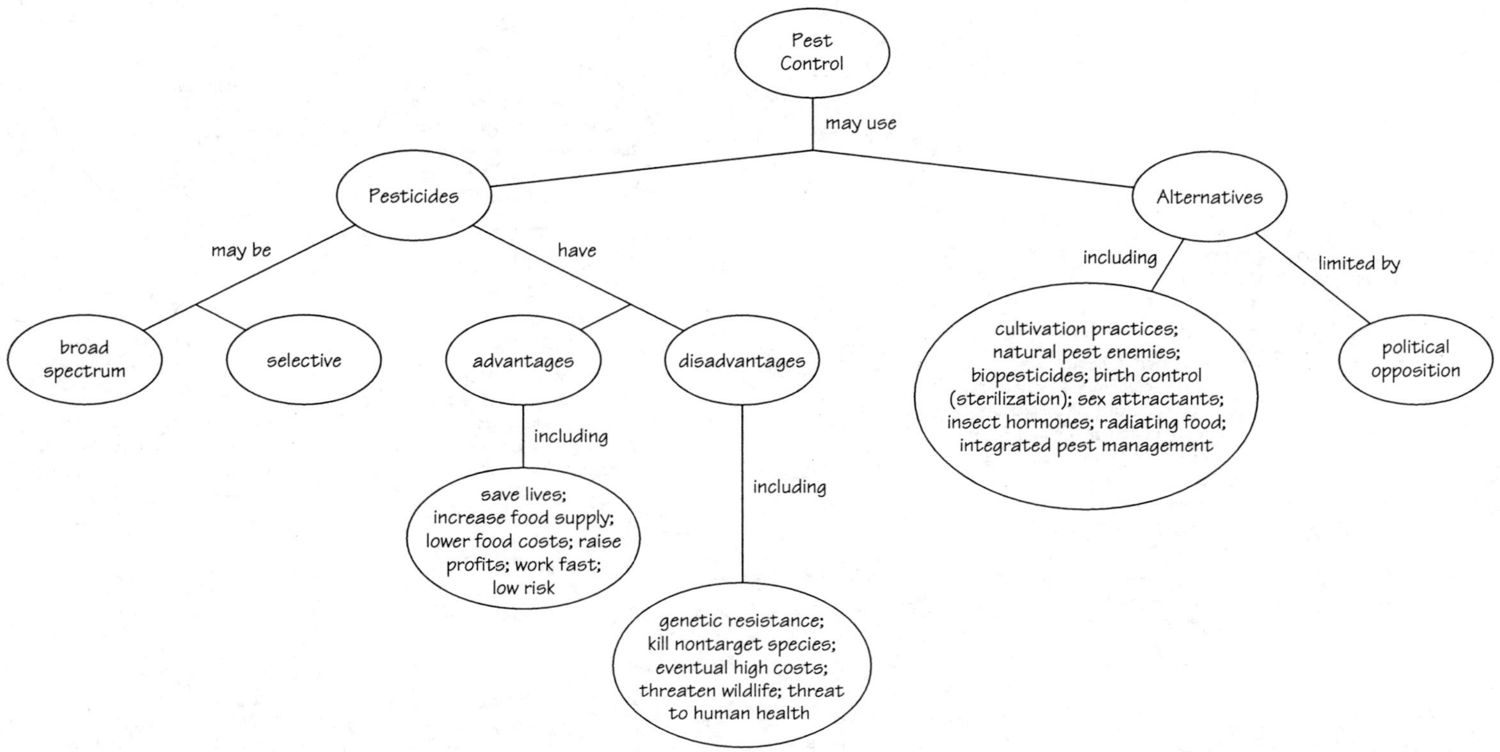

Map 21. Pest Control

Pest Control
— may use —

Alternatives
— limited by — political opposition
— including — cultivation practices; natural pest enemies; biopesticides; birth control (sterilization); sex attractants; insect hormones; radiating food; integrated pest management

Pesticides
— may be —
broad spectrum
selective

— have —

disadvantages
— including — genetic resistance; kill nontarget species; eventual high costs; threaten wildlife; threat to human health

advantages
— including — save lives; increase food supply; lower food costs; raise profits; work fast; low risk

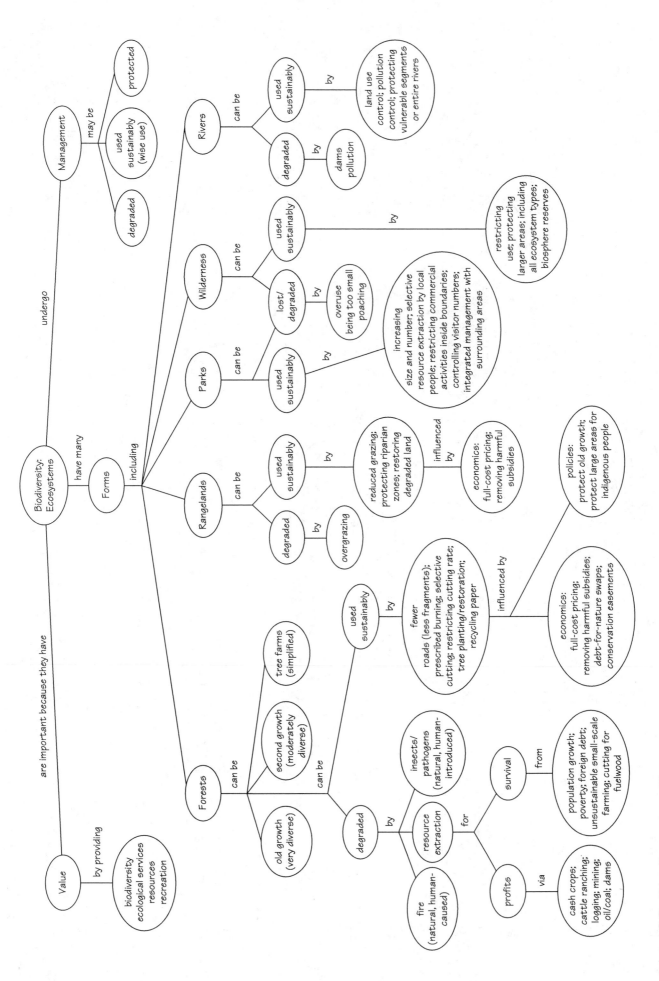

Map 22. Biodiversity: Ecosystems

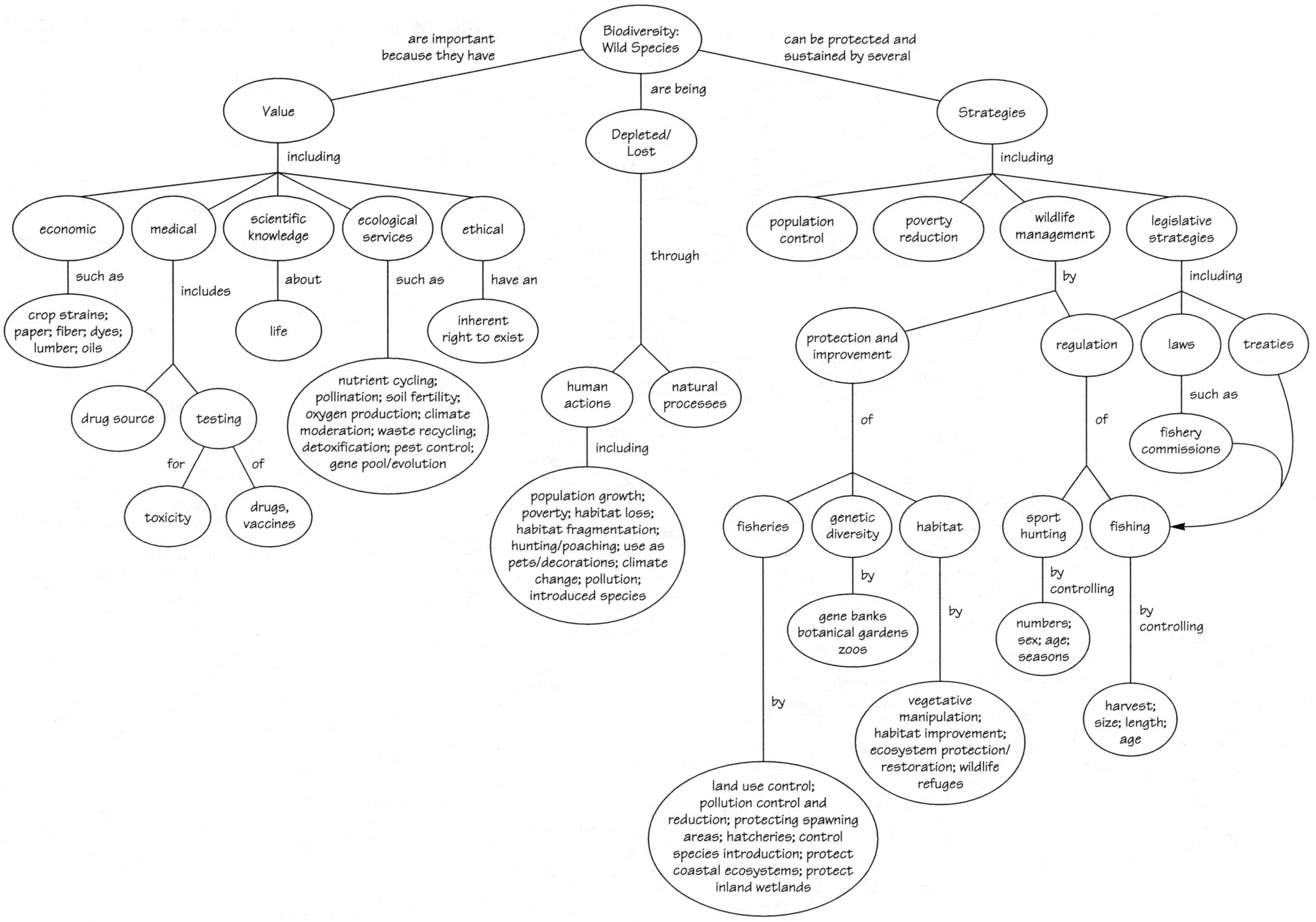

Map 23. Biodiversity: Wild Species

Biodiversity: Wild Species

are important because they have → **Value** including:
- **economic** such as → crop strains; paper; fiber; dyes; lumber; oils
- **medical** includes → drug source; testing — for → toxicity; of → drugs, vaccines
- **scientific knowledge** about → life
- **ecological services** such as → nutrient cycling; pollination; soil fertility; oxygen production; climate moderation; waste recycling; detoxification; pest control; gene pool/evolution
- **ethical** have an → inherent right to exist

are being **Depleted/Lost** through:
- **natural processes**
- **human actions** including → population growth; poverty; habitat loss; habitat fragmentation; hunting/poaching; use as pets/decorations; climate change; pollution; introduced species

can be protected and sustained by several **Strategies** including:
- **population control**
- **poverty reduction**
- **wildlife management** by:
  - **regulation** of:
    - **fishing** — by controlling → harvest; size; length; age
    - **sport hunting** — by controlling → numbers; sex; age; seasons
  - **legislative strategies** including:
    - **laws** such as → fishery commissions
    - **treaties**
- **protection and improvement** of:
  - **habitat** by → vegetative manipulation; habitat improvement; ecosystem protection/restoration; wildlife refuges
  - **genetic diversity** by → gene banks; botanical gardens; zoos
  - **fisheries** by → land use control; pollution control and reduction; protecting spawning areas; hatcheries; control species introduction; protect coastal ecosystems; protect inland wetlands

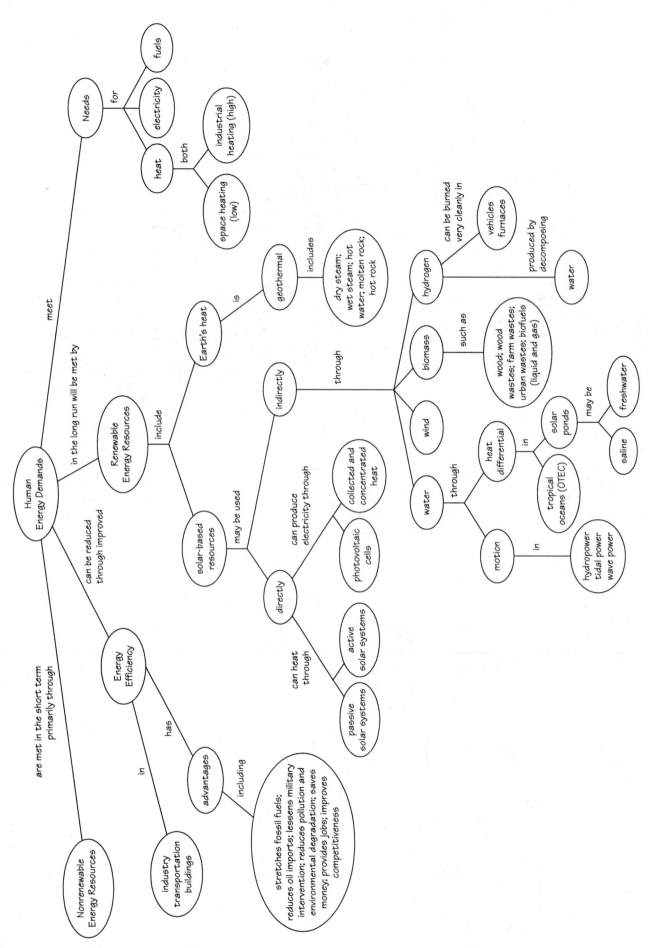

Map 24. Energy Efficiency and Renewable Energy

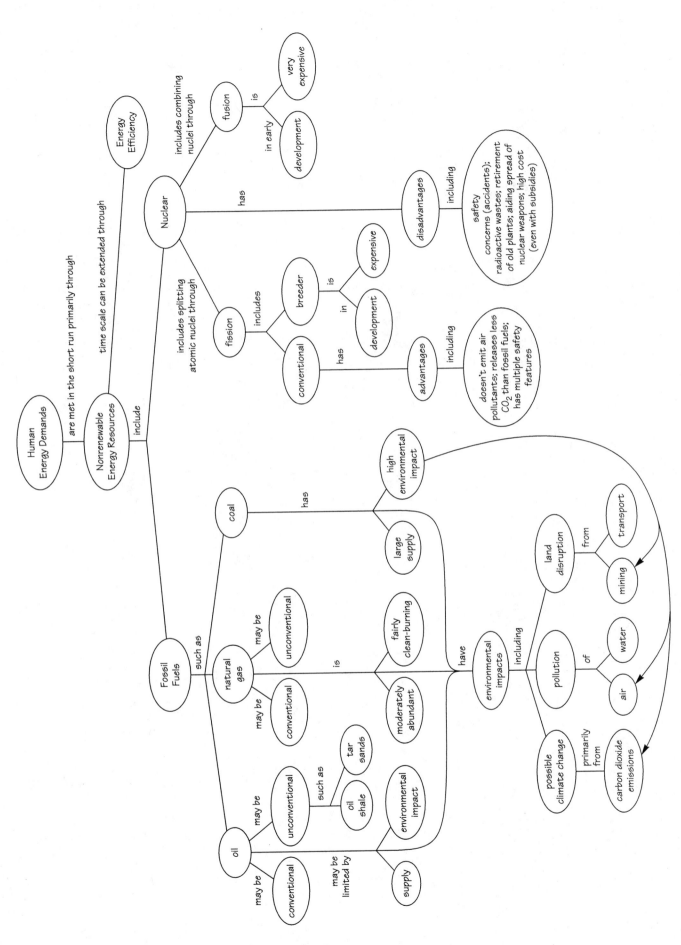

Map 25. Nonrenewable Energy Resources

APPENDIX B

MEDIA SOURCES AND DISTRIBUTORS

The following abbreviations were used in the <u>Multisensory Learning: Audiovisuals</u> subsections of the Goals and Activities material in this manual. Addresses and/or phone numbers of media sources are given. Some general suggestions about obtaining environmental films are given at the end of the abbreviations.

ACPB The Annenberg/CPB Project, 901 E Street, N.W., Washington, D.C. 20004-2006
  (800–LEARNER)
AHA American Hydrogen Association, Tempe, AZ (602-438-8005)
AMC American Mining Congress, 1920 North St., N.W., Washington, D.C. 20036 (202-861-2800)
AV American Visions (†800-553-8878)
BF Biosphere Films, 4 W. 105th St./6A, New York, NY 10025
BFF Bullfrog Films, Oley, PA 19547 (800-543-3764)
BP Brauer Productions, 402 Cass St., Traverse City, MI 49684
CBC Canadian Broadcasting Corporation Enterprises, P.O. Box 500, Station A,
  Toronto, M5W 1E6, Ontario, Canada
CBS Carolina Biological Supply Company, 2700 York Rd., Burlington, NC 27215
CF Commonwealth Films, 1500 Brook Rd., Richmond, VA 23220
CFS Conservation Film Service, 408 East Main St., P.O. Box 776, League City, TX 77574-0776
CMS Consolidated Media Services, 2565 Cloverdale Ave., Suite C, Concord, CA 94518-9955
CMTI Coronet Film and Video, 108 Wilmot Rd., Deerfield, IL 60015 (800-621-2131)
CNN Video Sales, P.O. Box 105366, Atlanta, GA 30348-5366
CP Centre Publications, Inc., 1327 Spruce St., Suite 3, Boulder, CO 80302
CPr Cinnamon Productions (212-431-4899)
CSWMB California Solid Waste Management Board, 1020 9th St., Suite 300, Sacramento, CA 95814
CTE Central Television Enterprises, 35-38 Portman Square, London W1A 2HZ, England
DC Direct Cinema (800-345-6748)
DWP David Weiss Productions, 1414 The Strand, Manhattan Beach, CA 90266
EBEC Encyclopedia Britannica Educational Corporation, 310 South Michigan Ave.,
  Chicago, IL 60604 (312-347-7000)
EP Energy Partners, West Palm Beach, Florida (407-688-0500)
FA Florida Audubon, Casselberry, FL (407-260-8300)
FDC Film Distribution Center, 13500 N.E. 124th St./2, Kirkland, WA 98034-8010 (206-820-2592)
FF Florida Films; P.O. Box 13712, Gainesville, FL 32604
FHS Films for the Humanities & Sciences, Box 2053, Princeton, NJ 08543
FI Films, Inc., 5547 North Ravenwood, Chicago, IL
FL Filmakers Library, 133 East 58th, New York, NY 10022
FLP Fine Line Productions (Film Library), P.O. Box 315, Franklin Lakes, NJ 07417
FP Fanlight Productions (617-524-0980)
FPL Florida Power & Light Company, Environmental Affairs, P.O. Box 14000,
  Juno Beach, FL 33408
FRF First Run Features (212-243-0600)
GF Griesinger Films, 7300 Old Mill Rd., Dept. IP, Gates Mills, OH 44040 (216-423-1601)
GMPF Green Mountain Post Films, P.O. Box 229/37 Ferry Rd., Turners Falls, MA 01376
GTC Global Tomorrow Coalition, 1325 G St., N.W., Suite 915, Washington, D.C. 20005-3104
HMDC Hackensack Meadowlands Development Commission, 1 DeKorte Park Plaza,
  Lyndhurst, NJ 07071
IFB International Film Bureau, 332 South Michigan Ave., Chicago, IL 60604
LF Landmark Films (703-241-2030)
LTS London Television Service, Hercules House, Hercules Road, London SE1 7DU, England

LW Leon Watson, University of Michigan, Ann Arbor, MI (517-353-9501)

MDC Missouri Department of Conservation, P.O. Box 180, Jefferson City, MO 65102
    ATTN: Film Library

MDNR Michigan Department of Natural Resources, P.O. Box 30028, Lansing, MI 48909

MM Michigan Media, 416 Fourth St., Ann Arbor, MI 48109 (313-764-8298)

NA National Audubon, New York, NY (212-832-3200)

NAP National Academy Press; 2101 Constitution Ave., N.W., Washington, D.C. 20418
    (202-334-2665)

NAVC National AudioVisual Center, 8700 Edgeworth Dr., Capital Heights, MD 20743-3701

NCAP Northwest Coalition for Alternatives to Pesticides, P.O. Box 1393, Eugene OR 97440
    (503-344-5044)

NFB National Film Board, 1241 Avenue of the Americas, 16th Floor, New York, NY 10020

NFBC National Film Board of Canada, 111 East Wacker Dr./313, Chicago, IL 60601

NG National Geographic Society Educational Services, Department 88, Washington, D.C. 20036
    (800-368-2728)

NI Noranda, Inc., P.O. Box 45, Commerce Court West, Toronto, Ontario M5L 2B6, Canada

NJN New Jersey Network, 1573 Parkside Ave., Trenton, NJ 08638

NREL National Renewable Energy Laboratory and VideoTakes for U.S. Department of Energy;
    free copies while supplies last (800-523-2929)

NWF National Wildlife Federation, 1400 16th St., N.W., Washington, D.C. 20036 (202-637-3700)

OFP Omni Film Productions, Ltd., #204-111 Water St., Vancouver, British Columbia
    V6B 1A7 Canada

PBS Public Broadcasting Service Video, 1320 Braddich Pl., Alexandria, VA 22314 (800-424-7963)

PF Partridge Films, 38 Mill Lane, London NW6 1NR England

PI Population Institute, 110 Maryland Avenue, N.E., Washington, D.C. 20036 (202-544-3300)

PRB Population Reference Bureau, 777 14th Street, N.W./800, Washington, D.C. 20005
    (202-639-8040)

PSU Pennsylvania State University, Audiovisual Services, Division of Media and Learning
    Resources, Special Services Building, University Park, PA 16802

SF Statens Filmcentral, Vestergrade 27, KD-1456, Copenhagen K, Denmark

SFWMD South Florida Water Management District, West Palm Beach, Florida (407-686-8800)

SITES Smithsonian Institute Traveling Exhibition Service, Washington, D.C. 20560
    (202-357-3168)

SLP Seven Locks Press, P.O. Box 27, Cabin John, MD 20818 (800-537-9359)

SLVP San Luis Video Publishing, P.O. Box 4604, San Luis Obispo, CA 93403

ST State of Tennessee, State Planning Office, 308 John Sevier Bldg., Nashville, TN 37219

SUNY State University of New York at Stony Brook, Department of Anatomical Sciences, Health
    Sciences Center, Stony Brook, NY 11794

TBS Turner Broadcasting System, CNN Center, P.O. Box 105366, Atlanta, GA 30348-5366
    (404-827-1700)

UF Umbrella Films, 60 Blake Rd., Brookline, MA 02146 (617-277-6639)

UFW United Farm Workers of America, AFL-CIO (805-822-5571)

USBM U.S. Bureau of Mines Motion Pictures, Cochrans Mill Rd., P.O. Box 18070,
    Pittsburgh, PA 15236 (412-675-4338)

VAP Video Active Productions, Rt. 2, Box 322, Canton, NY (315-386-8797)

VP The Video Project, 5332 College Ave., Suite 101, Oakland, CA 94618 (510-655-9050)

WBP World Bank Publications, Department 0552, Washington, D.C. 20073-0552

WETA WETA-TV Educational Distribution, P.O. Box 2626, Washington, D.C. 20013

WMM Women Make Movies (212-925-0606)

WN World Neighbors, 5116 North Portland Ave., Oklahoma City, OK 73112 (800-242-6387)

WRCF Wild Resource Conservation Fund, Room A1-85, Evan Press Bldg., 3rd & Reily Sts.,
    Harrisburg, PA 17120

ZPG Zero Population Growth, Washington, D.C. (202-332-2200)

From the sources listed above, I strongly recommend catalogues from the Video Project, Bullfrog Films, Carolina Biological Supply, the Race to Save the Planet Series, and National Geographic .

Contact the Cooperative Extension Service of the land grant university of your state and ask for a publications and videotapes list. You may be able to tailor your course to your own state's situation!

Shows from Public TV now available on video: Write for a catalog: *A Field Guide to Environmental Media* to Environmental Media, P.O. Box 1016, Chapel Hill, NC 27514 or call (800-ENV-EDUC)

A special "Eco-Video Collection for Schools" offers over 60 affordable selections for all grade levels. The aim is to help students become "good stewards for the planet." Contact: The Video Project, 5332 College Avenue, Oakland, CA 94618 or call (800-4-PLANET).

North American Association of Environmental Education, P.O. Box 400, Troy, Ohio 45373. (513-698-6493) This professional organization offers a number of environmental education publications including a Festival of Film and Video Series and Curriculum and Resource Fair Catalogs.

Global Tomorrow Coalition. *The Global Ecology Handbook: What You Can Do About the Environmental Crisis*, edited by Walter H. Corson. Beacon Press, Boston, 1990. This book is an excellent source of teacher support materials as well as films and videos.

SAFE PLANET: The Guide to Environmental Film and Video offers a source of high-quality documentaries. It is available to individuals for $7.50 plus $3 postage and handling. Make checks payable to Media Network, 39 W. 14th St., Suite 403, New York, NY 10011 (212-929-2663)